Composites and
Their Applications

Composites and Their Applications

Editor

Alano Riley

Composites and Their Applications

Edited by **Alano Riley**

ISBN: 978-1-68117-204-0
Library of Congress Control Number: 2016934751

© 2017 by
SCITUS Academics LLC,
www.scitusacademics.com
Box No. 4766, 616 Corporate Way,
Suite 2, Valley Cottage,
NY 10989

Preface

A composite material (shortened to composite) is a material made from two or more constituent materials with significantly different physical or chemical properties that, when combined, produce a material with characteristics different from the individual components. The individual components remain separate and distinct within the finished structure. The new material may be preferred for many reasons: common examples include materials which are stronger, lighter, or less expensive when compared to traditional materials. Composites are generally used for buildings, bridges, and structures such as boat hulls, swimming pool panels, race car bodies, shower stalls, bathtubs, storage tanks, imitation granite and cultured marble sinks and countertops. Composite materials are saving lives and assets around the world, every day. The lightweight and strong properties of composites make them an ideal fit for armour applications.Composites are changing the way we travel and ship goods. The most advanced examples perform routinely on spacecraft and aircraft in demanding environments. Concrete is the most common artificial composite material of all and typically consists of loose stones (aggregate) held with a matrix of cement.

This book, Composites and Their Applications, deals with prospective applications and the associated properties of several composites aiming on the following numerous topics: health or integrity monitoring techniques of composites structures, bio-medical composites and their solicitations in dental or tissue materials, natural fibre or mineral filler reinforced composites and their property characterization, catalysts composites and their applications, and some other potential applications of fibres or composites as sensors, etc.

Table of Contents

CHAPTER 1

Titania-Silica Composites: A Review on the Photocatalytic Activity and Synthesis Methods

Yuri Hendrix, Alberto Lazaro, Qingliang Yu, Jos Brouwers

Department of the Built Environment, Eindhoven University of Technology, Eindhoven, The Netherlands

ABSTRACT

The photocatalyic activity of titania is a very promising mechanism that has many possible applications like purification of air and water [1] -[4] . To make it even more attractive, titania can be combined with silica to increase the photocatalytic efficiency and durability of the photocatalytic material, while lowering the production costs [1]. In this article, relevant literature is reviewed to obtain an overview about the chemistry and physics behind some of the different parameters that lead to cost-effective photocatalytic titania-silica composites. The first part of this review deals with the mechanisms involved in the photocatalytic activity, then the chemistry behind certain methods for the synthesis of the titania-silica composites is discussed, and in the last and third part of this review, the influence of silica supports on titania is discussed. These three sections represent three different fields of research that are combined in this review to obtain better insights on the photocatalytic titania-silica composites. While many research subjects in these fields have been well known for some time now, some subjects are only more recently resolved and some subjects are still under discussion (e.g. the cause for the increased hydrophilic surface of titania after illumination). This article aims to review the most important literature to give an overview of the current situation of the fundamentals of photocatalysis and synthesis of the cost-effective photocatalyic composites. It is found that the most cost-effective photocatalytic titania-silica composites are the ones that have a thin anatase layer coated on silica with a large specific surface area, and are prepared with the precipitation or sol-gel methods.

Keywords: Photocatalysis, Photocatalytic TiO_2, Titania-Silica Composites, Low Cost Synthesis

1. INTRODUCTION

Composites made out of silica and titania can have the photocatalytic properties from titania, the high stability from silica and extra properties coming from chemical bonds between the two materials [1] . Titania is photocatalytic because it is able to absorb energy from light, and then use that energy to catalyze the degradation of organic molecules and the oxidation of some inorganic pollutants like nitrogen oxides (NO_x) [1] - [12] . As the photocatalytic activity takes place only on the exposed surface area of titania, the amount of titania needed for the same photocatalytic efficiency can be reduced enormously by coating a thin layer of titania on silica [5] [13] - [42] . As the production of silica can be cheaper than that of titania, the costs of the photocatalyic material can then be significantly reduced. In addition to lower production costs, the durability increases as silica has a higher mechanical and thermal stability than titania. So when the composites are used instead of pure titania, the photocatalytic material can be used for a longer time with high photocatalytic efficiencies. In addition, because of the enhanced thermal stability, the photocatalytic material can be used in applications that require higher preparation temperatures. In addition to the lower costs and increased durability, the photocatalytic efficiency of the material can be increased with the addition of silica because silica can have a large specific surface area and is able to adsorb some pollutants and intermediates for a longer time than pure titania.

One promising application of the photocatalytic materials is the degradation of pollutants. The main reasons why using photocatalytic materials for air purification is promising include: lower costs of materials and energy needed than the other current purification methods, the ability of many photocatalytic materials to oxidize pollutants even if they are present in low concentrations, and the fact that the pollutants do not have to be stored but are converted into less harmful side-products (e.g. CO_2 from organic molecules and NO_3^- from NO_x after a complete photocatalytic oxidation (PCO)). The photocatalytic titania has been, and is being used in many other applications as well, including: photoelectrolysis of water, medical applications (where titania works as a disinfectant by destroying bacteria and viruses), municipal and industrial wastewater treatment, self-cleaning glass with anti-fogging abilities, and even in textiles that are self-cleaning.

A good method to have air purification is by the incorporation of photocatalytic material in building materials (including: concrete, wallpaper, gypsum and paint [2] -[4] [8] -[10] [43] -[51]), due to the large illuminated surfaces areas that many building materials have. Investigations into the photocatalytic building materials showed that the concentration of pollutants close to photocatalytic building materials indeed significantly decreased. Since large areas of building materials are often illuminated anyway with sun-light or indoor light and because these building materials become self-cleaning, the maintenance costs of these materials can be very low. Because of the large illuminated area and low maintenance cost, the potential of the photocatalytic building material for air purification is very promising.

However, in most of the research field of the applications of photocatalytic materials, only pure and doped titania are mentioned and not the titania-silica composites despite the large benefits these composites can have (e.g. lower costs, higher durability). An important reason for this absence of composites can be the complexity of the research field of the titania-silica composites. For the synthesis of the titania-silica composites alone, there are many different methods, each with their own parameters that can be changed in multiple ways. As many studies on the titania-silica composites have been done with different goals in mind, many different kinds of titania-silica composites have been produced [1] [5] [13] - [42] [52] - [84] , from which some are either not suitable for photocatalysis or have a very expensive production method. Since the photocatalytic activity of titania alone is already a complex system [3] [4] [6] [11] [85] -[89] , it can be understandable that adding more complexity to the system (for example, with silica) is not desirable. This review is written in order to provide insight on the low cost synthesis of titania-silica composites, and how each different parameter can be tuned to produce highly efficient photocatalytic material to show how the composites can be an attractive alternative to the titania for photocatalytic applications.

2. PHOTOCATALYTIC TITANIA

2.1. Mechanism of Photocatalysis

The process of photocatalysis in titania starts when a photon is absorbed by an electron in the valance band of titania [3] [4] [6] [11] [85] -[89] . This electron is then excited to the conduction band, and by doing so, leaves a hole behind in the valance band (reaction 1). The valance and conduction bands of titania have the right energy levels for many important redox reactions. After the excitation of electrons, holes in the valance band have a redox potential of +2.53 V, which is enough for the oxidation of hydroxyl ions into OH˙ (see reaction 2) or the oxidation of adsorbed organic molecules groups. The largest source of hydroxyl ions comes from the dissociation of water (see reaction 3). The redox potential of electrons in the conduction band is −0.52 V, which is strong enough to reduce oxygen to superoxide (see reaction 4). It is also possible that the excited electrons and holes will react with different adsorbed species depending on the environment. For example, if there is a high amount of adsorbed water, it is possible that more radical hydroxyls will form through the reaction of hydrogen peroxide as shown in reaction 5 and 6.

$$TiO_2 + light \rightarrow TiO_2 + e^- + h^+ \tag{1}$$

$$OH^- + h^+ \rightarrow OH^˙ \tag{2}$$

$$H_2O \rightleftarrows OH^- + H^+ \tag{3}$$

$$O_2 + e^- \rightarrow O_2^- \tag{4}$$

$$O_2 + 2H_2O + 2e^- \rightarrow 2H_2O_2 \tag{5}$$

$$H_2O_2 + e^- \rightarrow OH^- + OH^\bullet \tag{6}$$

where e^- is an excited electron in the conduction band, h^+ is a hole in the valance band, $OH^\bullet$ is a radical hydroxyl and O_2^- is a superoxide. Radical hydroxyls and superoxides are strong oxidants that can react with certain inorganic pollutants like NO_x and many organic molecules. In Figure 1, a schematic view is given for the photocatalytic mechanism.

An important property of titania, which influences the photocatalytic efficiency, is the amount of hydroxyl groups in its environment. In turn, the amount of hydroxyl groups is determined by the humidity in air, or the amount of water and its pH in liquids. This amount will determine how many hydroxyl groups will be chemically bonded to the surface of the titania. Bonded hydroxyl groups can either react with holes themselves and form radicals, or adsorb other hydroxyl groups and water molecules which can subsequently react with the holes and excited electrons [90] [91] . The photocatalytic activity in air can thus be higher with a higher humidity. However, it is also possible that a very high humidity will lower the photocatalytic activity by taking up more adsorption sides on the surface. For example, the photocatalytic oxidation of NO_x depends on the adsorption to titania and is thus lower with a very high humidity.

2.2. Oxygen Vacancies, Hydrophilicity and Self-Cleaning Surfaces

By reacting Ti^{4+} and O^{2-} into Ti^{3+} and O^-, excited electrons and holes can remain at the surface longer if there are no adsorbed species they can react with directly [3] . Because the difference in charge between titanium and oxygen atoms is then reduced, the oxygen atoms are much less stable and can, with relatively little energy, leave the crystal forming oxygen vacancies. These oxygen vacancies are important in titania for different mechanisms. For example, around an oxygen vacancy there is an excess of electrons, making titania a n-type semiconductor, which has a higher conductivity than when titania is an intrinsic semiconductor.

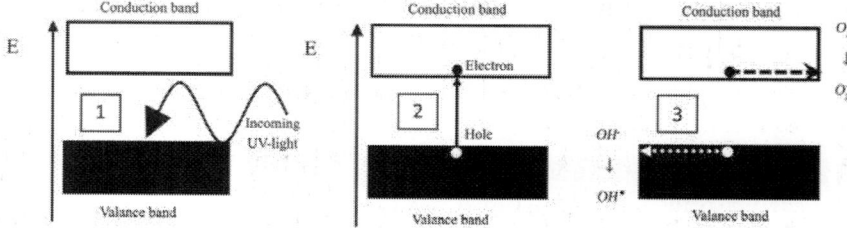

Figure 1. Schematic drawing of the photocatalytic activity of titania. 1: The absorption of a photon; 2: The excitation of an electron to the conduction band; 3: The transport of the electron and hole from the initial point to reach the surface of titania where the electron and hole can react with an adsorbed molecule.

Another reason why oxygen vacancies are important is because the surface of titania becomes more hydrophilic when water molecules occupy these oxygen vacancies. After a water molecule occupies the vacancy, one hydrogen atom of the water molecule can react with a neighboring oxygen atom forming two hydroxyl groups [3] . The increase in hydroxyl groups can lead to an increase in photocatalytic efficiency and to an increase in hydrophilicity of the surface. This increase in hydrophilicity, was first reported by Wang et al. [92] in 1997 with a titania coating on glass. By illuminating the coated glass with UV-light, the glass became transparent since the water fog that was present on the glass, defogged as the contact angle between the water droplets and the glass decreased to zero. They also showed that after keeping the hydrophilic surface away from any light source for some days, the glass became more hydrophobic, which means that the formation of a hydrophilic surface is a reversible process. However, it has been reported that oxygen vacancies are not solely responsible for the hydrophilicity, as some studies showed that hydrophilicity was independent in some cases on the number of oxygen vacancies [93] [94] . While it is possible that the degradation of organic materials on the surface can also play a role on the hydrophilicity increase, it nevertheless has been shown not to be a determining factor [95] .

The hydrophilicity and the degradation of organic materials on the surface of titania are two reasons why titania can be used for self-cleaning applications [3] . The degradation of organic materials through photocatalytic oxidation prevents organic substances to accumulate on the surface and can prevent the growth of bacteria and fungi. The hydrophilicity of the surface increases the water adsorption so that it can replace other adsorbed species and it lowers the energy required for water to slide over the surface so that contaminants can be washed off more easily.

2.3. Effect of Different Crystal Forms of Titania on the Photocatalytic Efficiency

Titania has several forms, but the two main crystal structures that most researchers focus on are rutile and anatase [3] . These two crystal forms are both tetragonal structures in which titanium atoms are 6-coordinated in an octahedral formation. The band gap of rutile is 3.0 eV and the band gap of anatase is 3.2 eV. For rutile and anatase to become photocatalytically active, they need to absorb electromagnetic radiation with wavelengths smaller than 413 nm and 387 nm respectively. While rutile is thus able to absorb more light in the visible range, anatase is more photocatalytically active. Luttrell et al. [96] showed this higher photocatalytic efficiency by studying the difference in PCO efficiencies of anatase and rutile thin films of different sizes for the PCO of methyl orange. They showed that for films thinner than 2.5 nm the difference was not significant between the two forms, but for thicker films, the anatase thin film had a higher efficiency. They measured that the maximum thickness, where the photocatalytic efficiency increases with increasing size, is 2.5 nm for rutile and 5 nm for anatase. Thus, from this study it can be concluded that excited electrons and holes in anatase can travel farther than in rutile so that more electrons and holes can reach the surface. The ability of exited electrons and holes to travel longer distances in anatase was contributed to a longer lifetime and higher conductivity [96] - [98].

An important reason why excited electrons and holes have a longer lifetime and higher conductivity in anatase is because of the differences between the oxygen vacancies that form in anatase and rutile. Oxygen vacancies cause extra energy levels within the band gap. Calculations done by Mattioli et al. [97] showed that oxygen vacancies in an anatase crystal can cause both shallow delocalized energy levels and deep localized energy levels in the band gap, while in rutile only deep localized levels can form. Since anatase has also shallow delocalized energy levels in its band gap, it has a higher conductivity and the excited electrons and holes have longer lifetimes than in rutile as they are less trapped in the deep localized energy levels where the chance of recombination is higher.

While anatase has a high photocatalytic efficiency, amorphous titania has the lowest efficiency [99]. The main reason for this lower efficiency is because, in amorphous titania, there are many spots where recombination of the electron-hole pair can happen. The recombination through defects is the most common way electrons and holes are lost. Thus, amorphous titania has a much higher recombination capacity. In addition, conductivity in amorphous materials is very low since energy levels in amorphous materials are much more localized. The high recombination rate and low conductivity means that only electrons which are excited directly at the surface play a part in the photocatalytic activity in amorphous titania.

Some researchers have measured higher PCO efficiencies in titania that contains both rutile and anatase than in titania with only anatase. Degussa P25 nanoparticles, which are commercial titania nanoparticles made out of around 80% anatase and 20% rutile, are well known for their high photocatalytic activity [100] and are often used as a reference material. The conduction band of rutile has been measured to start at a higher energy level than that of anatase even though its band gap is smaller, which is why titania with both crystal forms can have a higher photocatalytic efficiency [101]. Since electrons always go to a lower energy state if possible, excited electrons in rutile will go to the conduction band of anatase. As electron holes can be viewed as opposite electrons, electron holes in anatase will flow to the valance band of rutile because its top lies above the energy level of the valance band from anatase. Because holes move from anatase to rutile and excited electrons from rutile to anatase, the recombination chance is reduced and a difference in electron density is produced at the interface between the two forms, causing an increase in conductivity and lifetimes for the electrons and holes, resulting in a higher photocatalytic efficiency.

3. TITANIA-SILICA COMPOSITE SYNTHESIS

3.1. Silica Sources

Many different types of silica from different sources can be applied as a support for titania including: fumed silica, precipitation silica from alkali silicates, silica produced with the Stöber method, zeolites, clays, glass, silica from the dissolution of silica minerals and more [1] [5] [13] -[42] [52] -[84] [102] - [107] . Fumed silica is formed at high temperatures where silica compounds, like chlorosilanes, are transformed into silica [108] [109] . Silica from alkali silicates is formed by neutralization of the alkali solutions so that the silicate polymerizes to silica and precipitates from the solution. In both fumed silica and precipitated silica, amorphous aggregates are formed. These aggregates can have very large specific surface areas but also have complex undefined structures. Another silica which is often used because of the more defined structure, is silica made with a sol-gel method. The Stöber method [110] is the best known example of a sol-gel method for producing silica. During this method, Tetraethyl orthosilicate (TEOS) is slowly added to a solution of ethanol, water and ammonia. Depending on the composition of the solution, silica colloids of varies sizes and shapes can be formed. The advantage of this silica is that the resulting shape and size of the silica can be well controlled. However, the disadvantage is that this silica has a smaller specific surface area than fumed and precipitated silica. For both well-defined shapes and high specific surface areas, researchers have also used zeolites as support. However compared to the other mentioned supports, the zeolites are more expensive. For very low production costs and a high specific surface area, silica made during the dissolution of olivine has a great potential [111] - [114] but is still in its developing stage.

3.2. Titania-Silica Chemistry

The reaction of titania precursors with silica happens either directly with silanols or indirectly through hydrolysis into titania monomers ($Ti(OH)_4$) first and subsequently by condensation with silanols [1] [5] [13] -[42] [52] - [84] [109] [115] . Either way, the titania will form bonds with the silica through reaction 7.

$$\equiv Si - OH + R - Ti \equiv \rightarrow \equiv Si - O - Ti \equiv + HR \tag{7}$$

where R is a side group of a titania precursor or a hydroxyl group of a titania monomer. The Si-O-Ti bond can be measured by using techniques like infra-red/Raman spectrometry and X-ray photoelectron spectroscopy [14] - [16] [32] - [34] [52] [56] - [60] . This condensation reaction between the titania precursor and the silica surface depends mostly on the hydroxyl groups of the silica [16] [56] [59] [63] [66] [67] since the rest of the silica is very inert. In turn, the amount of hydroxyl groups on the silica is dependent on the temperature during the pretreatment, the method used, and the amount of water and its pH used [67] [109] . For example, if the silica undergoes pre-heating temperatures higher than

800°C and no water is used during the synthesis, there will be only a low amount of hydroxyl groups on the silica surface left so that only a few titanium atoms can be found on the silica after the reaction [67] . On the other hand, if lower temperatures are used, the density of hydroxyl groups will be high enough on the silica surface that hydrogen bonds between silanols can form. Titania precursors react more with these hydrogen bonded silanols than isolated silanols [66] [67] . These silanols are close enough to each other that a titania precursor can react with multiple hydroxyl groups, making the reaction of hydrogen bonded silanols favorable over the reaction of isolated silanol.

When water is used during the synthesis method of the titania-silica composites, the titania precursor undergoes hydrolysis first. During the hydrolysis, the side groups of the precursor are replaced by hydroxyl groups [116] - [119] . After a full hydrolysis at a neutral pH, $Ti(OH)_4$ is the most common product, as titanium has four valance electrons. Below a pH of 4, the ions $Ti(OH)_3^+$ and $Ti(OH)_2^{2+}$ can also be formed [116] [117] . It is also possible that, a double bonded oxygen atom which stays bonded during the hydrolysis, forms during the reaction of the precursor so that only two hydroxyl groups can bond to the titanium. Titanium hydroxides are titania monomers that can form larger titania molecules by polymerization through condensation with other monomers if their concentration is high enough.

3.3. Different Synthesis Methods

There are many different methods to synthesize the titania-silica composites. An indirect way to prepare them is by adding premade titania nanoparticles to a silica support [103] [120] [121] at a pH of around 3 - 4. At that pH, the titania and silica have opposite charges so that the titania and silica will have electrical attraction. However, for more stability and a better homogenous coating, direct methods are often more favorable. The vapor-deposi- tion methods (chemical vapor deposition (CVD) and physical vapor deposition (PVD)), for example, are such methods [5] [23] - [25] [66] - [70] . During the CVD method, the titania precursor is heated to the gas phase to react with dry silica in an inert environment and during the PVD method the titania is sputtered against a support surface for thin films. The impregnation [17] - [22] [52] [59] - [63] and the grafting [52] [64] [65] methods are also direct methods. During both these methods, the titania precursor is dissolved in an organic solvent like toluene or hexane. This solvent is then added to the silica support so that the precursor reacts with the silanols. During the grafting method, the solvent is removed through evaporation and during the impregnation method, the solvent is removed in some other way (e.g. filtration). During the vapor-deposition, the impregnation and the grafting methods, no water is used, which means that these methods do not have the option to form more than one layer of titania in one step, because no new hydroxyl groups can form on the coated titania during the reactions for further condensation. In addition, these methods are not optimal for low cost production since either very high temperatures or expensive organic solvents are required.

Methods that are more promising for low cost photocatalytic materials are the precipitation methods [13] - [16] [52] - [58] and the sol-gel methods [14]

[32] - [42] [72] - [84] . These methods are capable of forming more than one monolayer titania on silica, and do not require expensive solvents. During the precipitation method, the titania precursor is dissolved in an aqueous solution with a low pH and low temperatures, where titania does not form. After mixing the aqueous solution containing the precursor with the silica, the solution is either neutralized with an alkaline solution and/or heated up to a specific temperature. This specific temperature depends on the pH and solvent used. By increasing the pH and/or temperature, titania slowly forms, which can be on a silica support for a coating if the hydrolysis is slow enough so that the concentration of titania monomers does not reach the critical supersaturation. Titanium chlorides ($TiCl_3$, $TiCl_4$) and titanium oxysulfate ($TiOSO_4$) are the precursors, which are often used in the precipitation methods. During the sol-gel methods, titanium alkoxides (e.g. titanium isopropoxide, titanium n-butoxide) are often used. To form a titania coating, the precursor is slowly added to a silica dispersion in an organic solvent (ethanol, n-propanol) which contains a low amount of water or to which a low amount of water is added after the precursor is added.

An important parameter in the methods that involves hydrolysis is the pH. Below pH 6, part of the $Ti(OH)_4$ is replaced by $Ti(OH)_3^+$, and also by $Ti(OH)_2^{2+}$ below pH 4. With decreasing pH, more $Ti(OH)_4$ is replaced by the ions which lead to a higher solubility [116] . A higher solubility means that the equilibrium between monomers and condensed titania is then more to the side of the monomers. Thus, when a low pH is used, more precursor is needed for the same amount of the condensated titania, as some of the titania monomers stay dissolved [116] . Since it is mostly the removal of OH^- groups that lead to the formation of ions, hydrated, amorphous and small sized titania particles are more dissolvable than crystalline titania and titania bonded to larger particles like the silica [122] [123] . The peptizing method, which is a different kind of sol-gel method, uses this constant equilibrium between titania monomers and condensated titania in an aqueous solution and the difference in dissolvability. During this method, hydrated precipitates are first formed in an aqueous solution and then slowly dissolved by reducing the pH to around 2 - 4. Using the Ostwald ripening process, crystalline titania nanoparticles are then formed [122] [123] or coated on a silica support [15] .

Another way to use the sol-gel method for low cost photocatalytic material is by coating a support, like a glass plate, with a thin film using the dip-coating method [124] - [133] . During the dip-coating method, the support is dipped into a stable sol-gel mixture, and is then slowly pulled out of the mixture so that a thin layer of the mixture is adsorbed to the surface. During the drying, a thin titania film is then formed. Polymers (e.g. Poly (ethylene glycol)) can be used to obtain a higher porosity. By adding these large molecules in the sol-gel mixture, large pores are formed during the calcination step, when these molecules are removed.

Another method which is often used for the synthesis of titania-silica composites is the hydrothermal treatment [26] - [31] [71] [72] . The advantage of this method is that it can be used for both the coating step and the crystallization step (which will be discussed in 3.5). This method is done by adding a precursor, a solution containing some water and the silica (or silica source) to an autoclave. The solution is then heated up (e.g. to 200°C) for both the reaction and crystallization step.

3.4. Controlling the Hydrolysis Rate

For a homogenous coating with the seeded-growth process, the concentration of titanium monomers should not exceed the critical supersaturation, and thus the hydrolysis rate needs to be controlled. In aqueous solutions with a neutral pH, the hydrolysis of titania precursors happens so fast that the concentration of the monomers reaches the critical supersaturation point almost instantly, causing the titania to precipitate randomly in the solution instead of slowly forming on the silica surface.

The most important parameters on which the hydrolysis rate of titania precursors is dependent are: the pH, temperature, concentration of the precursor and of water, and type of precursor used. For example, in an aqueous solution with a pH below 1 and a temperature below 20°C, no titania will form [13] - [16] [52] - [58] . Having organic liquids (like n-propanol) [134] in the solvent increases the temperature and pH at which the titania is still soluble, because the dielectric constant of the solvent is then decreased. Having a low water content also prevents fast hydrolysis even when no acid is used [34] [81] [134] . However, as each hydrolysis-condensation reaction consumes a water molecule, enough water should be present, to add new hydroxyl groups on the surface of the forming silica-titania composites. Another way to slow down hydrolysis is by reacting the precursors first with molecules, like glycols, which are larger than the side-groups of the precursor. These molecules can replace the side-groups of the precursors [124] [135] if added in excess, so that new, less reactive titania precursors are formed. Depending on the exact method, another important variable is the speed at which a parameter is changed, for example, the change of pH during the neutralization method, the addition speed of a precursor during a sol-gel method and the speed at which the temperature increases during a hydrothermal treatment.

3.5. Transformation to Crystalline Titania

When the hydrolysis rate is very slow during the reaction, thermodynamics plays a more important role than kinetics. Since crystalline titania is more energetically favorable than amorphous titania, crystallization of the titania can then directly happen, especially at a low pH, where the solubility difference between amorphous titania and crystalline titania is larger [122] [134] [136] - [143] . However, the direct formation of crystalline titania is hard to control. If the hydrolysis is too slow, it can result in large rutile crystals with a low specific surface area, which is undesirable for the photocatalysis. In any other case, it is likely that most of the titania is amorphous titania after the reaction. Because amorphous titania has a much lower photocatalytic activity [96] [99] , it can be beneficial to either use calcination or hydrothermal treatment to transform it into anatase.

During the calcination of pure titania, the transformation of amorphous titania to anatase happens at a temperature of about 400°C and at temperatures above 600°C the transformation to rutile occurs [143] . At these high temperatures, chemically bonded hydroxyl groups condensate with each other so that more bonds are formed between the titanium and oxygen atoms. Through rearrangements, the crystal structures are then slowly formed. Once a crystal is large enough to be stable, it will further increase in size by taking up more titania atoms, either through more rearrangements, or by merging with other crystals.

Besides the calcination in dry air, it is also possible to use hydrothermal treatment for the formation of crystalline titania [25] - [31] [71] [72] [145] - [149] . Since the formation of crystalline titania takes place in an aqueous environment during a hydrothermal treatment hydroxyl groups are incorporated into the formed structure which can be helpful for the photocatalytic activity. Hydrothermal treatment works at lower temperatures than calcination because the water increases the mobility of the atoms, reduces surface tension of the titania and catalyzes nucleation of crystals [145] - [149] . Wang and Ying [147] showed that using a hydrothermal treatment on amorphous titania, smaller and more stable titania nanoparticles were produced than with calcination.

The exact temperature at which the transformation to either anatase or rutile happens during both calcination and hydrothermal treatment depends on the size of the particles (according to Banfield et al. [144] below a size of 14 nm, anatase is more thermodynamically stable than rutile), the pH and other chemicals (e.g. adsorbed polymers, salts) that can influence the mobility of the atoms [117] [134] [136] [140] [142] . The formation of anatase or rutile from amorphous titania does not start at a single point where all amorphous material crystallizes into anatase or into rutile. By using higher temperatures, more amorphous titania will transform into anatase. However, higher temperatures will also transform anatase into rutile and increase the growth rate of the crystals, which leads to a smaller specific surface area [72] [134] [136] - [143] .

When titania is chemically bonded to a substrate like silica, the substrate stabilizes the different structures of titania, and suppresses the transformation of amorphous titania to anatase and the transformation of anatase to rutile by decreasing the mobility of the titania atoms like an anchor [32] [33] [54] [72] [76] [80] [129] . Thus, higher temperatures are required to form anatase and rutile when titania is coated on silica. While more energy is needed for the formation of anatase from amorphous titania on a support, the anatase that is then formed has a higher thermal stability. It has even been reported that the anatase-rutile transformation only happens in some composites with a high temperature of 1000°C [54] [129] . The increase in temperature required for the crystalline transformations depends on the thickness of the titania, since a thicker layer is less influenced by the support [54] . For the titania-silica composites, the crystal growth by calcination can cause shrinkage stress when the titania structure shrinks due to the density increase and removal of chemically and physically adsorbed water. As the silica works like an anchor against the shrinkage, stress is produced on the structure which can lead to the breakage of some Ti-O-Si bonds [150] .

4. THE INFLUENCE OF SILICA ON PHOTOCATALYTIC TITANIA IN LOW TITANIA CONTENT COMPOSITES

Titania-silica composites have more different properties than pure titania than simply a higher stability and a higher specific surface area, especially when the titania content is very low. Many researchers have studied the low titania composites because of these different properties. Anpo et al. [5] were one of the first who studied them. Using the CVD method on a porous silica glass, they found some interesting results which include: 1) below three layers, no anatase could be measured with X-ray diffraction, while it could still be present; 2) the band-gap became larger (4.1 eV) for just a monolayer titania; 3) the titanium was 4-coordinated in a tetrahedral structure instead of the 6-coordinated octahedral structure in pure anatase or rutile; 4) the tetrahedral titania with a large band gap catalyzed different reactions like the decomposition of N_2O as will be explained in Section 4.1; and 5) the photocatalytic efficiency per amount of catalyst was much higher for low titania content composites, which will be explained in Section 4.3.

4.1. The Larger Band Gap and Its Influence on the Photocatalytic Activity

The band gap of titania increases when going from bulk anatase to the tetrahedral titania. The normal band gap for crystalline titania is around 3.0 - 3.2 eV, but the measureable band gap from a very low amount of titania on the surface of silica can be much larger [5] [21] [68] [69] [151] [152] . There are two effects responsible for this increase. The first is the quantum size effect, which increases the band gap with decreasing crystal size, when the size is below 2 nm [68] . The second effect is caused by the difference in energy levels of the energy bands from silica and the energy bands from titania, close to the titania-silica interface [1] [21] [68] . Band gaps up to 4.1 eV [5] [69] could be measured due to these two effects. When the band gap becomes larger, electrons require more energy to be excited to the conduction band. For the applications that use sun-light or normal indoor light as the light source, this larger band gap is a disadvantage, since even less of the light spectra can then be absorbed.

On the other hand, the energy that is absorbed is used more efficiently because of the larger band gap. A larger band gap lowers the chance for recombination and increases the redox potentials of the excited electrons and holes. This higher redox potential increases the efficiency of the formation of the radical hydroxyl and superoxides molecules and enables the titania to catalyze different reactions [5] [17] - [20] . For example, Yamashita et al. [18] measured that pure titania transformed NO mostly into oxidized species, while NO decomposed to N_2 and O_2 in the presence of composites prepared with an ion-exchange method, in which titanium ions replaced silicon ions. In the same system [19] and similar systems with other zeolites [20] , the same observation was made with the reaction of CO_2 and H_2O. With the ion-exchange composites, methanol was mostly produced while methane was produced by the titania samples. Another example of the difference in catalytic reactions taking place is from a study by Gao et al. [17] who observed that in the presence of tetrahedral titania, methanol reacted to methyl formate ($C_2H_4O_2$) and formaldehyde (CH_2O) while in the presence of octahedral titania, methanol reacted to dimethyl ether (C_2H_6O). While photocatalytic titania has some potential in reducing the amount of greenhouse gasses in air [153], these reactions show that the composites have an even greater potential to be useful against climate change.

4.2. Higher Density of Hydroxyl Groups

Binary metal oxides often have better catalytic properties than single metal oxides because they have extra acid sites on their surface in the form of hydroxyl groups [1] [154]. The titania-silica composites are one of those binary systems, and an increase in acid sites has been measured in several different studies [1] [18] [37] [56] [74] [80] [83]. The increase in hydroxyl groups is important for the photocatalytic activity and hydrophilicity as these depend on the amount of hydroxyl groups on the surface. Tanabe et al. [154] made the hypothesis that this increase in hydroxyl groups is caused by the difference in coordination numbers. The coordination number for silicon atoms in silica is 4 and for titanium atoms in crystalline titania it is 6. So when titanium atoms are introduced in, or on silica in low amounts and form the tetrahedral structure, an excess of negative -2 charge per titanium atom is created. This excess charge causes Brönsted acidity on the surface after absorbing enough protons to compensate the charge. Walter et al. [82] showed, using neutron diffraction, that the number of hydroxyl groups is indeed affected when titanium atoms are introduced into the silica structure by the difference in coordination number. They showed an increase in hydroxyl groups mainly caused by the increase in strain in the structure. Liu et al. [74] and Doolin et al. [80] both used the sol-gel method to make titania-rich and titania-poor composites and compared them to pure titania and silica. They measured indeed an increase in Brönsted acidity in the composites especially where there were Ti-O-Si bonds, which was in agreement with Tanabe. However, for the titania-poor composites, the increase in acidity was lower than for titania-rich composites, which is in disagreement with the model of Tanabe. So far, no model has been proposed yet, that explains the extra hydroxyl groups better than the model of Tanebe et al., but these studies about the extra hydroxyl groups do show that the mechanism is related with the Ti-O-Si bond [1].

4.3. Higher Photocatalytic Efficiency of Low Titania Content Composites

Other researchers [17] [35] [62] [67] [81] [83] found similar results as Anpo et al. [5] on different low titania content composites and these researchers often observed an increase in photocatalytic efficiency per amount of catalyst compared to pure titania. The high efficiencies of these low titania content composites, which do not even have enough titania for a full monolayer, are caused by: 1) the high specific surface area of the silica supports used; 2) the ability of the silica to adsorb many molecules for longer times than titania, especially with the extra hydroxyl groups; 3) the fact that the titania is used more efficiently since all the titania is at the surface; 4) the higher redox potentials of the electrons and holes; and 5) the fact that silica can scatter the light to the titania without being able to absorb its energy. In addition, during the photocatalytic measurements in these studies, UV-light was used. The measurements using UV-light might not represent the applications which use sun-light as the light source, since the decrease of possible light absorption caused by the increase in band gap is less with UV-light.

5. CONCLUSIONS

The titania-silica composites are interesting materials because they have the potential to make photocatalytic materials more cost-effective. For the same level of photocatalytic activity, fewer resources have to be invested with the titania-silica composites than with pure titania. The titania-silica composites can, with less and cheaper material, have the same photocatalytic efficiency as pure titania for a longer time since the composites can have a higher photocatalytic efficiency, lower production costs and increased durability. The applications of the photocatalytic material including the applications that degrade pollutants, become then more attractive for companies to produce on a larger scale which can eventually lead to an overall improvement of the quality of air and water.

To obtain this cost-effective photocatalytic material, the titania-silica composites need to be synthesized with a method that has low production costs but still produces composites which have a high photocatalytic efficiency.

- For a high efficiency, the method needs to deposit an anatase layer with thickness of maximum 5 nm on a large specific surface area. Any layer larger than 5 nm will have titania, which does not contribute to the photocatalysis, since it is too far away from the surface.
- If some of the crystal structure is rutile instead of anatase, it can have some increase in photocatalytic efficiency because of the separation of holes and electrons at the interface of the two forms. However, the amount of rutile should not be too high, as the lifetime and conductivity of excited electrons and holes in rutile are less favorable than in anatase.
- When the crystal size is too small, the titania may have an increase in band gap. While it has been reported that such an increase in band gap can cause higher photocatalytic efficiencies, it is important to note that it will make the titania absorb less visible light.
- The titania should be chemically bonded to a silica substrate which has a large specific surface area, high mechanical and thermal stability as well as low production costs.
- The most promising methods for low cost photocatalytic composites are the ones that involve hydrolysis (precipitation and sol-gel methods), as these methods ensure that more than one layer of titania can form without the need of expensive materials. However, the hydrolysis of titania precursors can be hard to control. The most important parameters on which the

hydrolysis rate is dependent are the pH, temperature, concentration of water and precursor, the speed at which these parameters are changed during the reaction (e.g. by addition of water) and the type of precursor used. How much influence each parameter has on the hydrolysis rate depends on the method used. It is important that the reaction speed of the hydrolysis should be slow enough so that the condensation of titania monomers on the substrate's surface is more likely to happen than polymerization between monomers.

- For the transformation of amorphous titania to anatase, calcination or hydrothermal treatment can be applied. The temperature and time required to obtain anatase crystals from amorphous titania depend on the mobility of the titania molecules, which can be influenced by: the chemical bonds to the silica, any nucleated crystals already present, and other chemicals present (e.g. adsorbed polymers, salts). While having more crystalline anatase is beneficial for the photocatalytic activity, crystallization does not always produce materials with a higher photocatalytic efficiency, since during the growth of the crystals, the specific surface area is reduced and anatase can transform into rutile at high temperatures.

When all these points are fulfilled, the resulting titania-silica composites will have the required properties to be a cost-effective material which can compete with pure titania in photocatalytic applications. Even with the increased complexity, the composites are an excellent alternative to pure titania nanoparticles.

REFERENCES

1. Gao, X. and Wachs, I.E. (1999) Titania-Silica as Catalysts: Molecular Structural Characteristics and Physico-Chemical Properties. Catalysis Today, 51, 233-254.
2. Hüsken, G., Hunger, M. and Brouwers, H.J.H. (2009) Experimental Study of Photocatalytic Concrete Products for Air Purification. Building and environment, 44, 2463-2474.
3. Fujishima, A., Zhang, X. and Tryk, D.A. (2008) TiO2 Photocatalysis and Related Surface Phenomena. Surface Science Reports, 63, 515-582.
4. Hashimoto, K., Irie, H. and Fujishima, A. (2005). TiO2 Photocatalysis: A Historical Overview and Future Prospects. Japanese Journal of Applied Physics, 44, 8269.
5. Anpo, M., Aikawa, N., Kubokawa, Y., Che, M., Louis, C. and Giamello, E. (1985) Photoluminescence and Photocatalytic Activity of Highly Dispersed Titanium Oxide Anchored onto Porous Vycor Glass. The Journal of Physical Chemistry, 89, 5017-5021.

6. Fujishima, A., and Zhang, X. (2006) Titanium Dioxide Photocatalysis: Present Situation and Future Approaches. Comptes Rendus Chimie, 9, 750-760.

7. Turchi, C.S. and Ollis, D.F. (1990) Photocatalytic Degradation of Organic Water Contaminants: Mechanisms Involving Hydroxyl Radical Attack. Journal of catalysis, 122, 178-192.

8. Ballari, M.M. and Brouwers, H.J.H. (2013) Full Scale Demonstration of Air-Purifying Pavement. Journal of hazardous materials, 254, 406-414.

9. Hüsken, G., Hunger, M., and Brouwers, H.J. (2007) Comparative Study on Cementitious Products Containing Titanium Dioxide as Photo-Catalyst. Proceedings of the International RILEM Symposium on Photocatalysis, Environment and Construction Materials—TDP, Florence, 8-9 October 2007, 147-154.

10. Hunger, M., Hüsken, G. and Brouwers, H.J.H. (2010) Photocatalytic Degradation of Air Pollutants—From Modeling to Large Scale Application. Cement and Concrete Research, 40, 313-320.

11. Linsebigler, A.L., Lu, G. and Yates Jr., J.T. (1995) Photocatalysis on TiO2 Surfaces: Principles, Mechanisms, and Selected Results. Chemical Reviews, 95, 735-758.

12. Anpo, M., Aikawa, N., Kodama, S. and Kubokawa, Y. (1984) Photocatalytic Hydrogenation of Alkynes and Alkenes with Water over Titanium Dioxide. Hydrogenation Accompanied by Bond Fission. The Journal of Physical Chemistry, 88, 2569-2572.

13. Sirikawinkobkul, N., Kalambaheti, C., Jiemsirilers, S., Kashima, D.P. and Jinawath, S. (2009) Synthesis, Characterization and Photocatalytic Activity of Visible-Light Titania/Silica Photocatalyst. 18th International Conference on Composite Materials, Edinburgh, 27-31 July 2009.

14. Montes, M., Getton, F.P., Vong, M.S.W. and Sermon, P.A. (1997) Titania on Silica. A Comparison of Sol-Gel Routes and Traditional Methods. Journal of Sol-Gel Science and Technology, 8, 131-137.

15. Huang, C.H., Bai, H., Liu, S.L., Huang, Y.L. and Tseng, Y.H. (2011) Synthesis of Neutral SiO2/TiO2 Hydrosol and Its Photocatalytic Degradation of Nitric Oxide Gas. Micro & Nano Letters, 6, 646-649.

16. Ding, Z., Lu, G.Q. and Greenfield, P.F. (2000) A Kinetic Study on Photocatalytic Oxidation of Phenol in Water by Silica-Dispersed Titania Nanoparticles. Journal of Colloid and Interface Science, 232, 1-9.

17. Gao, X., Bare, S.R., Fierro, J.L.G., Banares, M.A. and Wachs, I.E. (1998) Preparation and In-Situ Spectroscopic Characterization of Molecularly Dispersed Titanium Oxide on Silica. The Journal of Physical Chemistry B, 102, 5653-5666.

18. Yamashita, H., Ichihashi, Y., Anpo, M., Hashimoto, M., Louis, C. and Che, M. (1996) Photocatalytic Decomposition of NO at 275 K on Titanium Oxides Included within Y-Zeolite Cavities: The Structure and Role of the Active Sites. The Journal of Physical Chemistry, 100, 16041-16044.

19. Anpo, M., Yamashita, H., Ichihashi, Y., Fujii, Y. and Honda, M. (1997) Photocatalytic Reduction of CO2 with H2O on Titanium Oxides Anchored within Micropores of Zeolites: Effects of the Structure of the Active Sites and the Addition of Pt. The Journal of Physical Chemistry B, 101, 2632-2636.

20. Anpo, M., Yamashita, H., Ikeue, K., Fujii, Y., Zhang, S.G., Ichihashi, Y. and Tatsumi, T. (1998) Photocatalytic Reduction of CO2 with H2O on Ti-MCM-41 and Ti-MCM-48 Mesoporous Zeolite Catalysts. Catalysis Today, 44, 327-332.

21. Fernández, A., Caballero, A. and González-Elipe, A.R. (1992) Size and Support Effects in the Photoelectron Spectra of Small TiO2 Particles. Surface and Interface Analysis, 18, 392-396.

22. Anpo, M., Yamashita, H., Ichihashi, Y. and Ehara, S. (1995) Photocatalytic Reduction of CO2 with H2O on Various Titanium Oxide Catalysts. Journal of Electroanalytical Chemistry, 396, 21-26.

23. Ding, Z., Hu, X., Lu, G.Q., Yue, P.L. and Greenfield, P.F. (2000) Novel Silica Gel Supported TiO2 Photocatalyst Synthesized by CVD Method. Langmuir, 16, 6216-6222.

24. Yamashita, H., Ichihashi, Y., Harada, M., Stewart, G., Fox, M.A. and Anpo, M. (1996) Photocatalytic Degradation of 1-Octanol on Anchored Titanium Oxide and on TiO2 Powder Catalysts. Journal of Catalysis, 158, 97-101.

25. Sayilkan, F., Asilturk, M., Sener, S., Erdemoglu, S., Erdemoglu, M. and Sayilkan, H. (2007) Hydrothermal Synthesis, Characterization and Photocatalytic Activity of Nanosized TiO2 Based Catalysts for Rhodamine B Degradation. Turkish Journal of Chemistry, 31, 211-221.

26. Chuan, X.Y., Hirano, M. and Inagaki, M. (2004) Preparation and Photocatalytic Performance of Anatase-Mounted Natural Porous Silica, Pumice, by Hydrolysis under Hydrothermal Conditions. Applied Catalysis B: Environmental, 51, 255-260.

27. Hirano, M. and Ota, K. (2004) Preparation of Photoactive Anatase-Type TiO2/Silica Gel by Direct Loading Anatase-Type TiO2 Nanoparticles in Acidic Aqueous Solutions by Thermal Hydrolysis. Journal of Materials Science, 39, 1841-1844.

28. Hirano, M. and Ota, K. (2004) Direct Formation and Photocatalytic Performance of Anatase (TiO2)/Silica (SiO2) Composite Nanoparticles. Journal of the American Ceramic Society, 87, 1567-1570.

29. Kim, E.Y., Whang, C.M., Lee, W.I. and Kim, Y.H. (2006) Photocatalytic Property of SiO2/TiO2 Nanoparticles Prepared by Sol-Hydrothermal Process. Journal of Electroceramics, 17, 899-902.

30. Fu, X., Clark, L.A., Yang, Q. and Anderson, M.A. (1996) Enhanced Photocatalytic Performance of Titania-Based Binary Metal Oxides: TiO2/SiO2 and TiO2/ZrO2. Environmental Science & Technology, 30, 647-653.

31. Anderson, C. and Bard, A.J. (1995) An Improved Photocatalyst of TiO2/SiO2 Prepared by a Sol-Gel Synthesis. The Journal of Physical Chemistry, 99, 9882-9885.

32. Cheng, P., Zheng, M.P., Huang, Q., Jin, Y.P. and Gu, M.Y. (2003) Enhanced Photoactivity of Silica-Titania Binary Oxides Prepared by Sol-Gel Method. Journal of Materials Science Letters, 22, 1165-1168.

33. Smitha, V.S., Manjumol, K.A., Baiju, K.V., Ghosh, S., Perumal, P. and Warrier, K.G.K. (2010) Sol-Gel Route to Synthesize Titania-Silica Nano Precursors for Photoactive Particulates and Coatings. Journal of Sol-Gel Science and Technology, 54, 203-211.

34. Guo, X.C. and Dong, P. (1999) Multistep Coating of Thick Titania Layers on Monodisperse Silica Nanospheres. Langmuir, 15, 5535-5540.

35. Kamaruddin, S. and Stephan, D. (2014) Sol-Gel Mediated Coating and Characterization of Photocatalytic Sand and Fumed Silica for Environmental Remediation. Water, Air, & Soil Pollution, 225, 1948.

36. Shan, A.Y., Ghazi, T.I.M. and Rashid, S.A. (2010) Immobilisation of Titanium Dioxide onto Supporting Materials in Heterogeneous Photocatalysis: A Review. Applied Catalysis A: General, 389, 1-8.

37. Guan, K. (2005) Relationship between Photocatalytic Activity, Hydrophilicity and Self-Cleaning Effect of TiO2/SiO2 Films. Surface and Coatings Technology, 191, 155-160.

38. Jung, K.Y. and Park, S.B. (2000) Enhanced Photoactivity of Silica-Embedded Titania Particles Prepared by Sol-Gel Process for the Decomposition of Trichloroethylene. Applied Catalysis B: Environmental, 25, 249-256.

39. Ismail, A.A., Ibrahim, I.A., Ahmed, M.S., Mohamed, R.M. and El-Shall, H. (2004) Sol-Gel Synthesis of Titania-Silica Photocatalyst for Cyanide Photodegradation. Journal of Photochemistry and Photobiology A: Chemistry, 163, 445-451.

40. Xie, C., Xu, Z., Yang, Q., Xue, B., Du, Y. and Zhang, J. (2004) Enhanced Photocatalytic Activity of Titania-Silica Mixed Oxide Prepared via Basic Hydrolyzation. Materials Science and Engineering: B, 112, 34-41.

41. Zhang, X., Zhang, F. and Chan, K.Y. (2005) Synthesis of Titania-Silica Mixed Oxide Mesoporous Materials, Characterization and Photocatalytic Properties. Applied Catalysis A: General, 284, 193-198.
42. Yang, J., Zhang, J., Zhu, L., Chen, S., Zhang, Y., Tang, Y. and Li, Y. (2006) Synthesis of Nano Titania Particles Embedded in Mesoporous SBA-15: Characterization and Photocatalytic Activity. Journal of Hazardous Materials, 137, 952-958.
43. Maggos, T., Plassais, A., Bartzis, J.G., Vasilakos, C., Moussiopoulos, N. and Bonafous, L. (2008) Photocatalytic Degradation of NOx in a Pilot Street Canyon Configuration Using TiO2-Mortar Panels. Environmental Monitoring and Assessment, 136, 35-44.
44. Strini, A., Cassese, S. and Schiavi, L. (2005) Measurement of Benzene, Toluene, Ethylbenzene and o-Xylene Gas Phase Photodegradation by Titanium Dioxide Dispersed in Cementitious Materials Using a Mixed Flow Reactor. Applied Catalysis B: Environmental, 61, 90-97.
45. &Acircngelo, J., Andrade, L. and Mendes, A. (2014) Highly Active Photocatalytic Paint for NOx Abatement under Real-Outdoor Conditions. Applied Catalysis A: General, 484, 17-25.
46. Chen, J. and Poon, C.S. (2009) Photocatalytic Construction and Building Materials: From Fundamentals to Applications. Building and Environment, 44, 1899-1906.
47. Paz, Y. (2010) Application of TiO2 Photocatalysis for Air Treatment: Patents' Overview. Applied Catalysis B: Environmental, 99, 448-460.
48. Yu, Q.L. and Brouwers, H.J.H. (2009) Indoor Air Purification Using Heterogeneous Photocatalytic Oxidation. Part I: Experimental Study. Applied Catalysis B: Environmental, 92, 454-461.
49. Yu, Q.L. and Brouwers, H.J.H. (2013) Design of a Novel Photocatalytic Gypsum Plaster: With the Indoor Air Purification Property. Advanced Materials Research, 651, 751-756.
50. Poon, C.S. and Cheung, E. (2007) NO Removal Efficiency of Photocatalytic Paving Blocks Prepared with Recycled Materials. Construction and Building Materials, 21, 1746-1753.
51. de Melo, J.V.S., Trichês, G., Gleize, P.J.P. and Villena, J. (2012) Development and Evaluation of the Efficiency of Photocatalytic Pavement Blocks in the Laboratory and after One Year in the Field. Construction and Building Materials, 37, 310-319.
52. Castillo, R., Koch, B., Ruiz, P. and Delmon, B. (1994) Influence of Preparation Methods on the Texture and Structure of Titania Supported on Silica. Journal of Materials Chemistry, 4, 903-906.
53. Morrison, C. and Kiwi, J. (1989) Preparation and Characterization of TiO2-SiO2 Aerosil Colloidal Mixed Dispersions. Journal of the Chemical

Society, Faraday Transactions 1: Physical Chemistry in Condensed Phases, 85, 1043-1048.

54. Hsu, W.P., Yu, R. and Matijevic, E. (1993) Paper Whiteners: I. Titania Coated Silica. Journal of Colloid and Interface Science, 156, 56-65.

55. Galan-Fereres, M., Mariscal, R., Alemany, L.J., Fierro, J.L.G. and Anderson, J.A. (1994) Ternary V-Ti-Si Catalysts and Their Behaviour in the CO + NO Reaction. Journal of the Chemical Society, Faraday Transactions, 90, 3711-3718.

56. Galan-Fereres, M., Alemany, L.J., Mariscal, R., Banares, M.A., Anderson, J.A. and Fierro, J.L. (1995) Surface Acidity and Properties of Titania-Silica Catalysts. Chemistry of Materials, 7, 1342-1348.

57. Choi, H.H., Park, J. and Singh, R.K. (2005) Nanosized Titania Encapsulated Silica Particles Using an Aqueous TiCl4 Solution. Applied Surface Science, 240, 7-12.

58. Sun, Z., Bai, C., Zheng, S., Yang, X. and Frost, R.L. (2013) A Comparative Study of Different Porous Amorphous Silica Minerals Supported TiO2 Catalysts. Applied Catalysis A: General, 458, 103-110.

59. Srinivasan, S., Datye, A.K., Hampden-Smith, M., Wachs, I.E., Deo, G., Jehng, J.M., Turek, A.M. and Peden, C.H.F. (1991) The Formation of Titanium Oxide Monolayer Coatings on Silica Surfaces. Journal of Catalysis, 131, 260-275.

60. Srinivasan, S., Datye, A.K., Smith, M.H. and Peden, C.H.F. (1994) Interaction of Titanium Isopropoxide with Surface Hydroxyls on Silica. Journal of Catalysis, 145, 565-573.

61. Mariscal, R., Palacios, J.M., Galan-Fereres, M. and Fierro, J.L.G. (1994) Incorporation of Titania into Preshaped Silica Monolith Structures. Applied Catalysis A: General, 116, 205-219.

62. Salama, T.M., Tanaka, T., Yamaguchi, T. and Tanabe, K. (1990) EXAFS/XANES Study of Titanium Oxide Supported on SiO2: A Structural Consideration on the Amorphous State. Surface Science, 227, L100-L104.

63. Ellestad, O.H. and Blindheim, U. (1985) Reactions of Titanium Tetrachloride with Silica Gel Surfaces. Journal of Molecular Catalysis, 33, 275-287.

64. Aronson, B.J., Blanford, C.F. and Stein, A. (1997) Solution-Phase Grafting of Titanium Dioxide onto the Pore Surface of Mesoporous Silicates: Synthesis and Structural Characterization. Chemistry of Materials, 9, 2842-2851.

65. Huang, Y.Y., Zhao, B.Y. and Xie, Y.C. (1998) A Novel Way to Prepare Silica Supported Sulfated Titania. Applied Catalysis A: General, 171, 65-73.

66. Morrow, B.A. and McFarlan, A.J. (1990) Chemical Reactions at Silica Surfaces. Journal of Non-Crystalline Solids, 120, 61-71.

67. Haukka, S., Lakomaa, E.L. and Root, A. (1993) An IR and NMR Study of the Chemisorption of Titanium Tetrachloride on Silica. The Journal of Physical Chemistry, 97, 5085-5094.

68. Nakayama, T., Onisawa, K., Fuyama, M. and Hanazono, M. (1992) TiO2/SiO2 Multilayer Insulating Films for ELDs. Journal of the Electrochemical Society, 139, 1204-1206.

69. Lassaletta, G., Fernandez, A., Espinos, J.P. and Gonzalez-Elipe, A.R. (1995) Spectroscopic Characterization of Quantum-Sized TiO2 Supported on Silica: Influence of Size and TiO2-SiO2 Interface Composition. The Journal of Physical Chemistry, 99, 1484-1490.

70. Nakayama, T. (1994) Structure of TiO2/SiO2 Multilayer Films. Journal of the Electrochemical Society, 141, 237-241.

71. Hayashi, T., Yamada, T. and Saito, H. (1983) Preparation of Titania-Silica Glasses by the Gel Method. Journal of Materials Science, 18, 3137-3142.

72. Li, Z., Hou, B., Xu, Y., Wu, D., Sun, Y., Hu, W. and Deng, F. (2005) Comparative Study of Sol-Gel-Hydrothermal and Sol-Gel Synthesis of Titania-Silica Composite Nanoparticles. Journal of Solid State Chemistry, 178, 1395-1405.

73. Fu, X. and Qutubuddin, S. (2001) Synthesis of Titania-Coated Silica Nanoparticles Using a Nonionic Water-in-Oil Microemulsion. Colloids and Surfaces A: Physicochemical and Engineering Aspects, 179, 65-70.

74. Liu, Z.F., Tabora, J. and Davis, R.J. (1994) Relationships between Microstructure and Surface Acidity of Ti-Si Mixed Oxide Catalysts. Journal of Catalysis, 149, 117-126.

75. Mine, E., Hirose, M., Kubo, M., Kobayashi, Y., Nagao, D. and Konno, M. (2006) Synthesis of Submicron-Sized Titania-Coated Silica Particles with a Sol-Gel Method and Their Application to Colloidal Photonic Crystals. Journal of Sol-Gel Science and Technology, 38, 91-95.

76. Lee, D.W., Ihm, S.K. and Lee, K.H. (2005) Mesostructure Control Using a Titania-Coated Silica Nanosphere Framework with Extremely High Thermal Stability. Chemistry of Materials, 17, 4461-4467.

77. Lee, J.W., Kong, S., Kim, W.S. and Kim, J. (2007) Preparation and Characterization of SiO2/TiO2 Core-Shell Particles with Controlled Shell Thickness. Materials Chemistry and Physics, 106, 39-44.

78. Do Kim, K., Bae, H.J. and Kim, H.T. (2003) Synthesis and Characterization of Titania-Coated Silica Fine Particles by Semi-Batch Process. Colloids and Surfaces A: Physicochemical and Engineering Aspects, 224, 119-126.

79. Rupp, W., Hüsing, N. and Schubert, U. (2002) Preparation of Silica-Titania Xerogels and Aerogels by Sol-Gel Processing of New Single-Source Precursors. Journal of Materials Chemistry, 12, 2594-2596.

80. Doolin, P.K., Alerasool, S., Zalewski, D.J. and Hoffman, J.F. (1994) Acidity Studies of Titania-Silica Mixed Oxides. Catalysis Letters, 25, 209-223.

81. Hanprasopwattana, A., Srinivasan, S., Sault, A.G. and Datye, A.K. (1996) Titania Coatings on Monodisperse Silica Spheres (Characterization Using 2-Propanol Dehydration and TEM). Langmuir, 12, 3173-3179.

82. Walters, J.K., Rigden, J.S., Dirken, P.J., Smith, M.E., Howells, W.S. and Newport, R.J. (1997) An Atomic-Scale Study of the Role of Titanium in TiO2:SiO2 Sol-Gel Materials. Chemical Physics Letters, 264, 539-544.

83. Klein, S., Weckhuysen, B.M., Martens, J.A., Maier, W.F. and Jacobs, P.A. (1996) Homogeneity of Titania-Silica Mixed Oxides: On UV-DRS Studies as a Function of Titania Content. Journal of Catalysis, 163, 489-491.

84. Liu, G., Liu, Y., Yang, G., Li, S., Zu, Y., Zhang, W. and Jia, M. (2009) Preparation of Titania-Silica Mixed Oxides by a Sol-Gel Route in the Presence of Citric Acid. The Journal of Physical Chemistry C, 113, 9345-9351.

85. Fujishima, A., Rao, T.N. and Tryk, D.A. (2000) Titanium Dioxide Photocatalysis. Journal of Photochemistry and Photobiology C: Photochemistry Reviews, 1, 1-21.

86. Carp, O., Huisman, C.L. and Reller, A. (2004) Photoinduced Reactivity of Titanium Dioxide. Progress in Solid State Chemistry, 32, 33-177.

87. Kitano, M., Matsuoka, M., Ueshima, M. and Anpo, M. (2007) Recent Developments in Titanium Oxide-Based Photocatalysts. Applied Catalysis A: General, 325, 1-14.

88. Gaya, U.I. and Abdullah, A.H. (2008) Heterogeneous Photocatalytic Degradation of Organic Contaminants over Titanium Dioxide: A Review of Fundamentals, Progress and Problems. Journal of Photochemistry and Photobiology C: Photochemistry Reviews, 9, 1-12.

89. Macwan, D.P., Dave, P.N. and Chaturvedi, S. (2011) A Review on Nano-TiO2 Sol-Gel Type Syntheses and Its Applications. Journal of Materials Science, 46, 3669-3686.

90. Simonsen, M.E., Li, Z. and S&oslashgaard, E.G. (2009) Influence of the OH Groups on the Photocatalytic Activity and Photoinduced Hydrophilicity of Microwave Assisted Sol-Gel TiO2 Film. Applied Surface Science, 255, 8054-8062.

91. Yu, J., Jimmy, C.Y., Ho, W. and Jiang, Z. (2002) Effects of Calcination Temperature on the Photocatalytic Activity and Photo-Induced Super-Hydrophilicity of Mesoporous TiO2 Thin Films. New Journal of Chemistry, 26, 607-613.

92. Wang, R., Hashimoto, K., Fujishima, A., Chikuni, M., Kojima, E., Kitamura, A. and Watanabe, T. (1997) Light-Induced Amphiphilic Surfaces. Nature, 388, 431-432.

93. Mezhenny, S., Maksymovych, P., Thompson, T.L., Diwald, O., Stahl, D., Walck, S.D. and Yates, J.T. (2003) STM Studies of Defect Production on the TiO2(110)-(1×1) and TiO2(110)-(1×2) Surfaces Induced by UV Irradiation. Chemical Physics Letters, 369, 152-158.

94. White, J.M., Szanyi, J. and Henderson, M.A. (2003) The Photon-Driven Hydrophilicity of Titania: A Model Study Using TiO2(110) and Adsorbed Trimethyl Acetate. The Journal of Physical Chemistry B, 107, 9029-9033.

95. Miyauchi, M., Nakajima, A., Fujishima, A., Hashimoto, K. and Watanabe, T. (2000) Photoinduced Surface Reactions on TiO2 and SrTiO3 Films: Photocatalytic Oxidation and Photoinduced Hydrophilicity. Chemistry of Materials, 12, 3-5.

96. Luttrell, T., Halpegamage, S., Tao, J., Kramer, A., Sutter, E. and Batzill, M. (2014) Why Is Anatase a Better Photocatalyst than Rutile?—Model Studies on Epitaxial TiO2 Films. Scientific Reports, 4, Article No.: 4043.

97. Mattioli, G., Filippone, F., Alippi, P. and Bonapasta, A.A. (2008) Ab Initio Study of the Electronic States Induced by Oxygen Vacancies in Rutile and Anatase TiO2. Physical Review B, 78, Article ID: 241201.

98. Xu, M., Gao, Y., Moreno, E.M., Kunst, M., Muhler, M., Wang, Y., Idriss, H. and W&oumlll, C. (2011) Photocatalytic Activity of Bulk TiO2 Anatase and Rutile Single Crystals Using Infrared Absorption Spectroscopy. Physical Review Letters, 106, Article ID: 138302.

99. Ohtani, B., Ogawa, Y. and Nishimoto, S.I. (1997) Photocatalytic Activity of Amorphous-Anatase Mixture of Titanium (IV) Oxide Particles Suspended in Aqueous Solutions. The Journal of Physical Chemistry B, 101, 3746-3752.

100. Bickley, R.I., Gonzalez-Carreno, T., Lees, J.S., Palmisano, L. and Tilley, R.J. (1991) A Structural Investigation of Titanium Dioxide Photocatalysts. Journal of Solid State Chemistry, 92, 178-190.

101. Scanlon, D.O., Dunnill, C.W., Buckeridge, J., Shevlin, S.A., Logsdail, A.J., Woodley, S.M. and Sokol, A.A. (2013) Band Alignment of Rutile and Anatase TiO2. Nature Materials, 12, 798-801.

102. Mogyorósi, K., Farkas, A., Dékány, I., Ilisz, I. and Dombi, A. (2002) TiO2-Based Photocatalytic Degradation of 2-Chlorophenol Adsorbed on Hydrophobic Clay. Environmental Science & Technology, 36, 3618-3624.

103. Mogyorosi, K., Dekany, I. and Fendler, J.H. (2003) Preparation and Characterization of Clay Mineral Intercalated Titanium Dioxide Nanoparticles. Langmuir, 19, 2938-2946.

104. Kun, R., Mogyorósi, K. and Dékány, I. (2006) Synthesis and Structural and Photocatalytic Properties of TiO2/Mont-morillonite Nanocomposites. Applied Clay Science, 32, 99-110.

105. Kibanova, D., Trejo, M., Destaillats, H. and Cervini-Silva, J. (2009) Synthesis of Hectorite-TiO2 and Kaolinite-TiO2 Nanocomposites with

Photocatalytic Activity for the Degradation of Model Air Pollutants. Applied Clay Science, 42, 563-568.

106. Matthews, R.W. (1991) Photooxidative Degradation of Coloured Organics in Water Using Supported Catalysts. TiO2 on Sand. Water Research, 25, 1169-1176.

107. Matthews, R.W. and McEvoy, S.R. (1992) Photocatalytic Degradation of Phenol in the Presence of Near-UV Illuminated Titanium Dioxide. Journal of Photochemistry and Photobiology A: Chemistry, 64, 231-246.

108. European Commission (2007) Reference Document on Best Available Techniques for the Manufacture of Large Volume Inorganic Chemicals— Solids and Other Industry

109. Iler, R.K. (1979) The Chemistry of Silica: Solubility, Polymerization, Colloid and Surface Properties, and Biochemistry. Wiley, New York.

110. St&oumlber, W., Fink, A. and Bohn, E. (1968) Controlled Growth of Monodisperse Silica Spheres in the Micron Size Range. Journal of Colloid and Interface Science, 26, 62-69.

111. Lazaro, A., Brouwers, H.J.H., Quercia, G. and Geus, J.W. (2012) The Properties of Amorphous Nano-Silica Synthesized by the Dissolution of Olivine. Chemical Engineering Journal, 211, 112-121.

112. Lazaro, A., Van de Griend, M.C., Brouwers, H.J.H. and Geus, J.W. (2013) The Influence of Process Conditions and Ostwald Ripening on the Specific Surface Area of Olivine Nano-Silica. Microporous and Mesoporous Materials, 181, 254-261.

113. Lazaro, A., Quercia, G., Brouwers, H. and Geus, J. (2013) Synthesis of a Green Nano-Silica Material Using Beneficiated Waste Dunites and Its Application in Concrete. World Journal of Nano Science and Engineering, 3, 41-51.

114. Lazaro, A., Benac-Vegas, L., Brouwers, H.J.H., Geus, J.W. and Bastida, J. (2015) The Kinetics of the Olivine Dissolution under the Extreme Conditions of Nano-Silica Production. Applied Geochemistry, 52, 1-15.

115. Gu, W. and Tripp, C.P. (2005) Role of Water in the Atomic Layer Deposition of TiO2 on SiO2. Langmuir, 21, 211-216.

116. Sugimoto, T., Zhou, X. and Muramatsu, A. (2002) Synthesis of Uniform Anatase TiO2 Nanoparticles by Gel-Sol Method: 1. Solution Chemistry of $Ti(OH)_n^{(4-n)+}$ Complexes. Journal of Colloid and Interface Science, 252, 339-346.

117. Lee, G.H. and Zuo, J.M. (2004) Growth and Phase Transformation of Nanometer-Sized Titanium Oxide Powders Produced by the Precipitation Method. Journal of the American Ceramic Society, 87, 473-479.

118. Wang, T.H., Navarrete-López, A.M., Li, S., Dixon, D.A. and Gole, J.L. (2010) Hydrolysis of TiCl4: Initial Steps in the Production of TiO2. The Journal of Physical Chemistry A, 114, 7561-7570.

119. Yoldas, B.E. (1986) Hydrolysis of Titanium Alkoxide and Effects of Hydrolytic Pclycondensation Parameters. Journal of Materials Science, 21, 1087-1092.

120. Zhang, X.T., Sato, O., Taguchi, M., Einaga, Y., Murakami, T. and Fujishima, A. (2005) Self-Cleaning Particle Coating with Antireflection Properties. Chemistry of Materials, 17, 696-700.

121. Ryu, D.H., Kim, S.C., Koo, S.M. and Kim, D.P. (2003) Deposition of Titania Nanoparticles on Spherical Silica. Journal of Sol-Gel Science and Technology, 26, 489-493.

122. Bischoff, B.L. and Anderson, M.A. (1995) Peptization Process in the Sol-Gel Preparation of Porous Anatase (TiO2). Chemistry of Materials, 7, 1772-1778.

123. Mahshid, S, Askari, M. and Ghamsari, M.S. (2007) Synthesis of TiO2 Nanoparticles by Hydrolysis and Peptization of Titanium Isopropoxide Solution. Journal of Materials Processing Technology, 189, 296-300.

124. Takahashi, Y. and Matsuoka, Y. (1988) Dip-Coating of TiO2 Films Using a Sol Derived from Ti(O-i-Pr)4-Diethanolamine-H2O-i-PrOH System. Journal of Materials Science, 23, 2259-2266.

125. Kato, K., Tsuzuki, A., Taoda, H., Torii, Y., Kato, T. and Butsugan, Y. (1994) Crystal Structures of TiO2 Thin Coatings Prepared from the Alkoxide Solution via the Dip-Coating Technique Affecting the Photocatalytic Decomposition of Aqueous Acetic Acid. Journal of Materials Science, 29, 5911-5915.

126. Imoberdorf, G.E., Irazoqui, H.A., Cassano, A.E. and Alfano, O.M. (2005) Photocatalytic Degradation of Tetrachloroethylene in Gas Phase on TiO2 Films: A Kinetic Study. Industrial & Engineering Chemistry Research, 44, 6075-6085.

127. Negishi, N., Takeuchi, K. and Ibusuki, T. (1998) Preparation of the TiO2 Thin Film Photocatalyst by the Dip-Coating Process. Journal of Sol-Gel Science and Technology, 13, 691-694.

128. Negishi, N. and Takeuchi, K. (2001) Preparation of TiO2 Thin Film Photocatalysts by Dip Coating Using a Highly Viscous Solvent. Journal of Sol-Gel Science and Technology, 22, 23-31.

129. Kim, D.J., Hahn, S.H., Oh, S.H. and Kim, E.J. (2002) Influence of Calcination Temperature on Structural and Optical Properties of TiO2 Thin Films Prepared by Sol-Gel Dip Coating. Materials Letters, 57, 355-360.

130. Sonawane, R.S., Hegde, S.G. and Dongare, M.K. (2003) Preparation of Titanium (IV) Oxide Thin Film Photocatalyst by Sol-Gel Dip Coating. Materials Chemistry and Physics, 77, 744-750.

131. Daoud, W.A. and Xin, J.H. (2004) Low Temperature Sol-Gel Processed Photocatalytic Titania Coating. Journal of Sol-Gel Science and Technology, 29, 25-29.

132. Crepaldi, E.L., Soler-Illia, G.J.D.A., Grosso, D., Cagnol, F., Ribot, F. and Sanchez, C. (2003) Controlled Formation of Highly Organized Mesoporous Titania Thin Films: From Mesostructured Hybrids to Mesoporous Nanoanatase TiO2. Journal of the American Chemical Society, 125, 9770-9786.

133. Kajihara, K., Nakanishi, K., Tanaka, K., Hirao, K. and Soga, N. (1998) Preparation of Macroporous Titania Films by a Sol-Gel Dip-Coating Method from the System Containing Poly(ethylene glycol). Journal of the American Ceramic Society, 81, 2670-2676.

134. Park, H.K., Kim, D.K. and Kim, C.H. (1997) Effect of Solvent on Titania Particle Formation and Morphology in Thermal Hydrolysis of TiCl4. Journal of the American Ceramic Society, 80, 743-749.

135. Jiang, X., Herricks, T. and Xia, Y. (2003) Monodispersed Spherical Colloids of Titania: Synthesis, Characterization, and Crystallization. Advanced Materials, 15, 1205-1209.

136. Sun, J. and Gao, L. (2002) pH Effect on Titania-Phase Transformation of Precipitates from Titanium Tetrachloride Solutions. Journal of the American Ceramic Society, 85, 2382-2384.

137. Matthews, A. (1976) The Crystallization of Anatase and Rutile from Amorphous Titanium Dioxide under Hydrothermal Conditions. American Mineralogist, 61, 419-424.

138. Nam, H.D., Lee, B.H., Kim, S.J., Jung, C.H., Lee, J.H. and Park, S. (1998) Preparation of Ultrafine Crystalline TiO2 Powders from Aqueous TiCl4 Solution by Precipitation. Japanese Journal of Applied Physics, 37, 4603-4608.

139. Pedraza, F. and Vazquez, A. (1999) Obtention of TiO2 Rutile at Room Temperature through Direct Oxidation of TiCl3. Journal of Physics and Chemistry of Solids, 60, 445-448.

140. Terabe, K., Kato, K., Miyazaki, H., Yamaguchi, S., Imai, A. and Iguchi, Y. (1994) Microstructure and Crystallization Behaviour of TiO2 Precursor Prepared by the Sol-Gel Method Using Metal Alkoxide. Journal of Materials Science, 29, 1617-1622.

141. Sun, J., Gao, L. and Zhang, Q. (2003) Synthesizing and Comparing the Photocatalytic Properties of High Surface Area Rutile and Anatase Titania Nanoparticles. Journal of the American Ceramic Society, 86, 1677-1682.

142. Zhang, Q., Gao, L. and Guo, J. (2000) Effects of Calcination on the Photocatalytic Properties of Nanosized TiO2 Powders Prepared by TiCl4 Hydrolysis. Applied Catalysis B: Environmental, 26, 207-215.

143. Chen, J., Gao, L., Huang, J. and Yan, D. (1996) Preparation of Nanosized Titania Powder via the Controlled Hydrolysis of Titanium Alkoxide. Journal of Materials Science, 31, 3497-3500.

144. Banfield, J. (1998) Thermodynamic Analysis of Phase Stability of Nanocrystalline Titania. Journal of Materials Chemistry, 8, 2073-2076.

145. Yanagisawa, K. and Ovenstone, J. (1999) Crystallization of Anatase from Amorphous Titania Using the Hydrothermal Technique: Effects of Starting Material and Temperature. The Journal of Physical Chemistry B, 103, 7781-7787.

146. Bavykin, D.V., Dubovitskaya, V.P., Vorontsov, A.V. and Parmon, V.N. (2007) Effect of TiOSO4 Hydrothermal Hydrolysis Conditions on TiO2 Morphology and Gas-Phase Oxidative Activity. Research on Chemical Intermediates, 33, 449-464.

147. Wang, C.C. and Ying, J.Y. (1999) Sol-Gel Synthesis and Hydrothermal Processing of Anatase and Rutile Titania Nanocrystals. Chemistry of Materials, 11, 3113-3120.

148. Penn, R.L. and Banfield, J.F. (1999) Morphology Development and Crystal Growth in Nanocrystalline Aggregates under Hydrothermal Conditions: Insights from Titania. Geochimica et Cosmochimica Acta, 63, 1549-1557.

149. Zhang, H. and Banfield, J.F. (2002) Kinetics of Crystallization and Crystal Growth of Nanocrystalline Anatase in Nanometer-Sized Amorphous Titania. Chemistry of Materials, 14, 4145-4154.

150. Kumar, S.R., Suresh, C., Vasudevan, A.K., Suja, N.R., Mukundan, P. and Warrier, K.G.K. (1999) Phase Transformation in Sol-Gel Titania Containing Silica. Materials Letters, 38, 161-166.

151. Zhang, J., Hu, Y., Matsuoka, M., Yamashita, H., Minagawa, M., Hidaka, H. and Anpo, M. (2001) Relationship between the Local Structures of Titanium Oxide Photocatalysts and Their Reactivities in the Decomposition of NO. The Journal of Physical Chemistry B, 105, 8395-8398.

152. Yoneyama, H., Haga, S. and Yamanaka, S. (1989) Photocatalytic Activities of Microcrystalline Titania Incorporated in Sheet Silicates of Clay. The Journal of Physical Chemistry, 93, 4833-4837.

153. Caillol, S. (2011) Fighting Global Warming: The Potential of Photocatalysis against CO2, CH4, N2O, CFCs, Tropospheric O3, BC and Other Major Contributors to Climate Change. Journal of Photochemistry and Photobiology C: Photochemistry Reviews, 12, 1-19.

154. Tanabe, K., Sumiyoshi, T., Shibata, K., Kiyoura, T. and Kitagawa, J. (1974) A New Hypothesis Regarding the Surface Acidity of Binary Metal Oxides. Bulletin of the Chemical Society of Japan, 47, 1064-1066.

CHAPTER 2

Studies on Properties of Bio-Composites from Ecoflex/Ramie Fabric-Mechanical and Barrier Properties

K. A. Ajith Kumar[1], M. S. Sreekala[1], S. Arun[2]

[1]Post Graduate Department of Chemistry, Sree Sankara College, Kalady, India; [2]Fatima Mata National College, Kollam, India.

ABSTRACT

Nowadays, utilization of biodegradable materials has become necessary in order to maintain global environmental and ecological balance. "Green" composites offered the possible solution to waste disposal problems associated with traditional petroleum derived plastics. The use of plastics based on removable resources is enormous now a day for the development of true bio-composites. Fully biodegradable "Green" textile composites have been prepared from Ecoflex and ramie fabric. Textile composites were fabricated from the Ecoflex polymer and the ramie fabric by hot compression molding technique. Interactions at the fiber-matrix interface and the compatibility between ramie fabric and Ecoflex polymer will affect the properties of the system. The mechanical property and barrier property of the composites were investigated. Static mechanical properties such as tensile strength, tensile modulus, and elongation at break of the textile bio-composites were analyzed. Sorption characteristics of water, oil and diesel in the textile composites were analyzed in order to determine its outdoor applications and the influence of macro fibers on the transport phenomena was investigated. The kinetics of sorption-diffusion process was investigated. Kinetic parameters such as n, k, diffusion coefficient, permeability, solubility parameter, % swelling index, etc., were analyzed. The water sorption mechanism in the textile composites was found to exhibit slight deviation from Fickian mode.

Keywords: Biodegradable Composites; Mechanical Properties; Water Sorption; Oil Sorption; Diesel Sorption

1. INTRODUCTION

This template, Plastic materials are commonly used in agricultural practices for a variety of applications that include mulch films, greenhouse construction materials, packaging materials etc. [1]. Conventionally, such plastics are manufactured from petroleum derivatives that are not degradable and persist in the environment long after their useful life. As a result, interest in the use of naturally degradable and/or biodegradable polymers for plastic manufacturing, particularly for use in agriculture, has grown considerably in recent years [2,3]. The development of bio-composites started in the late 1980s and most of the biodegradable polymers which are now available in the market do not yet satisfy each of the requirements for bio-composites. The commercially available biodegradable polymers like Biopol, polycaprolacton, Bioceta, Mater Bi, Scona cell, etc. have been tested in order to examine their properties with special emphasis on the suitability of such polymers for use as matrix material for the fabrication of bio-composites especially with natural fibers [4]. The low viscosity of these matrix polymers at the processing temperature, good mechanical properties of both the matrix and reinforcing fiber as well as good matrix-fiber adhesion are required to obtain high quality bio-composites [5]. The degradation characteristics of short sisal/Mater Bi-Y bio-composites by soil burial were investigated by Alvarez et al. [6]. They observed a drop in mechanical properties of the composite as a function of exposure time.

Efforts have been made to develop environmentally compatible plastic products by incorporating renewable polymers as an alternative to petroleum-derived chemicals [7-11]. This fact leads to the use of eco-composites derived from bio-fibers and biodegradable polymers which not only can reduce the plastic pollution but can also reduce the wide spread dependence on fossil resources [5]. The renewable polymers are relatively inexpensive, environmentally friendly, and also naturally biodegradable. Particularly, plant material derived from renewable crops, byproducts or their industrially processed wastes, offer a good source of fiber for applications [12-14]. Recently, attention is increasingly devoted to biodegradable and plant-derived composites, which we designate as "green" composites [15-17], because of the strong demand for creating a resource-circulating society that poses no resource-shortage-related problems. In the case of the "green" composites, natural fibers derived from bamboo [15], hemp [16], or flax [17] are added to biodegradable resins to reinforce polymer matrix materials and improve the mechanical properties of the resultant composites.

Development of bio-composites is still in its preliminary stage. More data on properties of bio-composites are required to establish confidence in their use. Training must have priority to accelerate the acceptance of biocomposites for various applications. The Life Cycle Analysis (LCA) of environment friendly composites has to be carried out to make it suitable for a wide range of applications. A brief review of the most suitable and commonly used biodegradable polymer matrices and nanofiber (NF) reinforcements in eco-composites, as well as some of the already produced and commercialized NF eco-composites were recently reported by Bogoeva Gaceva et al. [18]. The nanofiber reinforcement can offer better barrier properties to the resultant composite. Studies have been reported in various bio-composite systems [19-21]. A theoretical approach to the moisture sorption in natural fiber plastic composite has been done by Wang et al. [19]. In another study Liu et al. [20] investigated the influence of processing methods and fiber length on physical properties of kenaf fiber reinforced soy based bio-composites. They found that the fractured fiber length on the impact fracture surface increased with increasing fiber length and fiber content. Torres et al. [21] studied processing and mechanical properties of natural fiber reinforced thermoplastic starch bio-composites. A recent report by Sorrentino et al. on the potential perspectives of bio-nanocomposites for food packaging applications reviews different types of new bio-based materials, such as edible and biodegradable nanocomposite films, their commercial applications as packaging materials, regulations and future trends [22].

The aim of the present work is to prepare a fully biodegradable "Green" composite—Ecoflex/ramie fabric composite. All the raw materials used in this study are fully biodegradable and hence the composite prepared will not cause any deleterious ecological impact. Cellulose rich ramie fibers can effectively reinforce Ecoflex neat. Ramie fiber is one of the strongest fine textile fibers. They are obtained from the stem of plants from the Utricacea family namely Boehmeria nivea chinensis and Boehmeria nivea indica. Ramie is one of the older fibers used by man, its use being reported in many ancient texts. Nowadays it is cultivated mainly in Asia, Taiwan, Korea, Philippines and Brazil, which are the main producers of ramie. Chemical composition and structural parameters of ramie fiber is as follows: cellulose—68.6 - 76.2 wt%; lignin—0.6 - 0.7 wt%; hemicellulose—13.1 - 16.7 wt%; pectin—1.9 wt%; wax—0.3 wt%; microfibrillar/spiral angle (.)—7.5 wt%; moisture content—8 wt%; density — 1.50 g/cm; tensile strength—400 - 938 MPa; Young's modulus—61.4 - 128 GPa; elongation at break—1.2% - 3.8% [5]. This fiber has also a strong potential as reinforcement of the green composites, as shown in literature [23]. A study on the moisture sorption characteristics in woven sisal fabric reinforced natural rubber bio-composites were reported elsewhere [24]. The authors have found that the moisture uptake of the textile composite depends upon fiber content as well as architecture. In another study they have evaluated water absorption characteristics of hybrid bio-fiber (sisal and oil palm)-reinforced natural rubber bio-composites [25]. Moisture uptake was found to be dependent on the properties of the bio-fibers.

The main objective of the present study is the investigation of mechanical and the swelling behaviour of the prepared composites in liquids like water, naphthenic oil and diesel. Generally, polymer-based materials are not soluble in water but they are capable of absorbing various amount of water depending on their chemical nature, formulation and on the humidity and temperature of the environment to which they are exposed [26]. Liquid swelling experiments of composites are important for the following reasons: as a method of analyzing their service performance in liquid environment, and to study the characteristics of matrix-fiber interface [27]. Articles made from composites come in contact with different liquids during their service performance. They come in contact with liquids or vapours, either aqueous or organic affecting the immediate and long-term performance of the material. This can happen either as a part of the service requirement, like in the case of oil seals, marine articles, etc., or by accidental splashing of oils and greases that occurs with automobile components. Applications of composites ranging from civil structures to medical implants require long-term prediction of material in a humid environment. Diffusion experiment is very useful to study the polymer-matrix interaction. Moisture penetration into composite materials is conducted by one major mechanism—diffusion. This mechanism involves direct diffusion of water molecules into the matrix and in some cases into the fibers. The common mechanism of moisture penetration into composite is capillary flow along the fiber-matrix interface, followed by diffusion from the interface into the bulk resin and transport by micro cracks. Each of these mechanisms becomes active only after the occurrence of specific damage to the composite. Often that damage that enhances moisture penetration by activating those additional mechanisms is in itself is a direct consequence of the exposure of the composite to moisture. Hence these two damage-dependent mechanisms are increasing both the rate and the maximum capacity of moisture absorption in an auto accelerative manner [27]. The absorbed moisture results in more detrimental effects in the mechanical properties of composite materials, since the water not only interacts with polymer matrices physically, i.e. plasticization and/or chemically, but it also attacks the fiber-matrix interface. In fact, the fibermatrix interface is the determining factor of the reinforcement mechanism, especially under wet conditions. Moisture absorption may induce irreversible changes to polymers and composites such as chemical degradation and cracking. This damage to the material will also change the weight gain behaviour of the material correspondingly. For example, cracking and blistering can cause exceptionally high uptake while the leaching of small molecule components results in gradually decreasing weight gain. The rate of ageing will be affected when liquid molecules diffuse into the polymeric system. In the present study, investigation on mechanical and the transport phenomenon of water, oil and diesel in Ecoflex neat/ramie fabric textile composites was carried out in detail.

2. EXPERIMENTAL

2.1. Materials

Ecoflex polymer is procured from BASF Co. A plain woven fabric of ramie fibers (supplied from Tosco Co. No. 25, 44 warps per inch and 46 wefts per inch) was used as reinforcement. Each warp fiber posses alternately under and over each weft fiber. The fabric is symmetrical with good stability and reasonable porosity. However it is the most difficult of the weaves to drape and high level of fiber crimp imparts relatively low mechanical properties compared with other weave styles. With large fibers (high tex) this weave style gives excessive crimp and therefore it is not generally used for very heavy fabrics.

2.2. Composite Preparation

Ecoflex polymer is fabricated by compression moulding at 120°C in hydraulic press for 15 minute and cooled at room temperature under the same pressure. The Ramie mat is sandwiched between two Ecoflex sheets by compression moulding at 120°C. Thus forms the Ecoflex/ ramie mat composite. Standard tensile test specimens were cut from the prepared sheets.

2.3. Tensile Properties

Tensile elongation and tensile modulus measurements are all among the most important indications of strength in a material and all the most widely specified properties cf plastic materials. Tensile test is a measurement of the ability of material to withstand forces that tend to pull it apart and to determine to what extend the material stretches before breaking. Tensile modulus is indication of the relative stiffness of a material can be determined from a stress-strain diagram. The elongation of a specimen continued until a rapture of the specimen is observed. Load value at break is also recorded. The tensile strength at yield and tensile strength at break is calculated.

$$\text{Tensile strength} = \frac{\text{Force (load)(N)}}{\text{Cross section area (mm}^2)}$$

$$\text{Tensile strength at yield (MPa)}$$

$$= \frac{\text{Max. load recorded (N)}}{\text{Cross section area (mm}^2)}$$

$$\text{Tensile strength at break (MPa)}$$

$$= \frac{\text{Load recorded at break (N)}}{\text{Cross section area (mm}^2)}$$

$$\text{Tensile modulus}$$

$$= \frac{\text{Difference in stress}}{\text{Difference in corresponding strain}}$$

$$\text{Elongation at yield, strain}$$

$$= \frac{\text{Change in length (elongation)}}{\text{Original length (gauge length)}}$$

$$\text{i.e.} \quad \sum = \frac{\Delta L}{1}$$

$$\Delta L = \sum \times l$$

$$S = \frac{3\,PL}{2bd^2}$$

where S = stress (MPa), P = load (N), L = length of span (mm), b = width in specimen (mm), d = thickness of specimen (mm).

The maximum strain in the outer fibres which also occurs at midspan is calculated using the following equation

$$R = \frac{6Dd}{L^2}$$

where R = strain, D = deflection (mm), L = Length of span (mm), d = thickness of specimen.

Yield point is the point on the stress-strain curve at which an increase strain occurs without increase in stress. Yield strength is the stress at which a material exhibits as specified limiting deviation from the proportionality of stress to strain.

Stress-strain measurements of the samples were determined according to ASTMD D 638. The testing was done on a Schimadzu model AG1 universal testing machine. All the tests were carried out in room temperature. Samples were punched out from the cured sheets and the dimensions were noted. Aluminum plates of 0.8 mm thickness were attached with specimen. In the universal testing machine, the sample was held tight by the two grips of the testing machines. Tests were carried out a grip separation speed 50 mm/min. A gauge length of 50 mm was used for both types of specimen.

2.4. Diffusion Experiment

Diffusion experiments were carried out using circular sample of 17 mm diameter and 0.80 mm thickness were punched from the composite sheets. The dried sample was weighed on a Sartonus BP 210 S balances that is accurate to 0.0001 g. After the first weighing, the samples were immersed in about 20 ml of the liquid 1) water, 2) Naphthenic oil, 3) Diesel in diffusion bottles, which were kept at constant room temperature. The sample were periodically removed from the liquid, dried of any liquid on the surface using a tissue paper and weighed again. During the initial stages of diffusion experiment, the liquid is sorbed by the samples at relatively fast pace and so it is necessary to weigh them every 10 min for the first hour, 20 min for the second hour, 30 min. for the third hour.

Then as the experiment progresses, it becomes necessary to weigh the samples only once an hour, and then once a day followed by once in a few days. Thus weighing was continued at frequent intervals until equilibrium swelling was reached. The result of the diffusion experiments is expressed as diffusion curves, where Q_t is the number of moles of solvent uptake by 100 g of polymer sample and is expressed as Q_t (mol·g^{-1} × 100).

The transport of liquids through composite follows different mechanisms depending upon many factors, such as chemical nature of polymer matrix and that of fiber, matrix-fiber compatibility and interfacial adhesion. Diffusion behaviour is classified into three categories depending upon the relative rates of penetrant mobility and polymer segmental relaxation. They include: 1) Fickian behaviour: penetrant mobility is much less than the polymer segmental relaxation rates; 2) Anomalous behaviour: penetrant mobility and polymer segmental relaxation rates are comparable; 3) Non-Fickian behaviour: penetrant mobility is much greater than polymer segmental relaxation rates.

In the present study, the mechanism of transport was analyzed using the empirical relation.

$$\log\left(\frac{Q_t}{Q_\alpha}\right) = \log k + n \log t$$

where, Q_t and Q_α are the number of moles of liquid absorbed by 100 g of sample at time t and at equilibrium swelling respectively. The constants n and k vary with the nature of materials and interfacial adhesion. The values of n and k were found out from the slope and y-intercept of the plots of log Q_t/Q_α vs. logt. The value of n determines the type of transport mechanism. For Fickian behavior, the value of n = 0.5; if n = 1, this indicates Non-Fickian behaviour, i.e. relaxation controlled transport. If its value is in between 0.5 and 1, the transport behaviour is termed as anomalous. The factor k is a constant that varies with the structure of composite and provides an idea about the interaction between the composite and solvent. Lower values of k indicate that there is less interaction between composite and solvent and also there is less absorption of solvent. If the composite and solvent both are either polar or non-polar, the solubility increases; this is explained on the basis of the principle that "like dissolves like" and hence the value of k is also high. If both composite and solvent are of different types, solubility decreases and hence the value of k also decreases.

$$Q_t \text{ sorption} = \frac{\dfrac{\text{Weight of solvent sorbed at time } t}{\text{Molecular weight of solvent}}}{\text{Weight of polymer compound}} \times 100$$

Diffusion experiment is very useful to study the polymer/fibre interaction. The transport of small liquid molecules through macromolecules proceeds via a two stages sorption diffusion process. At first the penetrate molecules sorbed by the polymer followed by diffusion (intermingling of the two systems) through the polymer. Hence the overall transport is determined by the difference in amount of the penetrant molecules between the two phases. The liquid absorption tends towards an equilibrium value, which depends on the nature of the material. The permeability "P" is given by equation P = DS where, D and S are diffusion coefficient and solubility parameter; respectively. The kinetic parameter, diffusion coefficient D is obtained by

$$D = \Pi \left[\frac{h\theta}{4Q_\alpha}\right]^2$$

where "h" is the sample thickness, q in the slope of the initial linear portion (before attaining of 50% of the equilibrium) of the Q_t verses $(time)^{1/2}$ curve. The thermodynamic solubility parameter "S" estimated by the number of grams of liquid sorbed per gram of polymer compound [27].

The swelling parameters like swelling index and swelling coefficient were evaluated to assess the extent of swelling behaviour of composites in solvents like oil and diesel [28]. Swelling index is calculated by the equation,

$$S = \frac{\text{Weight of solvent uptake at equilibrium}}{\text{Initial weight of sample}}$$

Swelling index % = $\dfrac{W_2 - W_1}{W_1} \times 100$

where, W_1 and W_2 are the initial and final (swollen) weights of the sample. Dimensional changes of the sample before and after sorption were measured. The swelling behaviour of the composites can also be analyzed from the swelling coefficient values. It is an index of the ability with which the sample swells and is determined by the equationSwelling coefficient, $\alpha = (A_s/m) \times (1/d)$ where, A_s is the weight of the solvent absorbed at equilibrium swelling, m the mass of the sample before swelling and d is the density of the solvent used.

3. RESULT AND DISCUSSION

3.1. Tensile Properties of Ecoflex Neat and Ecoflex Ramie Mat Composites

The Properties under tension are the most important indication of strength in a material and are one of the most widely specified properties of all plastic materials. Tensile property of Ecoflex mat and Ecoflex/ramie mat composites are given in **Table 1**. Tensile strength has shown very good improvement by textile ramie reinforcement. Ramie fibres have great potential as fibre reinforcement in resin matrix composite materials. Textile ramie mat reinforced Ecoflex composite result in dramatic increase in properties of the system. Tensile strength has been increased by 900% by ramie mat reinforcement.

Ecoflex neat sample exhibit very high extensibility. It is obvious from the strain value of 164% for the neat sample. By ramie mat reinforcement, the value decreased to 11%, good interface properties will lead to better stress-transfer between fibre and matrix and will lead to high strength properties. Strong ramie fibres will exert good reinforcing effect in the matrix and will increase the strength of the composite. Ecoflex neat has got inherent property of very high drawability. The inherent ductile behavior of Ecoflex sample is considerably reduced by the warp and weft of ramie fibre in the textile composite. In ramie mat composite, the warp and weft ramie fibres will effectively help the stress transfer upon tensile loading.

Tensile elongation and tensile modulus measurements are important indications of strength and stiffness of polymers respectively. The young's modulus value of Ecoflex/Ramie mat composite is given in **Table 1**. Ecoflex/Ramie mat composite has more strength and stiffness than Ecoflex neat. Typical stress-strain behaviour of Ecoflex neat and Ecoflex/ramie mat composite shown in **Figure 1**. Ecoflex neat sample shows plastic deformation and considerable necking effect. For a long duration upon tensile loading, the strain increases with constant stress due to the inherent ductile behaviour of Ecoflex neat. For Ecoflex/ramie mat, the elastic behaviour is observed. The elasticity and brittle fracture of the textile composite is clearly understood from the stress strain curve shown in **Figure 1**. From stress-strain graph we can see that upon tensile loading partial failure of the ramie fibres occur gradually and will create flaws in the composite. On further increase in loading more stress will be distributed among the remaining ramie fibre and finally brittle failure will occur to the composite. Up on textile ramie reinforcement in Ecoflex, high strength and high stiff composite material resulted.

Table 1. Tensile properties of Ecoflex neat and Ecoflex/ ramie mat composites.

Specimen	Tensile strength (MPa)	Elongation at break (%)	Young's modulus (MPa)
Ecoflex neat	9 SD (0.42)	164 SD (8.95)	41
Ecoflex/ramie mat	78 SD (2)	11 SD (3.7)	674

3.2. Water Sorption Characteristics

A systematic study on the water diffusion characteristics of Ecoflex neat and Ecoflex/ramie mat composite at room temperature were carried out. The mechanism of diffusion was analyzed. The mole percentage uptake of water with time by the Ecoflex neat and composite is shown in **Figure 2**. It was found that water diffusion is more for Ecoflex/ramie mat composite than Ecoflex neat. In the Ecoflex neat there is only one phase perfect polymer structure will be observed in Ecoflex neat.

But due to textile ramie reinforcement, it will interfere the three dimensional polymeric net work and the presence of fibre-matrix interface will affect the diffusion process. Water can penetrate and diffuse through the interface where as in the Ecoflex neat there is no possibility for this kind of process.

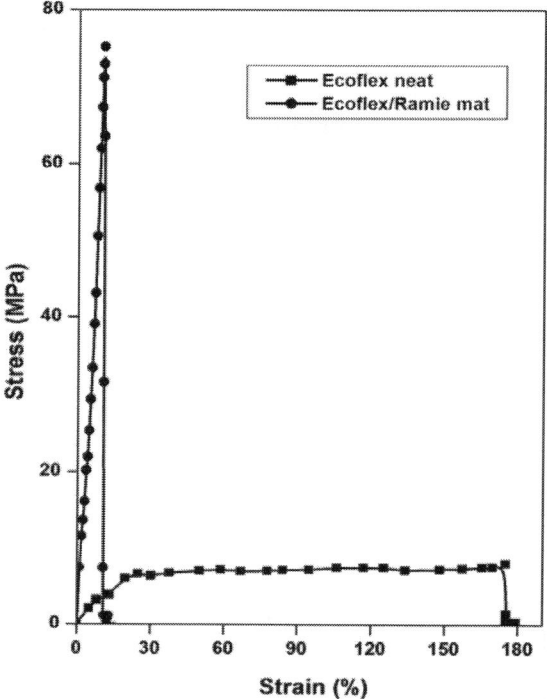

Figure 1. Stress-strain behaviour of Ecoflex neat and Ecoflex/ramie mat composite.

Since ramie fiber is lignocellulosic, it is more susceptible to water. Ramie mat reinforced Ecoflex show more affinity towards water due to its hydrophilicity arising from the exposed reinforcing fibres. This enable it to sorbs considerable amount of water than Ecoflex neat by forming H-bonds between water and hydroxyl groups of cellulose, lignin and hemi cellulose present in the cell wall. The fibre reinforcement causes subsequent reduction in the amount of matrix phase per unit volume of the composite and will result in increase of uptake of polar solvent. In the Ecoflex/ramie mat composite, fast water sorption is observed in the initial stage than Ecoflex neat in **Figure 2**. The difference will be attributed to the interface properties and presence of hydrophilic natural fibres in Ecoflex/Ramie mat composite system. After the initial sorption a slow increase in water sorption is observed for a long duration in the Ecoflex/ramie mat composite. This will be due to the good interface properties in the composite. The attainment of the equilibrium sorption occurred in long duration for the textile composite than that for Ecoflex neat in which the equilibrium is resulted at an early stage. Kinetic Parameters of water diffusion process is analyzed and were given in**Table 2.**

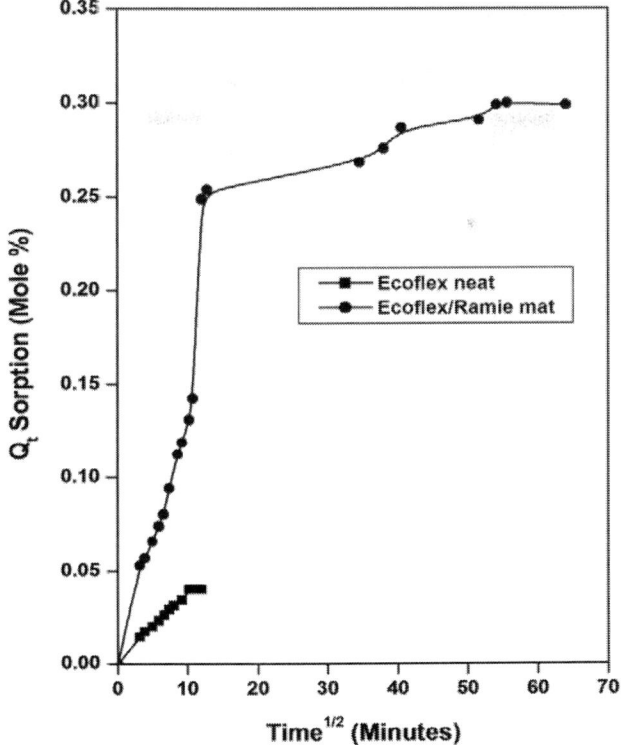

Figure 2. The water sorption behaviour of Ecoflex neat and Ecoflex/ramie mat composite.

From the **Table 2**, it is found that the value of n for Ecoflex neat and Ecoflex/Ramie mat composite is below 0.5. Hence transport behaviour slightly deviate from Fickian value. Deviation from Fickian behaviour is attributed to processes such as desorption, surface crazing, osmotic cracking, micro crack formation, moisture diffusion etc. The value of k is more for Ecoflex neat than for mat reinforced Ecoflex. The factor k is constant that vary with each polymer. Although there is more interaction of Ecoflex neat with water the absorption is less due to close packing of chains in this polymer and the water cannot penetrate easily in to the polymer.

Ecoflex/ramie mat composite has solubility parameter more than that for Ecoflex neat due to the presence of fibres. Percentage swelling index is also more for Ecoflex/ramie mat composite due to increase in number of hydroxyl group of cellulose fibre which makes hydrogen bonding possible with water molecules. D and P parameters vary with polymer and solvent. But by reinforcing mat, fibres help to retain water long time in composite and penetrating into the polymer. Hence sorption is high for Ecoflex/ramie mat composite.

Table 2. Kinetic parameters for Ecoflex neat and Ecoflex/ ramie mat composites in water sorption.

Specimen	n	k	S	D $(cm^2 \cdot S^{-1})$	P $(cm^2 \cdot S^{-1})$	Swelling index (%)
Ecoflex neat	0.396	0.138	0.0073	8.39×10^{-6}	6.14×10^{-6}	0.73
Ecoflex/ ramie mat	0.393	0.065	0.053	9.24×10^{-7}	4.90×10^{-8}	5.3

The dimensional stability of the unsorbed and sorbed resin and composite were investigated. The changes in thickness, diameter and colour of the sample upon water sorption were noted. The values are given in the **Table 3**. Diameter has no change before and after water sorption. Thickness of Ecoflex neat changed slightly only. This indicates the less absorption of water by Ecoflex. But in Ramie mat/Ecoflex composite, thickness is increased significantly due to increased uptake of water by the reinforcing fibers and matrix resin. There is no colour change observed for the textile composite and Ecoflex neat due to water sorption.

3.3. Oil Diffusion Characteristics

The weight percentage uptake of naphthenic oil with time by the Ecoflex neat and composites is shown in **Figure 3**. Here also Ecoflex/ramie mat composite has more oil sorption than Ecoflex neat. This is due to the presence of Ramie fiber mat. Percentage swelling index and solubility parameter S are calculated and given in **Table 4**.

Solubility parameter and swelling index is more for Ecoflex/ramie mat composite than Ecoflex neat. The presence of fibre content is the cause for increase in swelling index. Percentage swelling index is also more for Ecoflex/ramie mat composite due to increase in number of hydroxyl group of cellulose fibre. The dimensional stability of the unsorbed and sorbed Ecoflex neat and the composite were investigated. The changes in thickness, diameter and colour of the sample upon oil sorption were noted. The values are given in the **Table 5**.

Table 3. Thickness and diameter of Ecoflex neat and Ecoflex/ramie mat composite for water sorption.

Specimen	Before water sorption		After water sorption	
	Thickness (mm)	Diameter (mm)	Thickness (mm)	Diameter (mm)
Ecoflex neat	0.79	17	0.80	17
Ecoflex/ramie mat	0.51	17	0.56	17

Table 4. Solubility parameter (S) and swelling index percentage value for oil diffusion.

Specimen	Solubility parameter (S)	Swelling index (%)
Ecoflex neat	0.00459	0.459
Ecoflex/ramie-mat	0.0137	1.37

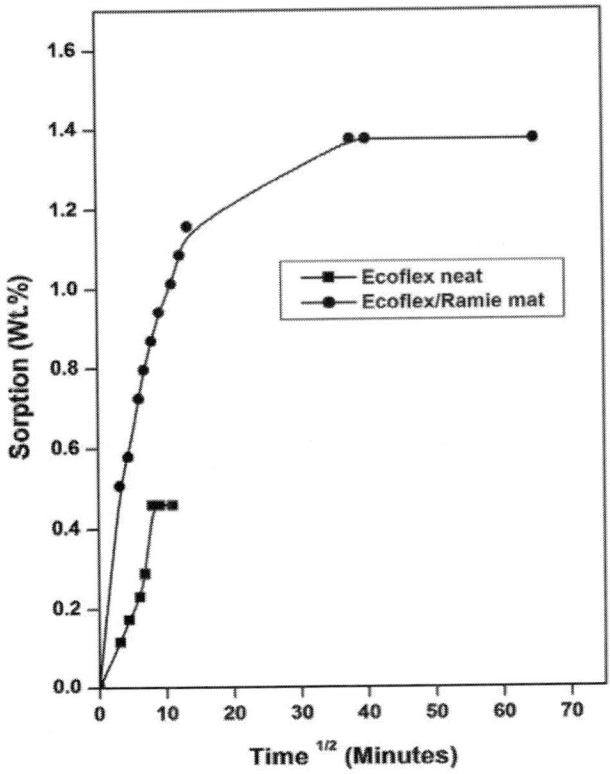

Figure 3. Oil sorption behaviour of Ecoflex neat and Ecoflex/ramie mat composite.

Table 5. Thickness and diameter of Ecoflex neat and Ecoflex/ramie mat composite for oil sorption.

Specimen	Before oil sorption		After oil sorption	
	Thickness (mm)	Diameter (mm)	Thickness (mm)	Diameter (mm)
Ecoflex neat	0.69	17	0.70	17
Ecflex/ramie mat	0.52	17	0.70	17

After oil absorption, diameter is not changed, but thickness is changed. In case of Ecoflex neat only slight increase in thickness observed. But Ecoflex/Ramie mat composite has notable increase is observed. This is due to presence of fibre content and more uptake of oil. After attaining equilibrium the colour of Ecoflex neat and composite changes to yellow.

3.4. Diesel Diffusion Characteristics

The weight percentage uptake of diesel with time by the Ecoflex neat and composites in shown in **Figure 4**. The Ecoflex neat shows more absorption of diesel than Ecoflex/ramie mat. The low molecular weight and molecular hydrocarbons present in Diesel can easily penetrate into neat polymer easily. Diesel molecules will have more interaction with the Ecoflex neat than the textile composite due to the aromatic hydrocarbon present in it. In Ecoflex/ramie mat composite, hydrocarbons diffuse through the fibre matrix interfaces of the composite. The hindrance exerted by the fibres restrict the movement of diesel within the composite and hence possibility of liquid diffusion decreased. In Ecoflex/ramie mat, there is decreased hydrophilicity of cellulose as a result of reaction of hydroxyl group and hydrocarbon. Also enhanced interfacial bonding between fibre and matrix prevent the fibre from absorbing diesel. One peculiarity observed in diesel sorption is the fast rate of initial absorption in both Ecoflex neat and mat composite. This will be due to the increased affinity of diesel to the matrix Ecoflex.

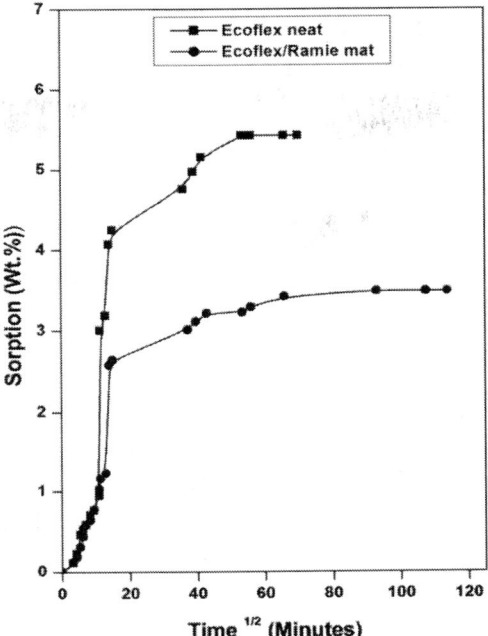

Figure 4. Diesel sorption behaviour of Ecoflex neat and Ecoflex/ramie mat composite.

Table 6. Solubility parameter (S) and swelling index percentage value for diesel diffusion.

Specimen	Solubility parameter (S)	Swelling index (%)
Ecoflex neat	0.0542	5.4
Ecoflex/ramie mat	0.035	3.5

Percentage swelling index and solubility parameter (S) were calculated in **Table 6**. Swelling index and solubility parameter (S) are more for textile composite than Ecoflex neat. This is due to high interaction of Ecoflex neat and solvent. High penetrating power of hydrocarbon contributes to the higher sorption of Ecoflex neat. But in Ecoflex/ramie mat composite, the hydrocarbon in diesel diffuse into the interfaces and voids if any present in composite which restrict the movement of diesel. Hence percentage swelling index and solubility parameter decreased in this case. The decreased amount of matrix due to fabric loading and the good fibre matrix interaction and effective binding of fibre in the composite contribute to the lower diesel uptake in the textile composite sample. Studies on the diesel uptake of natural fibres reinforced natural rubber composite were reported elsewhere. The study reported the effect of natural fibre loading upon the sorption behaviors. They have reported that the percentage swelling index and swelling coefficient of composite were found to decrease with increase in fibre loading. This is due to the increased hindrance exerted by the fibres at higher fibre loading and also due to the good fibre rubber interaction [28].

The dimensional stability of the unsorbed and sorbed Ecoflex neat and composite were investigated. The changes in thickness, diameter and colour of the sample up on diesel sorption were noted. The values are given in the **Table 7**. Up on diesel sorption the diameter of Ecoflex neat and composite does not change. The thickness of Ecoflex neat is increased only slightly up on diesel sorption. But for Ecoflex/Ramie mat composite more increase in thickness is observed. After attaining equilibrium, the colour of both composites changed from while to brown due to the possible chemical interaction. The attainment of equilibrium was of long duration for the textile composite than that for Ecoflex neat.

Table 7. Thickness and diameter of Ecoflex neat and Ecoflex/ramie mat composite for diesel sorption.

Specimen	Before diesel sorption		After diesel sorption	
	Thickness (mm)	Diameter (mm)	Thickness (mm)	Diameter (mm)
Ecoflex neat	0.67	17	0.69	17
Ecoflex/ ramie mat	0.58	17	0.62	17

4. CONCLUSION

Fully biodegradable "green" composites have been prepared from Ecoflex neat/ramie fabric. Mechanical properties such as tensile strength, tensile modulus, elongation at break and diffusion characteristics of the composites in water, naphthenic oil and diesel were studied. Tensile strength of Ecoflex/ramie mat composite is found to be more than that of Ecoflex neat. Young's modulus is also more for Ecoflex/ramie mat composite. Ecoflex neat shows considerable necking effect and plastic deformation. Water absorption is more for Ecoflex-Ramie mat composite than that of Ecoflex neat. The low molecular weight, small sized polar water molecule can easily penetrate into the textile composite which also contains polar cellulose nanofiber, ramie fiber. From the values of n it is clear that mechanism of transport of water through the textile composite shows slight deviation from Fickian mode of transport. The values of swelling parameters, like swelling index% during oil absorption, show that as fiber loading increases, oil sorption increases. But in the case of diesel sorption, as fiber loading increases, sorption decreases. Higher interaction of Ecoflex and diesel resulted in the higher diesel sorption of the matrix. Swelling index% values again supported this. The thickness of the composite is increased after swelling experiments. The diameter of the textile composite remains unchanged after swelling experiment in water, oil and diesel. Thus fully biodegradable "green" composite, Ecoflex/ramie fabric composite showed maximum uptake cf water than diesel and lubricating oil. The "green" composite prepared will be cost effective and will pose no health hazard. Having excellent mechanical and barrier properties it can be safely used in various environments of slight moisture and oil. The textile composite exhibits low oil sorption and reduced wear to the processing tools. Hence these "green" composites can find versatile applications due to their better properties.

5. ACKNOWLEDGEMENTS

One of the authors, M. S. Sreekala, is thankful to Council of Scientific and Industrial Research (CSIR), New Delhi for granting Senior Research Associateship under Scientist's Pool Scheme.

REFERENCES

1. C. L. McCormick, "Agricultural Applications," in: H. F. Mark, Ed., Encyclopedia of Polymer Science and Engineering, 2nd edition, John Wiley & Sons, New York, 1984, p. 611.

2. K. Fukuda, "An Overview of the Activities of the Biodegradable Plastic Society," in: M. Vert, et al., Eds., Biodegradable Polymers and Plastics, Royal Society of Chemistry, Cambridge, 1992, p. 169.

3. D. Briassoulis, "An Overview on the Mechanical Behaviour of Biodegradable Agricultural Films," Journal of Polymers and the Environment, Vol. 12, No. 2, 2004, pp. 65-81.

4. A. S. Herrmann, J. Nickel and U. Riedel, "Construction Materials Based upon Biologically Renewable Resources —From Components to Finished Parts," Polymer Degradation and Stability, Vol. 59, No. 1-3, 1998, pp. 251-261.

5. A. K. Mohanty, M. Misra and G. Hinrichsen, "Biofibers, Biodegradable Polymers and Biocomposites: An Overview," Macromolecular Materials and Engineering, Vol. 276-277, No. 1, 2000, pp. 1-24.

6. V. A. Alvarez, R. A. Ruseckaite and A. Vazquez, "Degradation of Sisal Fiber/Mater Bi-Y Biocomposites Buried in Soil," Polymer Degradation and Stability, Vol. 91, No. 12, 2006, pp. 3156-3162. doi:10.1016/j.polymdegradstab.2006.07.011

7. J. K. Pandey, et al., "An Overview on the Degradability of Polymer Nanocomposites," Polymer Degradation and Stability, Vol. 88, No. 2, 2005, pp. 234-250.

8. G. B. Kiran, K. N. S. Suman, N. M. Rao, R. Uma and M. Rao, "A Study on the Influence of Hot Press Forming Process Parameters on Mechanical

Properties of Green Composites Using Taguchi Experimental Design," International Journal of Engineering, Science and Technology, Vol. 3, No. 4, 2011, pp. 253-263.

9. M. M. Abd El-Latif, A. M. Ibrahim and M. F. El-Kady, "Adsorption Equilibrium, Kinetics and Thermodynamics of Methylene Blue from Aqueous Solutions Using Biopolymer Oak Sawdust Composite," Journal of American Science, Vol. 6, No. 6, 2010, pp. 267-283.

10. A. Krzan, S. Hemjinda, S. Miertus, A. Corti and E. Chiellini, "Standardization and Certification in the Area of Environmentally Degradable Plastics," Polymer Degradation and Stability, Vol. 91, No. 12, 2006, pp. 2819-2833.

11. S. N. Swain, S. M. Biswal, P. K. Nanda and P. L. Nayak, "Biodegradable Soy-Based Plastics: Opportunities and Challenges," Journal of Polymers and the Environment, Vol. 12, 2004, p. 35.

12. R. Jayasekara, et al., "Biodegradation by Composting of Surface Modified Starch and PVA Blended Films," Journal of Polymers and the Environment, Vol. 11, No. 2, 2003, pp. 49-56.

13. Y. X. Xu, et al., "Chitosan-Starch Composite Film: Preparation and Characterization," Industrial Crops and Products, Vol. 21, No. 2, 2005, pp. 185-192.

14. S. S. Joshi and A. M. Mebel, "Computational Modeling of Biodegradable Blends of Starch Amylase and Poly-Propylene Carbonate," Polymer, Vol. 48, No. 13, 2007, pp. 3893-3901.

15. H. Takagi and Y. Ichihara, "Effect of Fiber Length on Mechanical Properties of Green Composites Using a Starch-Based Resin and Short Bamboo Fibers," JSME International Journal, Vol. 47, No. 4, 2004, pp. 551-555.

16. H. Takagi, "Biodegradation Behavior of Starch-Based 'Green' Composites Reinforced by Manila Hemp Fibers," Proceedings of 3rd International Conference on EcoComposites, Stockholm, 20-21 June 2005, p. 14.

17. L. Jiang and G. Hinrichsen, "Flax and Cotton Fiber Reinforced Biodegradable Polyester Amide Composites, 2 Characterization of Biodegradation," Die Angewandte Makromolekulare Chemie, Vol. 268, No. 1, 1998, pp. 18-21.

18. G. Bogoeva-Gaceva, M. Avella, M. Malinconico, A. Buzarovska, A. Grozdanov, G. Gentile and M. E. Errico, "Natural Fiber Eco-Composites," Polymer Composites, Vol. 28, No. 1, 2007, pp. 98-107.

19. W. Wang, M. Sain and P. A. Cooper, "Study of Moisture Absorption in Natural Fiber Plastic Composites," Composites Science and Technology, Vol. 66, No. 3-4, 2006, pp. 379-386.

20. W. J. Liu, L. T. Drzal, A. K. Mohanty and M. Misra, "Influence of Processing Methods and Fiber Length on Physical Properties of Kenaf Fiber

Reinforced Soy Based Biocomposites," Composites Part B: Engineering, Vol. 38, No. 3, 2007, pp. 352-359.

21. F. G. Torres, O. H. Arroyo and C. Gomez, "Processing and Mechanical Properties of Natural Fiber Reinforced Thermoplastic Starch Biocomposites," Journal of Thermoplastic Composite Materials, Vol. 20, No. 2, 2007, pp. 207-223.

22. A. Sorrentino, G. Gorrasi and V. Vittoria, "Potential Perspectives of Bio-Nanocomposites for Food Packaging Applications," Trends in Food Science & Technology, Vol. 18, No. 2, 2007, pp. 84-95.

23. K. Goda, M. S. Sreekala, A. Gomes, T. Kaji and J. Ohgi, "Improvement of Plant Based Natural Fibers for Toughening Green Composites—Effect of Load Application During Mercerization of Ramie Fibers," Composites Part A: Applied Science and Manufacturing, Vol. 37, No. 12, 2006, pp. 2213-2220.

24. M. Jacob, K. T. Varughese and S. Thomas, "A Study on the Moisture Sorption Characteristics in Woven Sisal Fabric Reinforced Natural Rubber Biocomposites," Journal of Applied Polymer Science, Vol. 102, No. 1, 2006, pp. 416-423.

25. M. Jacob, K. T. Varughese and S. Thomas, "Water Sorption Studies of Hybrid Biofiber-Reinforced Natural Rubber Biocomposites," Biomacromolecules, Vol. 6, No. 6, 2005, pp. 2969-2979.

26. C. Z. Paiva Jr, L. H. de Carvalho, V. M. Fonseca, S. N. Monteiro and J. R. M. d'Almeida, "Analysis of the Tensile Strength of Polyester/Hybrid Ramie-Cotton Fabric Composites," Polymer Testing, Vol. 23, 2004, pp. 131-135.

27. V. G. Geethamama and S. Thomas, "Diffusion of Water and Artificial Seawater through Coir Fiber Reinforced Natural Rubber Composites," Polymer Composites, Vol. 26, No. 2, 2005, pp. 136-143.

28. L. Mathew, K. U. Joseph and R. Joseph, "Swelling Behaviour of Isora/Natural Rubber Composites in Oils Used in Automobiles," Chemistry and Materials Science, Vol. 29, No. 1, 2006, pp. 91-99.

CHAPTER 3

Functional Properties of Teff and Oat Composites

George E. Inglett, Diejun Chen, Sean X. Liu

Functional Foods Research Unit, National Center for Agricultural Utilization Research, Agricultural Research Service, United States Department of Agriculture, Peoria, USA

ABSTRACT

Teff-oat composites were developed using gluten free teff flour containing essential amino acids and minerals along with oat products containing β-glucan known for lowering blood cholesterol. Teff-oat composites were evaluated for their pasting and rheological properties by a Rapid Visco Analyzer (RVA) and an advanced rheometer. All teff-oat composites showed increased water holding and pasting viscosities with increasing oat contents compared to wheat flour. However, they were only significantly influenced by 80% oat products in teff-oat composites compared with teff flour alone. OBC (oat bran concentrate) had the highest elastic modulus G' among the starting materials. The elastic modulus G' for teff-Nutrim (oat bran hydrocolloid) composites were decreased with increasing Nutrim contents in composites. In contrast, the increasing content of OBC in composites significantly raised both G' and G". The elastic modulus G' and viscous modulus G" for all teff-OBC composites were higher than teff and wheat flour. All WOF composites showed similar rheological properties. All composites had shear thinning properties that are important to mouthfeel and industrial applications. These teff-oat composites were developed using feasible procedures. They have improved nutritional value and texture qualities for functional food applications.

Keywords: Teff, Oat, β-Glucan, Gluten-Free, Pasting, Rheology

1. INTRODUCTION

Teff (Eragrostis tef) is an important ancient grain finding resurgence in the modern age. The attention to teff has been recently increased because of its health benefits. Teff has an attractive nutritional profile having significant levels of minerals including calcium, iron, magnesium, phosphorus, potassium, and zinc. Also, teff is rich in vitamins, such as thiamin (B1), riboflavin (B2), vitamin A and K [1] . Furthermore, teff is high in proteins including all 8 essential amino acids that is superior in lysine than wheat or barley along with its high carbohydrates and fiber contents [2] . Teff is beneficial for those who are lactose intolerant since it is gluten-free.

In Egypt, red teff grains are greatly recommended for the improvement of osteoporosis and bone healing conditions. A chemical study of the red teff seeds reported the isolation of seven compounds from its ethanol extract, namely β-sitosterol (1), β-amyrin-3-O-(2¢-acetyl)-glucoside (2), β-sitosterol-3-O-β-D-glucoside (3), naringenin (4), naringenin-4¢-methoxy-7-O-α-L-rhamnoside (5), eriodictyol-30,7-dimethoxy-4¢-O-β-D-glucoside (6) and isorhamnetin-3-O-rhamnoglucoside (7). This was the first report on the isolation of compounds (2) and (6) in nature. A proximate analysis revealed the high nutritive value of the seeds: carbohydrates (57.27%), protein (20.9%), essential amino acids (8.15%) with major leucine and lysine (1.71% and 1.35%, respectively), vitamin B1 (1.56 mg/100g), potassium, and calcium. The seeds yielded 22% w/w of oil containing 72.46% of unsaturated fatty acids in which oleic acid was predominant (32.41%) following by linolenic acid (23.83%). The ethanolic extracted oil of the seeds exhibited anti-hyperlipedaemic and antihyperglycaemic activities. Oral administration of oil for 10 days resulted in a rise in serum calcium levels in rats [2] .

Although teff has excellent nutritional value, it has low pasting viscosity and water holding capacity. In contrast, some oat products have high viscosities and water holding capacities. Also, oat products, such as oat hydrocolloid (Nutrim), oat bran concentrate (OBC), and whole oat flour (WOF), contain β-glucan that has beneficial health effects on reducing serum cholesterol and postprandial serum glucose levels [3] . In addition, the phenolic and other antioxidant compounds in oat provide health benefits [4] . Oat hydrocolloid products containing β-glucan have numerous functional food applications to reduce fat content and calories in a variety of foods [5] ; control the rheology and texture of food products [6] ; modify starch gelatinization and retrogradation [7] [8] ; and also provide freezing/thawing stability [9] .

Therefore, teff flour and oat products complement each other in nutritional and textural qualities. There are some food products containing teff and oat, such as bread and cookies. However, teff and oat composites were not scientifically studied and reported. Thus, Nutrim, OBC, and WOF were used in this study to produce unique composites containing β-glucan in combination with teff's distinctive protein and other nutritional components along with its gluten free quality. The objective of our study was to find a feasible way to develop teff-oat composites and to evaluate their functional properties including water holding capacities, pasting, and rheological properties. The functional properties of teff-oat composites could provide useful information for future functional food product developments.

2. MATERIALS AND METHODS

2.1. Preparation of Teff-Oat Composites

Teff flour was purchased from Bob's Red Mill, Milwaukie, OR, USA. OBC (oat bran concentrate) was supplied by Quaker Oats, Chicago, IL, USA. Nutrim (oat bran hydrocolloid, 15 g/100g β-glucan) was provided by VDF FutureCeuticals (Momence, IL, USA). Nutrim was prepared by steam jet-cooking OBC, sieving, and drum- drying [10] [11] . Organic whole oat flour colloidal fine (WOF) was provided by Grain Millers (Eugene, OR, USA).

Teff flour was mixed with corresponding oat products separately using a mixer (KitchenAid, St Joseph, MI, USA) for 2 min. The mixtures were passed through a 20 mesh sieve followed by additional mixing for 1 min to obtain the desired consistency.

2.2. Water Holding Capacity

The water holding capacity (WHC) of the samples was determined according to a previous procedure with minor modifications [12] . Each sample (2 g, dry weight) was mixed with 25 g of distilled water and vigorously mixed for 1 min to a homogenous suspension using a Vortex stirrer, held for 2 h, and centrifuged at 1590 g for 10 min. Each treatment was replicated twice. Water capacity was calculated by the difference between the weight of water added and decanted on dry basis (g of water absorbed/100g of dry sample).

2.3. Pasting Property Measurement

The pasting properties of samples were evaluated using a Rapid Visco Analyzer (RVA-4, Perten Scientific, Springfield, IL). Samples (2.24 g, dry basis) were made up to a total weight of 28 g with distilled water in a RVA canister (8% solids, w/w). The viscosity of the suspensions was monitored during the following heating and cooling stages. Suspensions were equilibrated at 50°C for 1 min, heated to 95°C at a rate of 6.0°C/min, maintained at 95°C for 5 min, cooled to 50°C at rate of 6.0°C/min, and held at 50°C for 2 min. For all test measurements, a constant paddle rotating speed (160 rpm) was used throughout the entire analysis except for 920 rpm in the first 10 s to disperse sample. Each sample was analyzed in duplicate. The results were expressed in Rapid Visco Analyser units (RVU, 1 RVU = 12 centipoises).

2.4. Rheological Measurement

After samples from the RVA were cooled to 25°C and equilibrated, they were loaded on a rheometer (AR 2000, TA Instruments, New Castle, DE, USA) using a 4 cm diameter parallel stainless plate with 1 mm gap to the surface. The outer edge of the plate was sealed with a thin layer of mineral oil (Sigma Chemical Co., St Louis, MO, USA) to prevent dehydration during the test. All rheological measurements were carried out at 25°C using a circulation system within ± 0.1°C. A strain sweep experiment was conducted initially to determine the limits of linear viscoelasticity; then a frequency sweep test was carried out to obtain storage modulus (G') and loss modulus (G") at frequencies ranging from 0.1 to 10 rad s^{-1}. A strain of 0.5%, which was within the linear viscoelastic range, was used for the dynamic experiments. The steady shear viscosity of the paste was measured as a function of shear rates from 1 to 100 s^{-1}. The steady shear measurements apply varying steady shear deformation on sample material, with magnitude of each deformation depending on specified shear rates. All rheological measurements for samples were performed in duplicate.

2.5. Statistical Analysis

All data from replicated samples were analyzed with analysis of variance using Duncan's multiple comparison to determine significant differences ($P < 0.05$) between treatments [13] .

3. RESULTS AND DISCUSSION

3.1. Compositions of Teff, Oat, and Other Cereals

Teff has endured the ages from early civilizations as an important food of a nutritious gluten-free source. Both teff and oat are high in proteins, sugar, calcium, iron, magnesium, phosphorus, and zinc. In addition, teff and oat are rich in vitamins. Teff contains the highest potassium and vitamin B2 whereas oat has the highest thiamin (B1) among all products in Table 1. All B vitamins help the body to convert carbohydrates into energy (glucose). Vitamin B2 is important for normal vision along with other nutrients. Some early evidence shows that Vitamin B2 (riboflavin) might help prevent cataracts which can lead to cloudy vision [14] . Vitamin B1 helps the body metabolizes fat and protein and is required for healthy skin, hair, eyes, and liver. It also helps the nervous system function properly and good brain functions [14] . Remarkably, only teff and oat contain vitamin K among all products in Table 1. Vitamin K is a fat-soluble vitamin that the body stores it in fat tissue and the liver. It is best known for its role in helping blood clot and bone health [14] . Teff has endured the ages from early civilizations as an important highly nutritious food source with gluten-free quality. Oat was selected in this study because of its β-glucan content that is helpful for lowering blood cholesterol and improving food texture. Moreover, oat has high quality lipids including monounsaturated and polyunsaturated fatty acid contents. Therefore, the nutritional value and physical properties of gluten free products could be improved by adding the ancient grain teff and oat products to recipes.

Table 1. Compositions of Teff, oat, and other cereals.

Nutrient (per 100 g)	Unit	teff	rice	corn	whole wheat	Oat
Proximates						
Water	g	8.82	11.89	10.37	10.74	9.37
Energy	kcal	367	366	365	340	371
Protein	g	13.3	5.95	9.42	13.21	13.7
Total lipid (fat)	g	2.38	1.42	4.74	2.5	6.87
Carbohydrate, by difference	g	73.13	80.13	74.26	71.97	68.18
Fiber, total dietary	g	8	2.4	7.3	10.7	9.4
Sugars, total	g	1.84	0.12	0.64	0.41	1.42
Minerals						
Calcium, Ca	mg	180	10	7	34	47
Iron, Fe	mg	7.63	0.35	2.71	3.6	4.64
Magnesium, Mg	mg	184	35	127	137	270
Phosphorus, P	mg	429	98	210	357	458
Potassium, K	mg	427	76	287	363	358
Sodium, Na	mg	12	-	35	2	3
Zinc, Zn	mg	3.63	0.8	2.21	2.6	3.2
Vitamins						
Thiamin (vitamin B1)	mg	0.39	0.138	0.385	0.502	0.54
Riboflavin (vitamin B2)	mg	0.27	0.021	0.201	0.165	0.12
Niacin	mg	3.363	2.59	3.627	4.957	0.82
Vitamin B-6	mg	0.482	0.436	0.622	0.407	0.1
Folate, DFE	µg		4	19	44	32
Vitamin B-12	µg	-	-	-	-	-
Vitamin A, RAE	µg	-	-	11	-	-
Vitamin A, IU	IU	9	-	214	9	-
Vitamin E (alpha-tocopherol)	mg	0.08	0.11	0.49	0.71	0.7
Vitamin K (phylloquinone)	µg	1.9	-	-	-	3.2
Lipids						
Fatty acids, total saturated	g	0.449	0.386	0.667	0.43	1.11
Fatty acids, total monounsaturated	g	0.589	0.442	1.251	0.283	1.98
Fatty acids, total polyunsaturated	g	1.071	0.379	2.163	1.167	2.3

Data were selected from USDA nutrient data base. Teff: uncooked; rice: white, flour; corn: grain, yellow; whole wheat: whole wheat grain flour; oat: Quick Oats, dry.

3.2. Water Holding Capacities

Nutrim had the highest water holding capacity (602.99 g/100g) while wheat flour had the lowest water holding capacity (93.80 g/100g) among all the starting materials (Table 2 & Figure 1). Nutrim was produced by jet- cooking technology using thermal-shearing forces to promote molecular breakdown that probably contributed to its increased water holding capacity [10] [15] . The β-glucan contents for wheat flour, WOF, OBC, and Nutrim were 1.2 g/100g, 3.9 - 7.47 g/100g, 12.0 g/100g and 15.5 g/100g, respectively. The trend of wheat flour, WOF, OBC, and Nutrim water holding capacity (93.80 g/100g, 158.18 g/100g, 339.52 g/100g, 602.99 g/100g) (Table 2) appeared to be related to their β-glucan contents, suggesting β-glucan may be an important factor for WHC.

Data were selected from USDA nutrient data base. Teff: uncooked; rice: white, flour; corn: grain, yellow; whole wheat: whole wheat grain flour; oat: Quick Oats, dry.

At ratios 3:2 and 1:4, the WHC of teff-Nutrim composites (165.78/100g, 295.59 g/100g) were higher than that of teff-OBC composites (160.51 g/100g, 196.59 g/100g) (Table 2 & Figure 1). In contrast, the WHC values of teff-Nutrim composites (135.27 g/100g, 137.57 g/100g) were slightly higher than those of corresponding teff- OBC composites (139.26 g/100g, 154.43 g/100g) at ratios 4:1 and 3:2. The WHC of teff-WOF composites were the lowest (137.70 g/100g, 138.56 g/100g, 139.47 g/100g) compared to teff-Nutrim and teff-OBC composites at all ratios.

Table 2. Water holding capacities of starting materials and teff-oat composites.

Sample	Actual WHC %	Theoretical WHC %	Difference %
teff	134.92 ± 1.01 h		
Nutrim	602.99 ± 6.13 a		
OBC	339.52 ± 0.85 b		
WOF	158.18 ± 0.36 f		
wheat flour	93.80 ± 0.55 i		
teff:Nutrim 4:1	135.27 ± 0.27 h	228.53	93.26
teff:Nutrim 3:2	137.57 ± 1.95 h	322.15	184.58
teff:Nutrim 2:3	165.78 ± 2.08 e	415.76	249.99
teff:Nutrim 1:4	295.59 ± 7.64 c	509.38	213.79
teff:OBC 4:1	139.26 ± 0.67 h	175.84	36.58
teff:OBC 3:2	154.43 ± 8.10 fg	216.76	62.33
teff:OBC 2:3	160.51 ± 7.48 ef	257.68	97.17
teff:OBC 1:4	196.59 ± 2.36 d	298.60	102.01
teff:WOF 4:1	137.70 ± 1.80 h	139.57	1.87
teff:WOF 3:2	138.56 ± 2.05 h	144.22	5.67
teff:WOF 2:3	139.47 ± 1.57 h	148.88	9.41
teff:WOF 1:4	151.07 ± 0.92 g	153.53	2.46

Means ± standard deviation; n = 3; means followed by the same letter within the same column are not significantly different (P > 0.05).

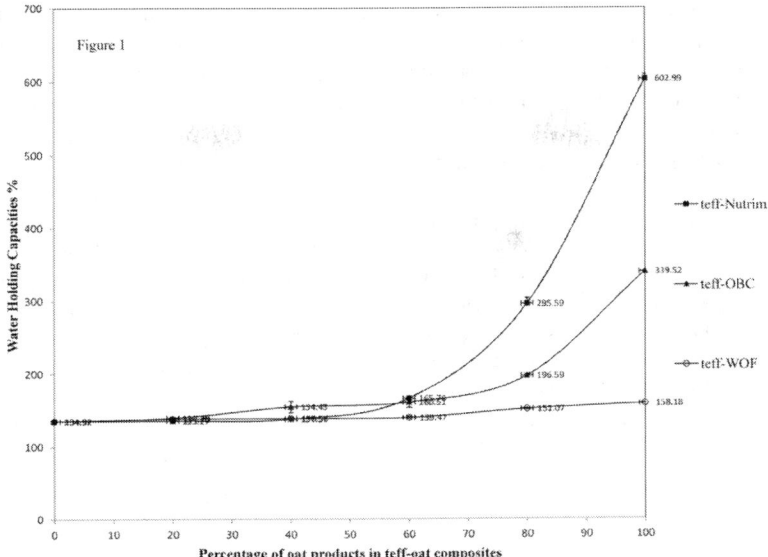

Figure 1. Water holding capacities of starting materials and teff-oat composites.

The actual WHC values of teff-OBC and teff-WOF composites are more close to theoretical WHC values (calculated using the actual WHC values of the starting materials) than teff-Nutrim composites (Table 2). In contrast, the actual WHC of teff-Nutrim composites were ~100% - 200% lower than the theoretical WHC as calculated using the WHC of starting materials (Table 2). Teff flour and oat contain about 13% protein, and teff is rich in calcium and magnesium (Table 1). It is possible that the calcium and magnesium in teff reacted with protein causing precipitation reducing WHC. It could be a similar mechanism to tofu preparation by coagulating proteins in soymilk with calcium or magnesium sulfate. The proteins coagulate when bonding occurs between the positively charged calcium ions and negatively charged anionic groups of the protein molecules. That reaction resulted in protein clumping and the removal of insoluble material from solution. Nutrim was produced by jet-cooking that may possibly produce anionic groups on the surface that could allow interactions with protein and calcium in teff. The data clearly showed that the teff-Nutrim composite with 20% Nutrim resulted in a significant reduction (93%) in WHC. However, the WHC was not decreased linearly when more than 20% Nutrim was replaced for teff (Figure 1). Those results suggested an interaction between Nutrim and teff. The interaction of teff with Nutrim appeared to reach equilibrium after 20% teff was added suggesting a possible saturated surface area.

In general, the WHC for teff-oat composites were all higher than teff flour or whole wheat flour alone. Also, the WHC of teff-oat composites were all apparently increased with the higher oat content (Table 2). Significant increases were found for all teff-oat products when the ratios of teff and oat were changed from 2:3 to 1:4.

Teff-oat composites could be widely used in different applications in the food industry because of improved WHC compared to teff or wheat flour alone. Also, they are notable for their thickening properties, syneresis control, and emulsion stabilization along with nutrients in teff.

3.3. RVA Pasting Properties of Composites

The pasting curves of all starting materials showed dissimilar patterns (Figure 2(a)). The Nutrim pasting viscosity curve increased sharply (~23 RVU/min) and showed a significantly high initial peak (~250 RVU) at 11 min (90°C), followed by a rapid decrease in viscosity to ~25 RVU during continued heating. The Nutrim viscosity slightly increased during cooling showing a considerably low final viscosity (~58 RVU) that was slightly higher than teff flour (~50 RVU). It is known that the viscosity of a completely gelatinized starch slurry decreases during heating [16]. These characteristics are common for pregelatinized flour [17] and typical for Nutrim since it had undergone jet-cooking during preparation where starch gelatinization occurred. The viscosity of OBC increased gradually (~7 RUV/min) to the initial peak (~100 RVU) at 95°C, remained almost constant viscosity during continued heating and shearing, and then increased sharply (~10 RVU/min) resulting in a considerably high final viscosity (~210 RVU) during cooling. This high viscosity could be due to starch gelatinization and interaction with β-glucan that resulted in an entanglement of molecules during cooling to form a matrix with greater stability under heating and shearing. WOF had a lower initial viscosity peak (~50 RVU) than Nutrim and OBC at 95°C, showing a small breakdown (peak viscosity minus the lowest point of viscosity after peak), and then slowly increased to a final viscosity (~89 RVU) that was lower than OBC but higher than the rest of the starting materials. The viscosity of teff showed a lower initial peak (~26 RVU) than all oat products but slightly higher than wheat flour (~16 RVU), and remained constant during heating and shearing, and reached a final viscosity (~50 RVU) similar to Nutrim during cooling. The wheat flour showed the lowest initial peak (~16 RVU) and final peak (~17 RVU) among all samples. The trend of initial peaks from wheat flour (~16 RVU), WOF (~48 RVU), OBC (~100 RVU), and Nutrim (~250 RVU) appeared to be related to their β-glucan contents (1.2 g/100g, 8 g/100g, 12 g/100g and 15 g/100g) with the same trends as their water holding capacities.

The initial and final viscosity peaks of teff-Nutrim 4:1 (~16 RVU), 3:2 (~20RVU), and 2:3 (~23 RVU) composites were increased gradually, and slightly lower than teff flour (~26 RVU, Figure 2(b)). Although the initial peak viscosity of teff-Nutrim 1:4 (~46 RVU) was significantly higher than the other teff-Nutrim composites and teff, it is much lower than Nutrim (~250 RVU). A similar trend was observed for teff-OBC composites as teff-Nutrim composites (Figure 2(c)). It suggested possible interactions between teff with Nutrim and OBC as showed for WHC. Teff-OBC 1:4 had significantly higher final viscosities than the other teff-OBC composites and teff suggesting great stability for the products requiring heating and shearing. All teff–WOF composites had higher initial peaks and final peaks compared to teff and wheat flour (Figure 2(d)). In general, the initial and final peak viscosities of all teff–WOF composites apparently increased with increasing amount of oat products in teff-WOF composites. Overall, all teff-oat composites had higher initial and final viscosities than wheat flour. It is probably attributed to the high protein content (30 g/100g) of teff and beta-glucan from oat products. Higher initial peak viscosity was related to starch gelatinization whereas higher final peak viscosities suggested stability after heating and shearing. Improvement in the textural properties of food using oat β-glucan hydrocolloids has been reported [15] . In general, all 20% teff-oat composites showed a similar viscosity, indicating they would have similar viscosity properties after shearing and heating. The RVA data were useful since it could provide information for food products. Composites having low viscosity may be suitable for products such as drinks or nutritional bars. Composites with high viscosities could be used for products such as breads, muffin, and cookies for improving the texture quality and health benefits.

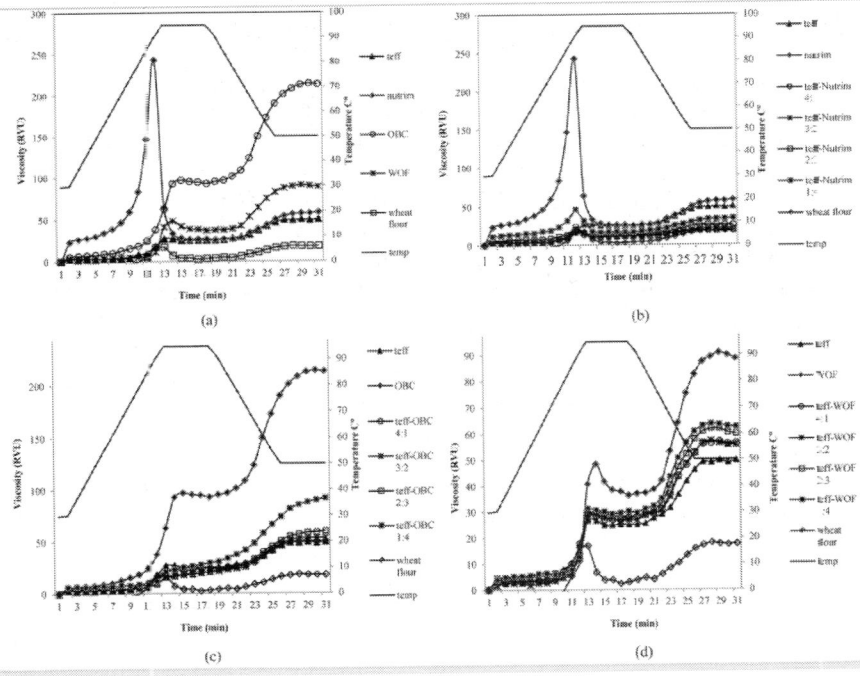

Figure 2. Rapid Visco-analyser pasting curve of starting ingredients and teff-oat composites. (a) Starting materials; (b) teff -Nutrim composites; (c) teff -OBC composites; (d) teff -WOF composites.

3.4. Rheological Properties of Composites

The elastic (storage) modulus G' and viscous (loss) G" with frequency for all the starting materials are displayed in Figure 3. The dynamic viscoelastic properties have been related to product quality [15] . Rheometric measurements are often performed to establish the elastic properties, such as gel strength and yield value, both important parameters affecting particle carrying ability and spreadability. The G', an elastic (storage) modulus, represents the non-dissipative component of the mechanical properties of a material and reflects its elastic characteristics. The viscous (loss) modulus (G") characterizes the dissipative part of the mechanical properties and represents the viscous flow of the material. Viscosity is usually meant resistance to flow or thickness whereas elasticity is related to stickiness or structure.

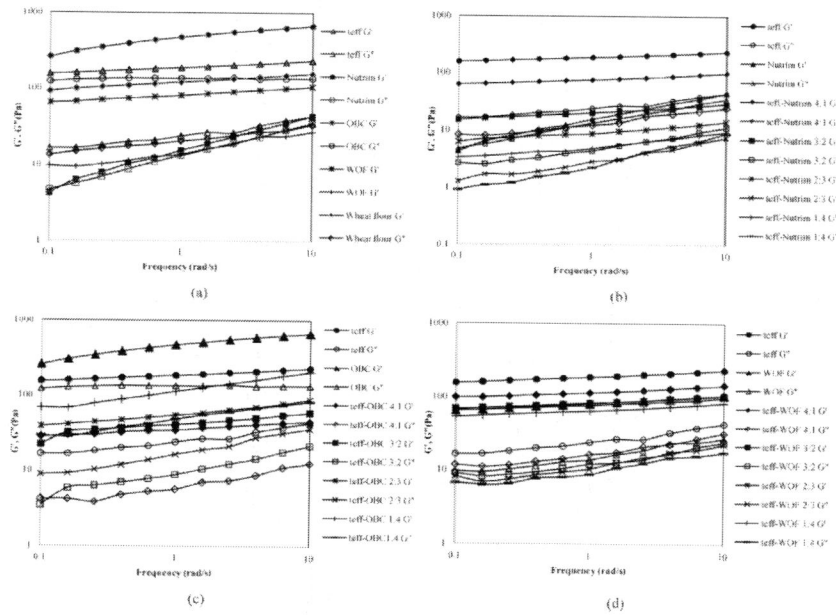

Figure 3. Dynamic viscoelastic properties of starting materials and teff-oat composites. (a) Starting materials; (b) teff- Nutrim composites; (c) teff-OBC composites; (d) teff-WOF composites.

OBC had the highest G' and G" moduli that may be contributed to the polysaccharides of OBC including amylose and amylopectin, followed by teff that maybe contributed to its high protein content. Both G' and G" for Nutrim were considerably lower than OBC and WOF perhaps because of their prior hydrothermal processing.

Elastic moduli G' were greater than viscous moduli G" throughout the frequency range for all starting materials with the exclusion of Nutrim (Figure 3(a)). Both moduli were almost frequency-independent throughout the frequency range tested for OBC, teff, WOF, and wheat flour. Also, the large differences between G' and G" were observed indicating that the composite gels could be classified rheologically as elastic gels [15] [18] . In contrast, both moduli G' and G" of Nutrim showed frequency dependence, suggesting colloidal gel properties of Nutrim. In general, a viscoelastic material behaves in a liquid-like manner at low frequencies when the viscoelastic moduli are considered as a function of frequency [19] . The frequency sweep gives information about the gel strength. A steep slope of the G' curve indicates low strength as Nutrim while a slight slope indicates high strength as OBC and teff flour.

The rheological trends of the starting materials were similar to the RVA data (Figure 2(a)). The molecular weight for Nutrim at its peak was 3.0×10^5 that is smaller than OBC (1.5×10^6) as measured by a size-exclusion chromatography instrument (Shiamdzu, VP series, Tokyo, Japan) [20] . Our rheological results showed a maximum value of dynamic storage modulus G' for Nutrim that was lower than OBC. It was in contrast with a previous report that the maximum value of dynamic storage modulus G' decreased with increasing molecular size of the polysaccharides [21] . It may be due to the heating effect on OBC during the RVA test that could change carbohydrate bonds and influence molecular weight.

In general, both moduli (G' and G") of teff-Nutrim composites are lower than that of teff (Figure 3(b)). The G' of teff-Nutrim 4:1 composites was significantly higher and showed less frequency dependency compared to Nutrim and the remaining teff-Nutrim composites. The elastic modulus G' for teff-Nutrim composites were decreased with the increasing Nutrim contents in composites, indicating that the elastic property of these materials were reduced by the addition of Nutrim. The elastic moduli G' of teff-Nutrim composites 2:3 and 1:4 were lower than Nutrim, suggesting interactions between teff and Nutrim. Such results implied that Nutrim contributed a more viscous property whereas teff contributed a more elastic property to the composite systems. All G' and G" from teff-OBC composites were lower than OBC and teff (Figure 3(c)). In contrast to the teff-Nutrim composites, the increasing content of OBC in composites significantly raised both G' and G". It suggests that the additional OBC increased the elastic properties of teff-OBC composites. The slope of teff-OBC composites 2:3 and 1:4 were increased compared to teff, indicating that the elastic gel property of these materials could be improved by the addition of OBC.

The elastic modulus G' for teff-WOF composite 4:1 was lower than that of teff but higher than WOF while the remaining teff-WOF composites were similar to WOF (Figure 3(d)). Despite the differences in the elastic modulus G' between teff-WOF 4:1 and other teff-WOF composites were noticeable, the viscosity curves of all composites were nearly identical showing similar rheological properties. Also, the elastic or storage modulus G' was greater than loss modulus G" for all teff-WOF composites. It is known that if G' > G", the material exhibited a solid behavior and the deformation in the linear range will be essentially elastic or recoverable [22] .

The rheological results were also explained by tan δ values (Figure 4). Loss tangent, tan d = loss modulus G"/storage modulus G', is a dimensionless value that compares the amount of energy lost during a test cycle to the amount of energy stored during this time [19] which indicates whether a material is solid (d = 0°), liquid (d = 90°), or between (0° < d < 90°). The values of tan d are from zero to infinity; and tan d = 1 means G' = G", tan d < 1 represents G' > G", and tan d > 1 indicates G' < G". The loss tangent indicates whether viscous (tan d >

1) or elastic (tan d < 1) properties predominate in a sample [23] . The tan δ has been used for food products to indicate the strong relationship between the viscous behavior and the degree of hydrolysis such as casein [22] . Of starting materials, Nutrim is the only one having higher tan δ value greater than 1 in the beginning and slightly decreased to 0.92 at the end of test indicating more viscous behavior compare with other starting materials (Figure 4(a)). OBC had a lower tan δ than Nutrim and a decreased trend with an increasing frequency suggesting more elastic properties that may be attributed to some components and structures of OBC (Figure 4(a)). The other starting materials (teff, WOF and wheat flour) had the tan δ value less than 1 indicating G' > G" or elastic properties.

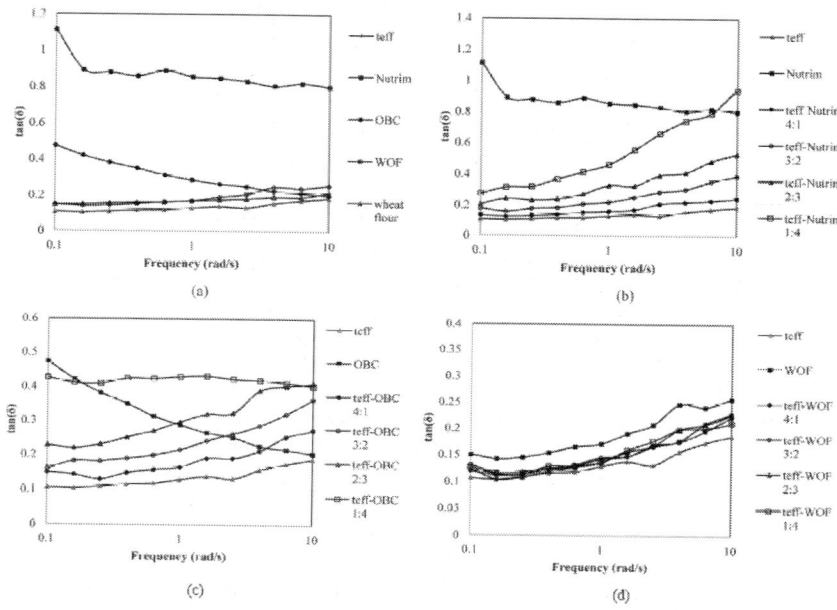

Figure 4. Curves of tan (δ) versus frequency (rad/s) for starting materials and teff-oat composites. (a) Starting materials; (b) teff-Nutrim composites; (c) teff-OBC composites; (d) teff-WOF composites.

Teff-Nutrim composites have tan δ values less than one suggesting that the elastic properties predominate. Also, tan δ values increased from 0.25 to 0.94 at frequency 10 (red/s) as the amount of Nutrim increased from 20% to 80% in composites indicating the loss moduli increased with increasing Nutrim contents in composites (Figure 4(b)). The tan δ values for all teff-OBC composites were higher than that of teff, and lower at low frequency than OBC but higher than OBC at higher frequencies (Figure 4(c)). The tan δ is a rheological parameter that shows the contributions of elastic and viscous components to the rheological properties of the materials. Cross points for tan δ of teff-OBC composites were observed reflecting possible interactions of teff and OBC. This enhanced elastic property could provide better shape retention during handling and cooking. The tan δ curve for teff-OBC 1:4 were nearly flat with frequencies showing great stability and elastic properties of teff- OBC. No dramatic changes were observed for WOF and teff-WOF composites indicating that adding WOF to teff had little effect on the viscosity properties of teff (Figure 4(d)).

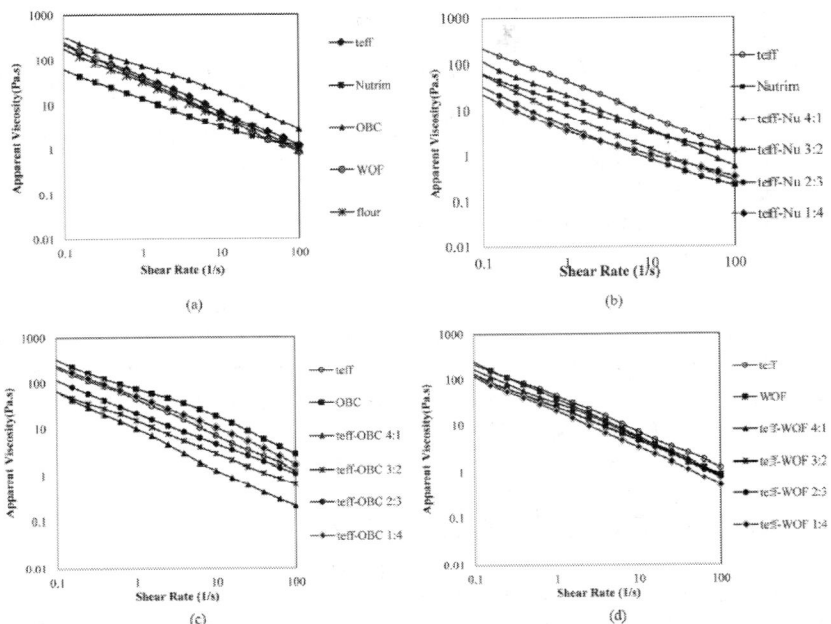

Figure 5. Apparent viscosities versus shear rates of starting materials and teff-oat composites. (a) Starting materials. (b) teff-Nutrim composites. (c) teff-OBC composites. (d) teff-WOF composites.

A viscometric measurement normally consists of a shear rate analysis. The apparent viscosity of the samples vs. shear rate was illustrated in Figure 5. Rheological properties of food products, especially apparent viscosity, have been used as references for predicting their performance during processing [24]. Most food processing and mastication occur in a shear rate range of 1 to 100/sec as used in this study [25]. All of the samples exhibited shear thinning behaviors over the entire measured shear rates at 25°C. Shear-thinning behavior can be observed in many food materials such as soy, rice, and wheat flour. This type of shear-thinning behavior is attributed to the disruption of random coil polymers and/or their parallel alignment with flow stream during shearing [26]. OBC seems to have a higher apparent viscosity, followed by teff, WOF, and wheat flour (Figure 5(a)). Interestingly, Nutrim had low apparent viscosity at the low shear rate, and the apparent viscosities increased gradually with increasing shear rate that were higher at high shear rate similar to teff. These properties could be attributed to the beta-glucan component of Nutrim with its high WHC.

The apparent viscosities of teff-Nutrim composites were decreased with the addition of Nutrim to teff, especially at high shear rate (Figure 5(b)). Teff-Nutrim 1:4 composites showed a similar tendency as Nutrim, which has low apparent viscosity at the low shear rate but increased gradually with increasing shear rate, and reached higher apparent viscosities at high shear rate. The apparent viscosities of the teff-OBC composites were increased with the increasing contents of OBC. Similar patterns between teff-OBC composites and teff were observed (Figure 5(c)). A similar trend was observed for WOF composites compared to Nutrim and OBC composites (Figure 5(b) &Figure 5(d)). The apparent viscosities of teff-WOF composites were not greatly affected by the addition of WOF (Figure 5(d)). The apparent viscosities of WOF composites suggested that those samples could have similar thickening qualities in food processing. Shear-thinning behavior of a material has several potential advantages in food applications since viscosity can be reduced with increasing shear rates. This behavior becomes important in industrial operations such as mixing and pumping. In addition, it was reported that polysaccharide solutions are easily and quickly swallowed where viscosity decreases rapidly at shear rates [27] . Hence, the shear-thinning behavior may contribute a light and nonslimy mouth feel to food products. Since the experimental conditions were similar to actual processing conditions, all our findings on rheological characteristics could be beneficial for processing and developing teff-oat composites for new food applications.

4. CONCLUSION

The pasting and rheological properties of teff-oat composites revealed some useful properties for their potential functional food applications. The teff-oat composites were innovated by a feasible procedure that maintains the original product quality and nutritional values of the starting materials. Teff-oat composites were distinct because the health benefits of the essential amino acids and gluten free uniqueness of teff along with the oat soluble fiber β-glucan for food texture and coronary heart disease prevention. The composites also have improved water holding capacities, texture, and viscoelastic qualities. The teff-oat composites could be valuable for developing new functional foods having improved nutritional value and desirable texture qualities for health concerned consumers.

REFERENCES

1. USDA Nutrient Database (2014)http://ndb.nal.usda.gov/ndb/search/list http://ndb.nal.usda.gov/ndb/foods/show/6238
2. Taha, S.E., Shahira M.E. and Amani A.S. (2012) Chemical and Biological Study of the Seeds of Eragrostis tef (Zucc.) Trotter. Natural Product Research: Formerly Natural Product Letters, 26, 619-629.

3. Klopfenstein, C.F. (1988) The Role of Cereal Beta-Glucans in Nutrition and Health. Cereal Food Worlds, 33, 865-869.

4. Madhujith, T. and Shahidi, F. (2007) Antioxidative and Antiproliferative Properties of Selected Barley (Hordeum vulgare L.) Cultivars and Their Potential for Inhibition of Low-Density Lipoprotein (LDL) Cholesterol Oxidation. Journal of Agricultural and Food Chemistry, 55, 5018-5024.

5. Lee, S., Inglett, G.E. and Carriere, C.J. (2004) Effect of Nutrim Oat Bran and Flaxseed on Rheological Properties of Cakes. Cereal Chemistry, 81, 637-642.

6. Rosell, C.M., Rojas, J.A. and de Barber, C.B. (2001) Influence of Hydrocolloids on Dough Rheology and Bread Quality. Food Hydrocolloids, 15, 75-81.

7. Rojas, J.A., Rosell, C.M. and de Barber, C.B. (1999) Pasting Properties of Different Wheat Flour-Hydrocolloid Systems. Food Hydrocolloids, 13, 27-33.

8. Lee, S., Warner, K. and Inglett, G.E. (2005) Rheological Properties and Baking Performance of New Oat β-Glucan-Rich Hydrocolloids. Journal of Agricultural and Food Chemistry, 53, 9805-9809.

9. Lee, M.H., Baek, M.H., Cha, D.S., Park, H.J. and Lim, S.T. (2002) Freeze-Thaw Stabilization of Sweet Potato Starch Gel by Polysaccharide Gums. Food Hydrocolloids, 16, 345-352.

10. Inglett, G.E. (2000) Soluble Hydrocolloid Food Additives and Method of Making. U. S. Patent Number 6,060,519.

11. Inglett, G.E. (2011) Low-Carbohydrate Digestible Hydrocolloidal Fiber Compositions. U. S. Patent Number 7,943,766B2.

12. Ade-Omowaye, B.I.O., Taiwo, K.A., Eshtiaghi, N.M., Angersbach, A. and Knorr, D. (2003) Comparative Evaluation of the Effects of Pulsed Electric Field and Freezing on Cell Membrane Permeabilisation and Mass Transfer during Dehydration of Red Bell Peppers. Innovative Food Science & Emerging Technologies, 4, 177-188.

13. SAS Institute Inc. (1999) The SAS® System for Windows®, Version 8e. Cary, NC.

14. University of Maryland Medical Center.http://umm.edu/health/medical/altmed/supplement/vitamin-b2-riboflavin#ixzz3YoxeKZsF

15. Lee, S. and Inglett, G.E. (2006) Rheological and Physical Evaluation of Jet-Cooked Oat Bran in Low Calorie Cookies. International Journal of Food Science and Technology, 41, 553-559.

16. Guha, M., Ali, S.Z. and Bhattacharya, S. (1998) Effect of Barrel Temperature and Screw Speed on Rapid Viscoanalyser Pasting Behaviour of Rice Extrudate. International Journal of Food Science and Technology, 3, 259-266.

17. Lai, H.-M. and Cheng, H.-H. (2004) Properties of Pregelatinized Rice Flour Made by Hot Air or Gum Puffing. International Journal of Food Science and Technology, 39, 201-212.

18. Lai, L.S. and Liao, C.-L. (2002) Steady and Dynamic Shear Rheological Properties of Starch and Decolorized Hsiantsao Leaf Gum Composite Systems. Cereal Chemistry, 79, 58-63.

19. Ferry, J.D. (1980) Viscoelastic Properties of Polymers. 3rd Edition, John Wiley, New York.

20. Kim, S., Inglett, G.E. and Liu, S.X. (2008) Content and Molecular Weight Distribution of Oat β-Glucan in Oatrim, Nutrim, and C-Trim Products. Cereal Chemistry, 85, 701-705.

21. Vaikousi, H., Biliaderis, C.G. and Izydorczyk, M.S. (2004) Solution Flow Behavior and Gelling Properties of Water-Soluble Barley $(1\rightarrow3,1\rightarrow4)$-$\beta$-Glucans Varying in Molecular Size. Journal of Cereal Science, 39, 119-137.

22. Gravier, N.G., Zaritzky, N.E. and Califano, A.N. (2004) Viscoelastic Behavior of Refrigerated and Frozen Low-Moisture Mozzarella Cheese. Journal of Food Science, 9, 123-128.

23. Sutheerawattananonda, M. and Bastian, E.D. (1998) Monitoring Process Cheese Meltability Using Dynamic Stress Rheometry. Journal of Texture Studies, 29, 169-183.

24. Salvador, A., Sanz, T. and Fiszman, S. (2002) Effect of Corn Flour, Salt, and Leavening on the Texture of Fried, Battered Squid Rings. Journal of Food Science, 76, 730-773.

25. Bloksma, A. (1998) Rheology of the Bread Making Process. Paper presented at 8th International Cereal and Bread Congress, Lausanne, PMid: 9772201.

26. Salamone, J.C. (1996) Polymeric Materials Encyclopedia. CRC Press, Boca Raton.

27. Szczesniak, A.S. and Farkas, E. (1962) Objective Characterization of the Mouthfeel of Gum Solutions. Journal of Food Science, 27, 381-385.

CHAPTER 4

Mineralogical Study of Polymer-Mortar Composites with PET Polymer by Means of Spectroscopic Analyses

Ahmed Soufiane Benosman[1,2]*, Mohamed Mouli[2], Hamed Taibi[1], Mohamed Belbachir[1]. Yassine Senhadji[2], Ilies Behlouli[1], David Houivet[3]

[1]Faculty of Science, Laboratory of Polymer Chemistry, University of Oran Es-Senia, Oran, Algeria;[2]Department of Civil Engineering, Laboratory of Materials, ENSET, Oran, Algeria; [3]Laboratory of LUSAC, University of Caen Basse-Normandie, CherbourgOcteville, France.

ABSTRACT

The sheer amount of disposable bottles being produced nowadays makes it imperative to identify alternative procedures for recycling them since they are non-biodegradable. Experimental investigation on the effects of polyethylene terephthalate (PET) polymer, which is a waste material obtained by crushing of used PET bottles, on the mineralogical composition of composites after 28 days of casting are presented in this paper. Various weight fractions of cement 2.5%, 5% and 7.5% were replaced by the same weight of PET plastic; they were then moulded into specimens and cured. The fine powder samples obtained from broken specimens were subjected to X-ray diffraction, FT-IR spectroscopy, differential thermal analysis, thermogravimetric analysis and the composites were also observed by optical microscope. Thermogravimetry (TG) and derivative thermogravimetry (DTG) were used to study the interaction between polymers and cements. Differential thermal analysis (DTA), X-ray diffraction and FT-IR were also used to investigate the cement hydration according to the additions. The results showed that an increase in polymer-cement ratio meets with a decrease in the quantity of $Ca(OH)_2$; in terms of bonding, the rough surface of particle favours greater contact between PET and cement matrix and doesn't seem to have chemical interaction between the mineral species and the organic molecules which could lead to the formation of new compounds. The present study highlights the capabilities of the different methods for the analysis of composites and opened new way for the recycling of PET in polymer-mortars.

Keywords: Composite; PET Polymer; Interaction; TG/DTG; DTA; X-Ray
Diffraction; FT-IR

1. INTRODUCTION

Polymer-modified mortars have been popular construction materials because of
their excellent properties in comparison with ordinary mortars. Polymers have
been used for improving mechanical properties, adhesion with substrates, or
waterproofing properties of mortars and concretes. The literature agrees that the
properties of polymer modified mortar and concrete depend significantly on the
polymer content or polymercement ratio, that is, the mass ratio of the amount of
polymer solids in a polymer-based admixture to the amount of cement in a
polymer-modified mortar or concrete [1-3].

Although extensive research has been done on recycled materials, there have
been very few studies concerning lightweight concrete which incorporates waste
products as aggregates. The self weight of concrete elements is high and can
represent a large proportion of the load on a structure [4-6]. Therefore, using
lightweight concrete with a lower density can result in significant benefits such
as superior load-bearing capacity of elements, smaller cross-sections and
reduced foundation sizes. A lightweight structure is also desirable in earthquake
prone areas.

Polyethylene terephthalate (PET) is one of the most common consumer
plastics used and is widely employed as a raw material to realize products such
as blown bottles for soft-drink use and containers for the packaging of food and
other consumer goods. PET bottles have taken the place of glass bottles as
storing vessel of beverage due to its lightweight and easiness of handling and
storage.

In 2007, it is reported a world's annual consumption of PET drink covers of
approximately 10 million tons, which presents perhaps 250 milliards bottles.
This number grows about up to 15% every year [7]. On the other hand, the
number of recycled or returned bottles is very low. Generally, the empty PET
packaging is discarded by the consumer after use and becomes PET waste
(WPET).

The major problems that this level of waste production generates initially
entail storage and elimination [8]. The recycling of PET bottles and the
preservation of natural resources are priority items but to date, the recycling of
PET bottles as a lightweight aggregate for concrete has not been studied because
of the high melting cost [9].

Researches into new and innovative uses of waste materials are continuously
advancing. These research efforts try to match society's need for safe and
economic disposal of waste materials. The use of recycled aggregates saves
natural resources and dumping spaces, and helps to maintain a clean
environment [10-16]. Thus, this current study concentrates on the waste
materials without any further transformation beyond crushing, in order to
minimize final material costs. Specifically plastics waste (PET) is used as

substitute for conventional materials, mainly cement, in polymer-mortar composites mixes.

Even though the main effects of some polymeric additives on the mechanical properties of mortars or concrete are well established, the mechanisms responsible for these effects are not yet fully understood.

Furthermore calcium hydroxide, $Ca(OH)_2$, being one of the major phases in hydrated Portland cement, occupies about 20% to 25%, by volume, of the total mass [17]. It may affect the physico-mechanical properties of the hydrated cement to a considerable extent [18]. Additionally, the role which $Ca(OH)_2$ plays towards the engineering properties and durability characteristics of the systems based on hydrated cement is still a matter of debate among researchers. While studying the durability aspects of cement systems, many researchers have found $Ca(OH)_2$ being much more unstable in resisting the adverse environmental conditions than calcium silicate hydrate (CSH) gel [19]. $Ca(OH)_2$ is produced by the hydration of tricalcium silicate (C_3S) and dicalcium silicate (C_2S) phases of the cement. In the set Portland cement, $Ca(OH)_2$ is formed with well-defined crystallinity having a definite stochiometry, which may affected by the presence of additives and admixtures [17].

In the previous work, the author studied the effects of PET polymer on the mortar properties, specifically to decrease the chloride ion penetration depth and apparent chloride ion diffusion coefficient of polymer-mortar composites. This may be explained due to the reduced volume of large-sized pores and the improved resistance to the absorption of the test solutions with an increase in polymer-cement ratio [20]. In addition, Benosman et al., [21] showed the improvement of the adherence strength and the resistance to aggressive solutions of composites using such additions. For this reason, aim of this work is to study the influence of such admixtures concerning the hydration of Portland cement by means of the mineralogical study of the composites with the same composition used in the mentioned works [20,21].

The interaction between polymers and cement Portland can be investigated through several techniques such as thermal analysis, X-ray diffraction and FT-IR. Thermogravimetry (TG), derivative thermogravimetry (DTG) and differential thermal analysis (DTA) are considered important tools for evaluating the nature of hydrated products according to different stages of cement hydration, in addition to quantifying the different phases [22- 25].

When cement is hydrated, its main components are transformed into hydration products, mainly calcium silicate hydrate (C-S-H gel) and portlandite. The hydration can be evaluated by measuring the mass loss of hydrated compounds up to 900°C. The following peaks and temperature ranges have been studied when hydrated cement is heated in thermobalance and they are interpreted as described below [23,24]:

- ~100°C: dehydration of pore water
- 100°C - 300°C: different stages of C-S-H dehydration
- ~500°C: dehydroxylation of $Ca(OH)_2$
- ~700°C: decarbonation of $CaCO_3$.

This study reports the results of investigations in which methods of thermal analysis, TG, DTG and DTA, were applied to investigate the effects of polymer modification on the process of hydration of portland cement by estimating $Ca(OH)_2$ content and calcium hydrate content. X-ray diffraction and FT-IR were carried out to study the hydrate products of cement. Certain key proportions are also studied, in contrast with what has been undertaken in previous work [16,26] in order to determine feasibility limits. Also, the analysis by the optical microscope was carried to view the arrangement of PET particles in the cement matrix. The present study highlights the capabilities of the different methods for the analysis of composites and has shown quite encouraging results and opened new way for the recycling of PET waste in polymer-mortars.

2. EXPERIMENTAL

The mortar and/or mortar-polymer composites mixtures were prepared in collaboration with two laboratories: the Laboratory of Materials, Civil Engineering Department, ENSET Oran (Algeria) and the Laboratory of Polymer Chemistry, University of Oran Es-Senia, using the following materials:

2.1. Cement

The cement used was a blended Portland cement type CPJ-CEM II/A (pouzzolanic cement) delivered from Zahana factory located in the western Algeria, chemical and physical properties of cement are shown in Tables 1 and 2, respectively, according to the manufactories. The chemical composition was obtained by using an X-ray fluorescence spectrometer analysis type OXFORD MDX^{1000}.

Table 1. The chemical composition of cement.

Chemical Composition	CPJ-CEM II/A (%)
Loss on Ignition	2.09
SiO_2	21.82
Al_2O_3	6.57
Fe_2O_3	4.01
CaO	63.43
MgO	0.21
SO_3	1.86
CaO free	0.24

2.2. Polymer

PET is a thermoplastic polyester with tensile and flexural modulus of elasticity of about 2.9 and 2.4 GPa, respectively, tensile strength up to 60 MPa and excellent chemical resistance. It is a semi-crystalline polymer, with a melting point of about ~260°C (**Figure 1**). Its density (specific gravity) is around 1.3 - 1.4 g/cm³. The PET powder is obtained by finely crushing the drink bottles (waste plastics). After preliminary tests, polymer particles of size lower than 1 mm were used in this study.**Figure 2** and **Table 3** show the infrared spectra of PET in KBr pellet.

The analysis of PET was performed in the Laboratory of Applied Sciences of Cherbourg, University of Caen Basse Normandie, France, using DTA/TGA 92 Setaram equipment. The experimental conditions were: N₂ gas dynamic atmosphere; heating rate (10 K/min); alumina top-opened crucible. The samples were heated in the range of 20°C - 350°C.

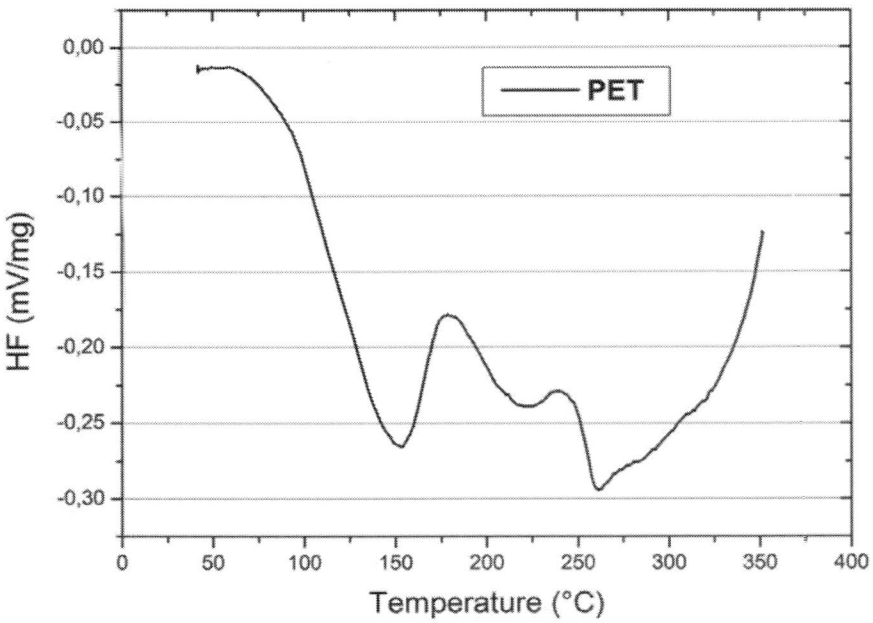

Figure 1. DTA curve at 10 K/min of the polyethylene terephthalate PET.

Table 2. Physical properties of cement.

Setting Time (min)		Blaine Surface Area (cm³/g)	Absolute Density (g/cm³)	Compressive Strength (MPa)
Initial	Final			28-days
120	200	2987	3.09	32.5

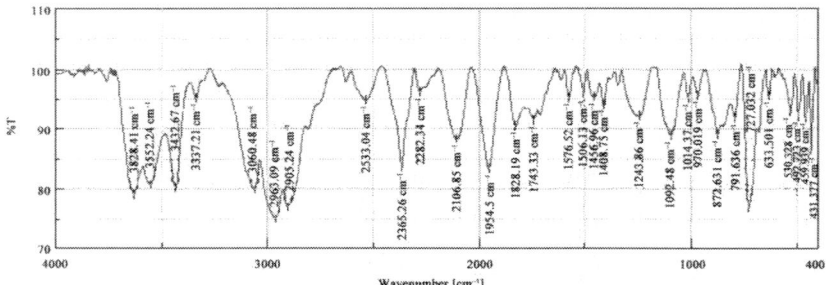

Figure 2. Fourier-transform infrared spectrum of PET polymer in KBr pellet.

Table 3. Fourier-transform infrared table of PET polymer in KBr pellet.

Bands identification	
(cm⁻¹)	Group
872.63	Alternating deformation of benzene, out of the plane (para-disubstituted)
1092.48	-C-O
1576.52 - 1660.2	C=C, aromatic character
1743.33	-COO-
2963.09 - 2905.24	-CH₂-
3060.48	CH, benzene group

Table 4. The chemical composition of sand.

Chemical Composition	Sand (%)
Loss on ignition	43.83
SiO₂	0.77
Al₂O₃	0.11
Fe₂O₃	0.36
CaO	54.71
MgO	0.21
SO₃	Nil
Chlorures	Nil

2.3. Sand

Crushed sand obtained from Kristel quarry in Oran, West Algeria was used. The chemical properties shown in Table 4 were obtained by using an X-ray fluorescence spectrometer analysis type OXFORD MDX^{1000}.

Table 5. Mix proportions and physical properties of polymer-mortar composites.

Mix Design	Polymer-Cement Ratios (%)	Water Demand for Standard Consistency (%)	Setting Time (min)		Density (g/cm³)
			Initial	Final	
PET 0	0	24.5	120	200	2.28
PET2.5	2.5	25	125	205	2.23
PET5.0	5.0	25.5	130	210	2.22
PET7.5	7.5	26	145	225	2.21

2.4. Experimental Program

Four mixtures were prepared as described in **Table 5**, which are the composites with the same proportions used in the previous works [20,21], as explained in the introduction. The materials were weighed and mixed in a planetary type mortar mixer. The mortar mixes had proportions of 1 binder: 3 Sand (by weight). The binder consisted of cement and polyethylene terephthalate. The water to binder ratio was kept constant at 0.5. The physical properties of the pastes of mortars were determined in accordance with EN 196-3 [27].

The preparation for TG/DTG, DTA, X-ray diffraction and FT-IR analysis was carried out using agate crucible, in which the mortar or composite was manually ground until the size of particles was lower than 0.160 mm. For the prevention of carbonation and maintenance of relative humidity, all specimens were stored in the vacuum up to the time when the test started.

The analyses were performed in the Laboratory of Polymer Chemistry, University of Oran, using Labsys TGA SETARAM Thermogravimetric Analyzer equipment, easy to use (ambient to 1600°C). The experimental conditions were: He gas dynamic atmosphere; heating rate (10°C·min⁻¹). The samples were heated in the range of 20°C - 1100°C at a constant rate. The $Ca(OH)_2$ and calcite were estimated from the weight losses measured in the TG curve between the initial and final temperature of the corresponding TG peak.

The composites were submitted to thermal analysis in a DTA equipment, type Linseis mark DTA L62, in the Laboratory of Materials, ENSET Oran. The experimental conditions were 1) continuous heating from 25°C to ~1100°C; 2) heating rate: 10°C/min; 3) alumina, topopened crucible. DTA curves were obtained.

XRD was used to identify the polycrystalline phases of cement and hardened cement paste by means of the recognition of the X-ray patterns that are unique for each of the crystalline phases. The qualitative XRD investigation was performed in a Philips PW 1710 X-ray diffractometer with Cu Kα radiation and 2θ scanning, ranging between 5° and 80° of 2θ.

The composites were analyzed by an optical microscope, type Keyence VH-5911 commissioned by Microvision software, available at the Laboratory LMDC of INSA, Toulouse, France.

A FT/IR-4200 Type A, Photometrec Fourier-transform infrared spectrometer was used, and the mortar/composite cement were analyzed in KBr pellets. The spectra were traced in the range of 4000 - 400 cm⁻¹ (wave number), and the band intensities were expressed in transmittance (%).

3. RESULTS AND DISCUSSION

3.1. Thermal Analysis (DTA, TG/DTG)

The DTA and TG curves obtained in all of the tests are typical of hydrated cement composites containing carbonate phases. Three major endothermic reactions occurred during the heating of the samples: 1) release of the evaporable and part of the adsorbed water (CSH and ettringite) at 110°C approximately; 2) Ca(OH)$_2$ dehydration, between 450°C and 510°C; 3) decomposition of the carbonate phases at ~850°C - 900°C (**Figure 3**). As shown in **Figure 4**, PET polymer strongly influences the DTA curve, causing 1) enlargement of the exothermic shoulder at ~200°C - 420°C; 2) decreasing of the endothermic peak intensity and weight loss on the dehydration of calcium hydroxide; 3) sharp changes in the curves at temperatures higher than 510°C (in most cases, an endothermic peak at 580°C and three exothermic shoulders at 820°C - 840°C, 620°C - 660°C and 540°C were detected); 4) enlargement of the endothermic peak intensity at 620°C - 780°C and decreasing of the peak temperature (from about ~937°C to ~926°C); and e) decreasing of the weight loss in the latter temperature range. These results comply with an early study of Silva et al. [28], in which EVA copolymer strongly influences the DTA/TG curves.

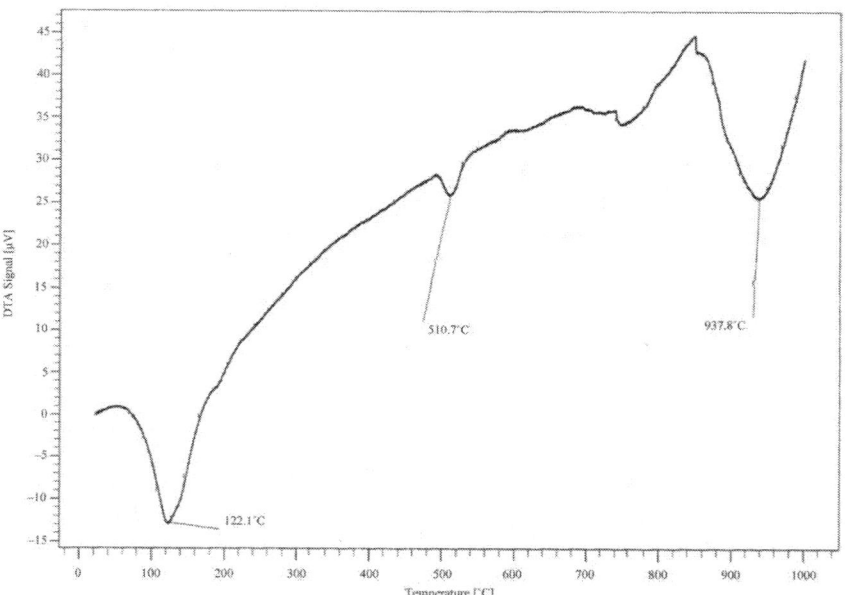

Figure 3. DTA curve at 10°C/min of an unmodified mortar PET0.

Figure 5(a) shows the TG curves of mortar without polymer PET0. It can be seen that TG/dTG curves for this mortar consist of four zones:

~40°C - 123.3°C: dehydration of pore water~123.3°C - 420°C: dehydration of calcium silicate hydrates~420°C - 510°C: dehydroxylation of calcium hydroxide~840°C: decarbonation of $CaCO_3$.

Figure 5(b), Figures 6(a) and 6(b) show TG curves of composites with PET polymer addition of 2.5%, 5% and 7.5% (polymeric solids). The TG curves obtained in these tests are typical of hydrated cement composites containing carbonate phases and PET polymeric admixtures influences. As it is shown, the curves can be divided into five major parts, according to different reactions:

~123.3°C: dehydration of pore water~123.3°C - 345°C: dehydration of calcium silicate hydrates~490°C - 517°C: dehydroxylation of calcium hydroxide~650°C - 780°C: Enlargement of the endothermic peak ~800°C: decarbonation of $CaCO_3$.

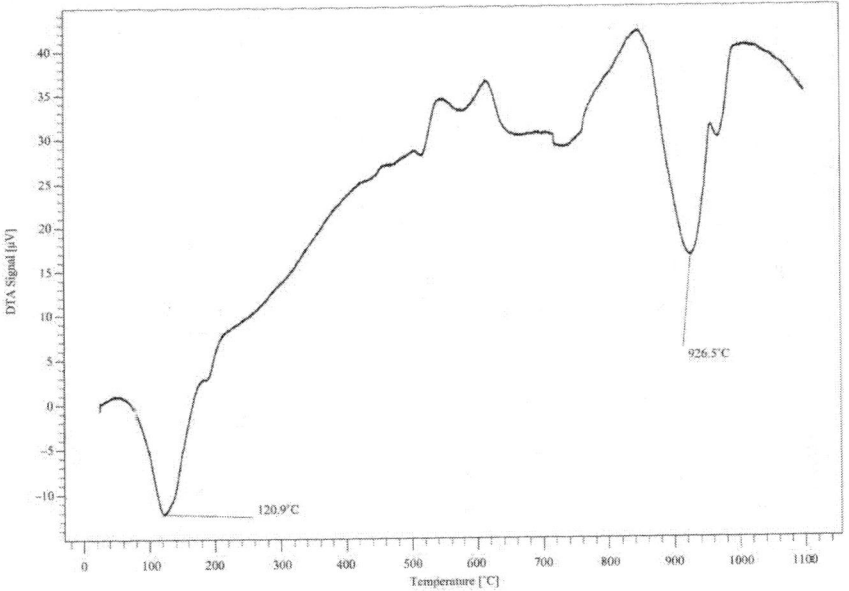

Figure 4. DTA curve at 10°C/min of the composite PET5.

All the weight loss data are expressed as a function of the ignited weight of the sample, as suggested by Taylor [29]. The calcium hydroxide content was determined from the following equation:

$$CH(\%) = WL_{CH}(\%) \times \frac{MW_{CH}}{MW_H}$$

(1)

where CH(%) is the content of $Ca(OH)_2$ (in weight basis), WL_{CH}(%) is the weight loss occurred during the dehydration of calcium hydroxide (in weight basis), MW_{CH} is the molar weight of calcium hydroxide and MW_H is the molar weight of water. Since the exact stoichiometry of decomposition reactions of the carbonate phases is not known, the results are expressed in function of the weight of CO_2 gas released during the decomposition, and not as carbonate phase's content.

When one adds the PET to the composite there is a diminution of the calcium hydroxide (CH) content and an increase of the dehydration temperature, as can be seen in Figures 7(a) and (b). Also, Silva et al. [28] demonstrated that EVA polymer strongly reduces the CH content. **Figure 8**shows that PET polymer sharply decreases the quantity of carbonate phases in the composites, measured by TG/DTG.

Because when the rate of substitution of PET increases, there is less formation of portlandite $Ca(OH)_2$, which is mainly the result of the following reaction:

$$C_3S \text{ and } C_2S + H_2O \quad \blacksquare\text{-S-H} + Ca(OH)_2 \qquad (2)$$

3.2. XRD Analyses

The XRD results show some qualitative differences in the hydration rate due to the incorporation of PET polymer. **Figure 9** shows the X-ray patterns of the composites with 7.5% of polyethylene terephthalate, and composites without polymer PET0. The main compounds observed are $Ca(OH)_2$in the form of portlandite, a big amount of $CaCO_3$ resulting from carbonation of $Ca(OH)_2$ and calcium silicate anhydrous. The peak intensity in the region $2q = 18° - 18.1°$ [30,31] has been considered as a measure of the quantity of $Ca(OH)_2$ [32].

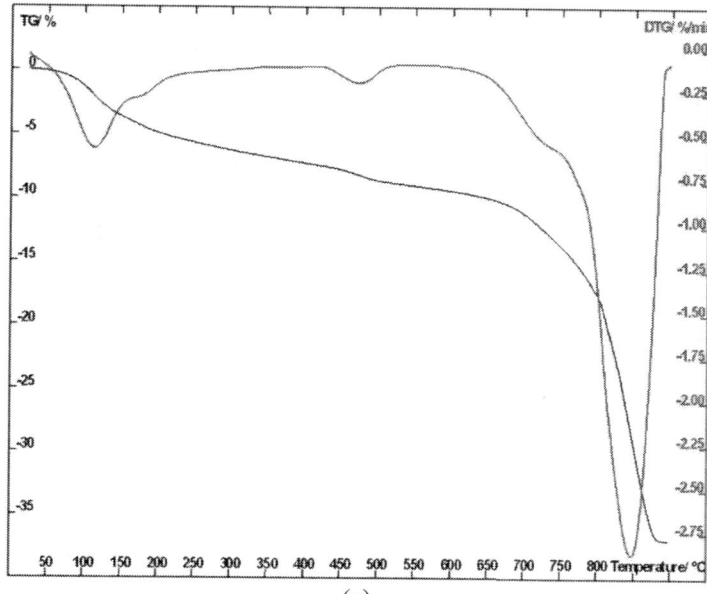

(a)

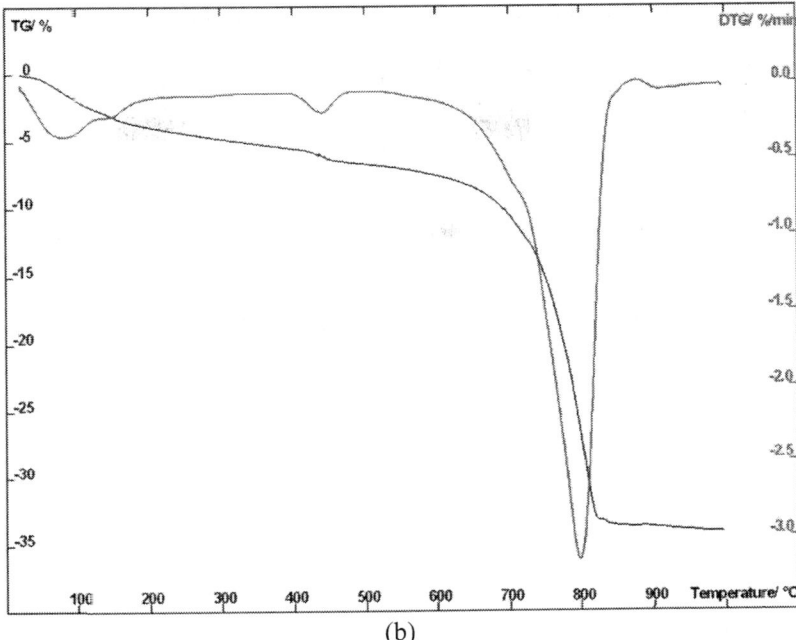

(b)

Figure 5. TG/DTG curves at 10°C•min–1 of unmodified mortar CPJ-PET0 (a) and the composite PET2.5 (b).

So, it is also noted here that at a polymer-cement ratio of 7.5% a slight increase in the peak intensity compared to the unmodified mortar is observed. Further, $Ca(OH)_2$ crystals may possibly produce sharper reflection in the presence of PET polymer due to a change in the orientation pattern of the crystals. This sharper reflection offsets the effect gained by the lowered quantity of $Ca(OH)_2$, and so the peak intensity at a polymer-cement ratio of 7.5% appears at an increased level compared to that of the unmodified mortar. These results are in line with those of some previous research's [19,26,31].

On the basis of the experience gained by DTA/TGA analysis, it is possible to explain this variation in terms of the fact that the PET addition causes a progressive decrease in the amount of portlandite $Ca(OH)_2$ in the composites compared to unmodified mortar.

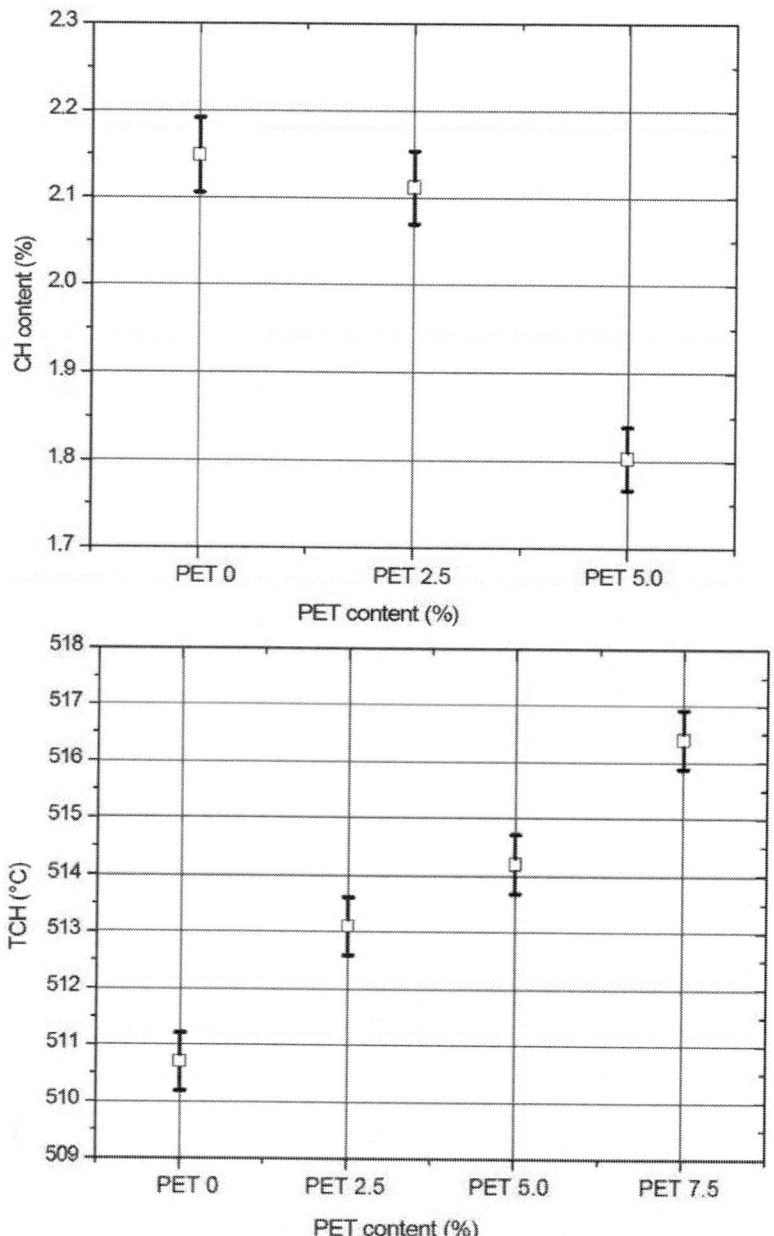

Figure 7. Effect of PET content on the calcium hydroxide content (CH) (a) and its dehydration temperature (TCH) (b).

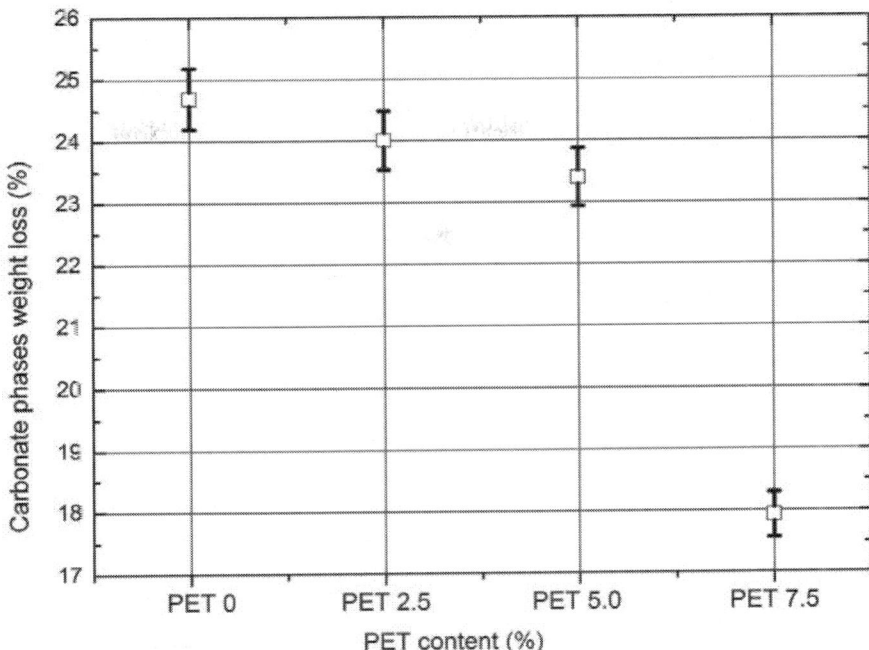

Figure 8. Effect of PET content on the weight loss due to carbonate phases' decomposition.

3.3. The Optical Microscope Results

Figure 10 shows optical micrograph of the composite containing the PET particles. In terms of bonding, the rough surface of particle favours greater contact between PET and cement matrix. The particles appear well covered by cement matrix. Furthermore, Benazzouk et al., [33] found that there is a good adherence between rubber particles and cement matrix. So, the introduction of the PET addition leads to a densification of the cement matrix, which serves to improve material strength and durability [21,34].

3.4. FT-IR Analyses of Hydrated Cement and PET Polymer

The FT-IR spectra of the PET0 and PET7.5 composite cement hydrated up to 28 days and PET polymer are presented in **Figure 11**. The major changes of the FTIR spectra in the hydrated cement are:

Calcium hydroxide bands (~3638 cm^{-1}) and also for the free OH groups, combined and adsorbed water of CS-H, AFm and AFt phases (3451.96 cm^{-1}), molecular water (3440 - 3446 and 1640 - 1654 cm^{-1}), carbonate phases (~1427, 874.56 and 713.5 cm^{-1}). The broad band at ~1020 - 1018 cm^{-1} arises from C-S-H vibrations, in agreement with those reported by Martinez-Ramirez [35]. As can be seen in Figures 2, 11 and **Table 3**, PET polymer doesn't cause any modifications in the spectra profile.

The PET0 and PET7.5 composite have the similar IR spectra (**Figure 11**). So, there isn't any chemical interaction between the mineral species and the organic molecules which could lead to the formation of new compounds, in agreement with XRD analysis.

4. CONCLUSIONS

The following conclusion can be drawn from the obtained experimental data:

From the thermogravimetric investigations performed, showed in the TG/DTG and DTA curves, it is possible to conclude that polymeric additions have influenced the cement hydration.

The qualitative XRD investigation revealed that a lower intensity of $Ca(OH)_2$ (in the region $2q = 18°$) was obtained in the presence of PET, compared to mortar without polymer. Similarly, we found a decrease in the $Ca(OH)_2$ content in the TG analyses for the modified mortar with polymer addition. As it can be seen, composite with 5% polymer content presented the lowest $Ca(OH)_2$ compared with the other composites.

The FT-IR analyses exhibited that there isn't any chemical interaction between the mineral species and the organic molecules which could lead to the formation of new compounds, in agreement with XRD analysis.

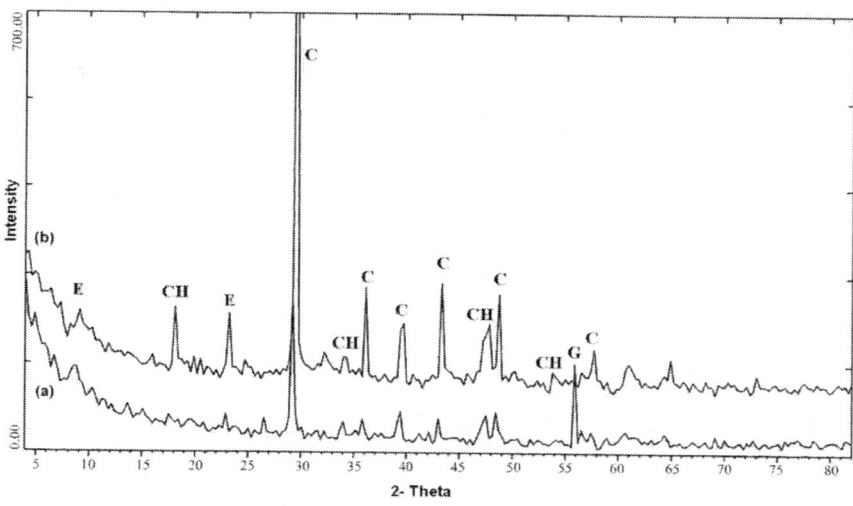

Figure 9. XRD patterns of the PET0 (a) and PET 7.5 (b) composites after 28 days of Hydration. CH, portlandite $(Ca(OH)2)$; C, calcium carbonate $(CaCO3)$; E, ettringite $(Ca6[Al(OH)6]2(SO4)3•26H2O)$; G, gypsum $(CaSO4•2H2O)$.

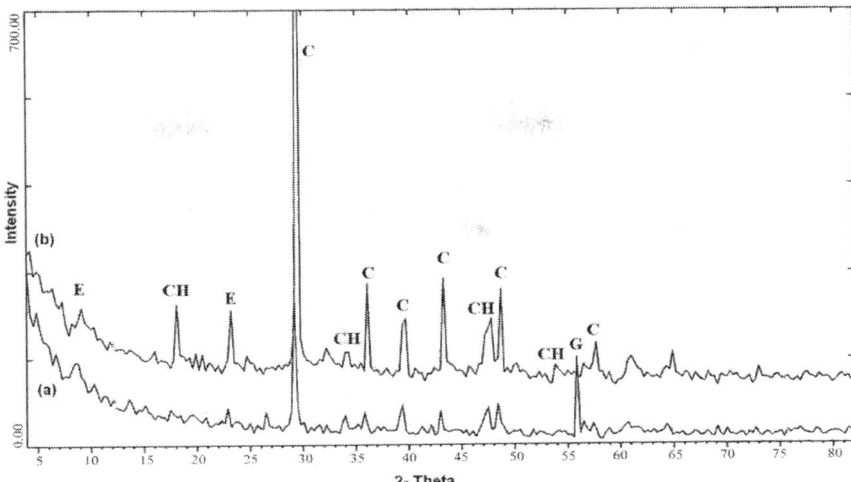

Figure 9. XRD patterns of the PET0 (a) and PET 7.5 (b) composites after 28 days of Hydration. CH, portlandite (Ca(OH)2); C, calcium carbonate (CaCO3); E, ettringite (Ca6[Al(OH)6]2(SO4)3•26H2O); G, gypsum (CaSO4•2H2O).

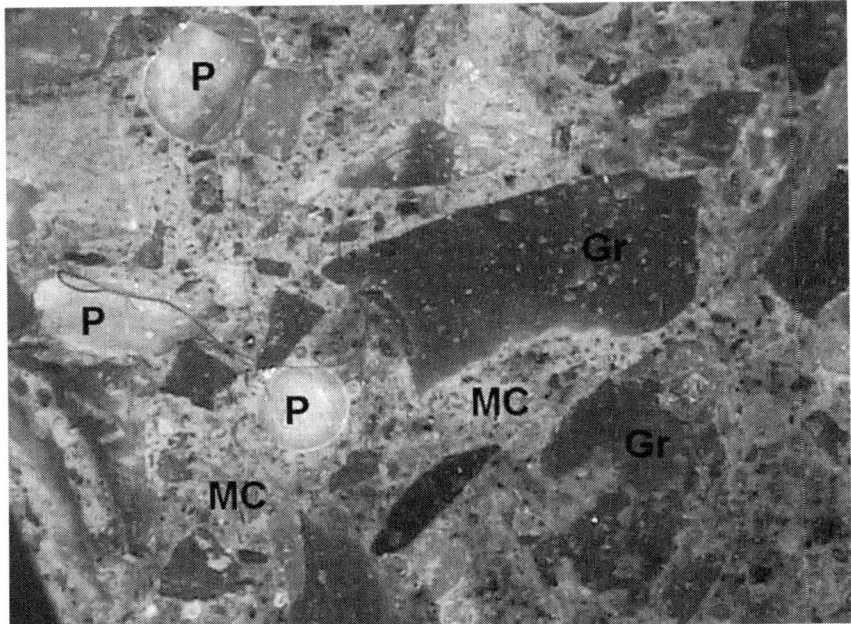

Figure 10. Optical micrograph of composite containing the PET particles at 360 days—adherence of particle additives to the matrix (magnification X25). P-PET, Gr-aggregate, MC-cement matrix.

In terms of bonding, the rough surface of particle favours greater contact between PET and cement matrix. The particles appear well covered by cement matrix. So, the introduction of the PET addition leads to a densification of the cement matrix, which serves to improve material strength and durability.

The application in civil construction of cement composite-based PET wastes appears to be feasible considering the results obtained from analysis of its properties. This study contributes toward the program of PET wastes recycling and pollution reduction. To conclude, studies on the durability of the composite when exposed to aggressive environment, like resistance to acid, sulfate attack and corrosion have started and displayed encourageing first results.

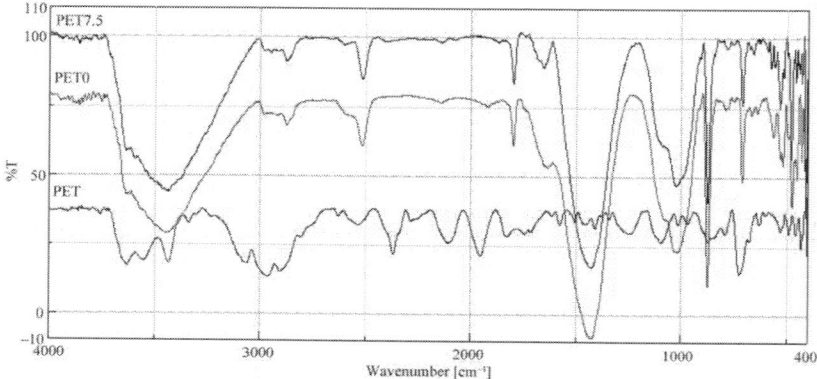

Figure 11. FT-IR spectra of PET0 and PET7.5 composite cured for 28 days and PET polymer.

The durability of the concrete is an important engineering property. It determines the service life of concrete structures very significantly. Some conclusions were stated by the author in his previous study [20,21]:

- The resistance to acids solutions explained by the mass loss of polymer-mortar composites exposed to HCl and CH_3COOH solution were lower than those of unmodified mortar.
- The basic solutions are harmless to composite materials.
- The resistance to chloride ion penetration was improved [20].

Such mortars can be recommended as effective materials for preventing the chloride-induced corrosion of reinforcing steel in various concrete structures and towards the chemical resistance to typical reagents for industrials uses.

5. ACKNOWLEDGEMENTS

The authors acknowledge the financial support from the Ministry of Higher Education and Scientific Research of Algeria, under the grants CNEPRU J0405520060009.

REFERENCES

1. Y. Ohama, "Polymer Based Admixtures," Cement and Concrete Composite, vol. 20, No. 2-3, 1998, pp. 189-212.
2. Y. Ohama, "Recent Progress in Concrete-Polymer Composites," Advanced Cement Based Materials, vol. 5, No. 2, 1997, pp. 31-40.
3. D. W. Fowler, "Polymers in Concrete: A Vision for the 21st century," Cement and Concrete Composites, vol. 21, No. 5-6, 1999, pp. 449-452.
4. Y. W. Choi, Y. J. Kim, H. C. Shin and H. Y. Moon, "An Experimental Research on the fluidity and Mechanical Properties of High-Strength Lightweight Self-Compacting Concrete," Cement and Concrete Research, vol. 36, No. 9, 2006, pp. 1595-1602.
5. E. Yasar, C. D. Atis, A. Kilic and H. Gulsen, "Strength properties of Lightweight Concrete Made with basaltic Pumice and Fly Ash," Materials Letters, vol. 57, No. 15, 2003, pp. 2267-2270.
6. J. A. Rossignolo, M. V. C. Agnesini and J. A. Morais, "Properties of High-Performance LWAC for Precast Structures with Brazilian Lightweight Aggregate," Cement and Concrete Composites, vol. 25, No. 1, 2003, pp. 77-82.
7. ECO PET, 2007. http://www.ecopet.eu/Domino_english/ecopet.htm
8. The Korea Institute of Resources Recycling, "The Korean Institute of Resources Recycling, Recycling Handbook," The Korea Institute of Resources Recycling, Seoul, 1999.
9. T. Ochi, S. Okubo and K. Fukui, "Development of Recycled PET Fiber and Its Application as Concrete-Reinforcing Fiber," Cement and Concrete Composites, vol. 29, No. 6, 2007, pp. 448-455.
10. M. Batayneh, M. Iqbal and A. Ibrahim, "Use of Selected Waste Materials in Concrete Mixes," Waste Management, vol. 27, No. 12, 2007, pp. 1870-1876.
11. O. Y. Marzouk, R. M. Dheilly and M. Queneudec, "Valorization of Post-Consumer Waste Plastic in Cementitious Concrete Composites," Waste Management, vol. 27, No. 2, 2007, pp. 310-318. doi:10.1016/j.wasman.2006.03.012
12. S. C. Kou, G. Lee, C. S. Poon and W. L. Lai, "Properties of Lightweight Aggregate Concrete Prepared with PVC Granules Derived from scraped PVC pipes," Waste Management, vol. 29, No. 2, 2009, pp. 621-628.
13. Y. W. Choi, D. J. Moon, Y. J. Kim and M. Lachemi, "Characteristics of mortar and Concrete Containing Fine Aggregate Manufactured from Recycled Waste Polyethylene Terephthalate Bottles," Construction and Building Materials, vol. 23, No. 8, 2009, pp. 2829-2835.

14. S. Akçaözoglu, C. D. Atis and K. Akçaözoglu, "An investigation on the use of Shredded Waste PET Bottles as aggregate in Lightweight Concrete," Waste Management, vol. 30, no. 2, 2010, pp. 285-290.

15. M. Frigione, "Recycling of PET bottles as fine aggregate in concrete," Waste Management, vol. 30, No. 6, 2010, pp. 1101-1106.

16. J. M. L. Reis, R. Chianelli Jr., J. L. Cardoso and F. J. V. Marinho, "Effect of recycled PET in the Fracture Mechanics of Polymer Mortar," Construction Building Materials, vol. 25, no. 6, 2011, pp. 2799-2804.

17. J. I. Bhatty, D. Dollimore, G. A. Gamlen, R. J. Mangabhai and H. Olmez, "Estimation of Calcium Hydroxide in OPC, OPC/PFA and OPC/PFA polymer Modified Systems," Thermochimica Acta, vol. 106, 1986, pp. 115-123.

18. V. S. Ramachandra, "Differential Thermal Method of Estimating Calcium Hydroxide in Calcium Silicate and Cement Pastes," Cement and Concrete Research, vol. 9, no. 6, 1979, pp. 677-684.

19. M. U. K. Afridi, Y. Ohama, M. Z. Iqbal and K. Demura, "Morphology of $Ca(OH)_2$ in Polymer-Modified Mortars and effect of freezing and Thawing Action on its Stability," Cement and Concrete Composites, vol. 12, No. 3, 1990, pp. 163-173.

20. A. S. Benosman, H. Taibi, M. Mouli, M. Belbachir and Y. Senhadji, "Diffusion of Chloride Ions in Polymer-Mortar Composites (PET)," Journal of Applied Polymer Science, vol. 110, no. 3, 2008, pp. 1600-1605.

21. A.S. Benosman, H. Taibi, M. Mouli, M. Belbachir and Y. Senhadji, "Resistance of Polymer (PET)—Mortar Composites to Aggressive Solutions," International Journal of Engineering Research in Africa, vol. 5, No. 1, 2011, pp. 1-15.

22. J. Dweck, P. M. Buchler, A. C. V. Coelho and F. K. Cartledge, "Hydration of a Portland Cement Blended with Calcium Carbonate," Thermochimica Acta, vol. 346, No. 1-2, 2000, pp. 105-113.

23. C. J. Fordham and I. J. Smalley, "A Simple Thermogravimetric study of Hydrated Cement," Cement and Concrete Research, vol. 15, No. 1, 1985, pp. 141-144.

24. S. Tsivilis, G. Kakali, E. Chaniotakis and A. Souvaridou, "A study on the hydration of Portland Limestone Cement by means of TG," Journal of Thermal Analysis and Calorimetry, vol. 52, No. 3, 1998, pp. 863-870.

25. R. Vedalakshmi, A. Sundara Raj, S. Srinivasan, and K. Ganesh Babu, "Quantification of Hydrated Cement Products of Blended Cements in low and Medium Strength Concrete Using TG and DTA technique," Thermochimica Acta, vol. 407, No. 1-2, 2003, pp. 49-60.

26. M. U. K. Afridi, Y. Ohama, M. Z. Iqbal and K. Demura, "Behavior of $Ca(OH)_2$ in Polymer-Modified Mortars," The International Journal Cement Composites and Lightweight Concrete, vol. 11, no. 4, 1989, pp. 235-244.

27. EN 196-3, "Methods of Testing Cement—Part 3: Determination of Setting Time and Soundness," Comité Européen de Normalisation, Brussels, 1995.

28. D. A. Silva, H. R. Roman and P. J. P. Gleize, "Evidences of Chemical Interaction between EVA and Hydrating Portland Cement," Cement and Concrete Research, vol. 32, No. 9, 2002, pp. 1383-1390.

29. H. F. W. Taylor, "Studies on the chemistry and microstructures of Cement Pastes," Proceedings of the British Ceramic Society, Vol. 35, 1984, pp. 65-82.

30. D. B. Kopil'skii, M. Yu. Butt and V. M. Kolbasov, "Question of the composition and properties of portlandite in Hydrated Portland Cements," Soviet Physics-Crystallography, vol. 13, no. 6, 1969, pp. 945-948.

31. L. Ben-Dor, C. Heitner-Wirguin and H. Diab, "The effect of Ionic Polymers on the hydratation of C_3S," Cement and Concrete Research, vol. 15, no. 4, 1985, pp. 681- 686.

32. A. E. F. de. S. Almeida and E. P. Sichieri, "Mineralogical study of Polymer Modified Mortar with Silica Fume," Construction and Building Materials, vol. 20, No. 10, 2006, pp. 882-887.

33. A. Benazzouk, O. Douzane, T. Langlet, K. Mezreb, J. M. Roucoult and M. Quéneudec, "Physico-Mechanical Properties and Water Absorption of Cement Composite Containing Shredded Rubber Wastes," Cement and Concrete Composites, vol. 29, No. 10, 2007, pp. 732-740.

34. A. S. Benosman, "Mechanical Performance and Durability of Cementitious Materials Modified by adding polymer (PET)," Ph.D. Thesis, University of Oran, Oran, 2010.

35. S. Martínez-Ramírez, "Influence of SO_2 deposition on Cement Mortar Hydration," Cement and Concrete Research, vol. 29, No. 1, 1999, pp. 107-111.

CHAPTER 5

Biocomposite Materials

Khaled R. Mohamed[1]

[1] Biomaterials Department, National Research Centre, Cairo, Egypt

1. INTRODUCTION

Composite materials may be restricted to emphasize those materials that contain a continuous matrix constituent that binds together and provides form to an array of a stronger, stiffer reinforcement constituent. The resulting composite material has a balance of structural properties that is superior to either constituent material alone. Combining the advantages of inorganic and organic components such as HA/organic composites that show good biocompatibility and favorable bonding ability with surrounding host tissues inherent from HA. Besides, the problems associated with HA ceramic, such as its intrinsic brittleness, poor formability and migration of HA particles from the implanted sites, can be circumvented by the integration of HA ceramic with biopolymers. In order to achieve controlled bioactivity and biodegradability, polymer-ceramic composites have been proposed. The composites of ceramics with natural degradable polymers have attracted much interest as bone filler. Several particle composites based on degradable biopolymers such as chitosan and gelatin with inorganic powders, were developed as bone filler.

2. HYDROXYAPATITE (HA)

Hydroxyapatite (HA) ($Ca_{10}(PO_4)_6(OH)_2$, is widely used in musculoskeletal procedures due to its chemical and crystallographic similarity to the carbonated apatite in human bones and teeth (Suchanek and Yoshimura, 1998). While sintered HA can be machined and used in pre-fabricated forms, several formulations of calcium phosphate cements can be molded as pastes and harden in situ. HA is the main component of teeth and bones in vertebrates. Good mechanical properties with superior biocompatibility of sintered HA make it well preferred bone and tooth implant material (Kim et al., 2008). Calcium phosphates, especially HA, are excellent candidates for bone repair and regeneration and have been used in bone tissue engineering for two decades.

Although HA is bioactive and osteoconductive, its mechanical properties are inadequate, making it unable to be used as a load bearing implant.

2.1. Properties

HA power (HAp), a major inorganic component of bone, has been used extensively for biomedical implant applications and bone regeneration due to its bioactive, biodegradable and osteoconductive properties (Gomez-Vega et al., 2000). HA is available in market in many forms like solids blocks, micro-porous blocks and as granules (Murgan and Ramakrishna, 2005). Nano-hydroxyapatite (n-HA) has been proven to be of great biological efficacy. nHA precipitates may have higher solubility and therefore affect the biological responses. It was able to promote the attachment and growth of human osteoblast-like cells (Huang et al., 2004). Clinical trials have shown that HA cement is both biocompatible and resistant to infection and that the HA coating improves the success rate of implants. It has also been demonstrated that HA ceramics support mesenchymal stem cell (MSC) attachment, proliferation, and differentiation (Zhao et al., 2006). Nano scaled HA with extraordinary properties such as high surface area to volume ratio and ultra fine structure similar to that of biological apatite, which is of great effect on cell-biomaterial interaction, has been reported to be used for the treatment of bone defects, and it could bond to living bone in implanted areas (Liuyun et al., 2008). HAp has been used in bone regeneration and as a substitute of bone and teeth because it is a biocompatible, bioactive, non-inflammatory, non-toxic, osteoconductive and non-immunogenic material (Grande et al., 2009).

Instability of the particulate nHA is often encountered when the particles are mixed with saline or patient's blood and hence migrate from the implanted site into surrounding tissues and causing damage to health tissue (Miyamato and Shikawa, 1998). Also, HA ceramic is difficult to shape in specific forms required for bone substitution, due to its hardness and brittleness. Therefore, composites of HAp and organic polymers have become of great interest to compensate the weak mechanical points of HAp (Furukawa et al., 2000). Mohamed and Mostafa, (2008) reported that HA has low fracture toughness, hardness and brittleness, therefore, HA cannot serve as a bulk implant material under the high physiological loading conditions traditionally associated with implants.

Since the natural bone is a composite mainly consisted of nano-sized, needle-like HAp crystals and collagen fibers, many efforts have been made to modify HAp by polymers, such as poly-lactic acid (Kasuga et al., 2001), chitosan (Viala et al., 1998), polyethylene (Wang and Bonfield, 2001) to compensate the weak mechanical points of HAp. Yamaguchi et al., (2003) reported that chitosan is flexible and has a high resistance upon heating due to the intra-molecular hydrogen bonds formed between hydroxyl and amino groups. A composite biomaterial of HAp and chitosan is, therefore, expected to show an increased osteoconductivity and biodegradation together with a sufficient mechanical strength for orthopedic use. A major disadvantage of current orthopedic implant materials such as sintered hydroxyapatite is that they exist in a hardened form, requiring the surgeon to drill the surgical site around the implant or to carve the graft to the desired shape. This can lead to increases in bone loss, trauma, and surgical time. Hence, the moldable, self-setting calcium phosphate cement (CPC)-chitosan composite is desirable for dental, craniofacial and orthopedic repairs, especially where shaping and contouring for esthetics are needed. The CPC powder can be mixed with the chitosan liquid to form a paste that can be applied in surgery via minimally invasive techniques such as injection, with fast-setting and anti-washout capabilities to form a scaffold in situ (Moreau and Xu, 2009).

2.2. Applications

Bone defects that are generated by tumor resection, trauma, and congenital abnormality have been clinically treated by the implantation of bioceramics or autogenous and allogenous bone grafts. Although autografting is a popular procedure for reconstructive surgery, it has several disadvantages, such as the shortage of donor supply, the persistence of pain, the nerve damage, fracture, and cosmetic disability at the donor site. On the other hand, there are no donor site problems for allografting, while allografting has some clinical risks including disease transmission and immunological reaction (Takahashi et al., 2005). HA has been incorporated into a wide variety of biomedical devices including dental implants, coatings on Ti based hip implants, biodegradable scaffolds, and other types of orthopedic implants (Wilson and Hullb, 2007). Synthetic HA, is a bioactive material that is chemically similar to biological apatite HA, has been used as a bioactive phase in the composites, coating on metal implants, and granular filler for direct incorporation into human tissues (Rehman et al., 1995). The characteristics of an ideal ceramic composite for bone tissue engineering should comply to the following parameters:(1) A biodegradability for bone remodeling, (2) A macroporosity for in-growth into the composite, (3) Mechanical stability/ease of handling, (4) Osteoconductivity to guide bone around/inside the implant, (5) Carrier for growth factors/cells. The use of these materials for tissue engineering purposes is still explored. Most researchers are aware that HA has low resorbability of sintered Ca P-ceramics. Because of the positive influence of ceramics on cell differentiation/proliferation, it is not surprising that bone forming cells are introduced into these ceramics to speed-up tissue in-growth. The surface of sintered ceramics is chemically stable and therefore a good substrate for seeding cells. In the field of bone tissue engineering most scaffolds are ceramic or ceramic-derivatives. Especially HA-based calcium phosphate compound is regarded as high-potential scaffolds due to their osteoconductive properties. Next to the scaffold material, in bone tissue engineering, cells and growth factors are also introduced to speed-up tissue in-growth. In most ceramic scaffolds, however, difficulties arise when cells are added, most probably due to a limited supply of nutrition at the inside of the implant or less than optimal cell-cell interactions. Growth factors like bone morphogenic protein-2 (BMP-2), transforming growth factor (TGF-β), basic fibroblast growth factor (b-FGF) and vascular endothelial growth factor (VEGF) are commonly introduced into these scaffolds due to their osteoinductive properties and vascularization (Habrakenb et al., (2007). HA has a composition and structure very close to natural bone mineral and therefore has been considered to be the ideal material to build bone tissue engineering scaffold due to its osteoconductivity and osteoinductivity (Wang et al., 2007). Some in vitro studies demonstrated that nano-phase HA (67 nm grain size) significantly enhanced osteoblast adhesion and strikingly inhibited competitive fibroblast adhesion compared to conventional, 179 nm grain size HA, after just 4 h of culture. Researchers believe they know why they have elucidated the highest adsorption of vitronectin (a protein well known to promote osteoblast adhesion) on nanophase ceramics, which may explain the subsequent enhanced osteoblast adhesion on these materials. In addition, enhanced osteoclast-like cell functions (such as the synthesis of tartrate-resistant acid phosphatase (TRAP) and the formation of resorption pits) have also been observed on nano-HA compared to conventional HA, nano-porous or nano-fibrous polymer matrices can be fabricated via electrospinning, phase separation, particulate leaching, chemical etching and 3-D printing techniques (Zhang et al., 2008).

3. CHITOSAN POLYMER (C)

3.1. Structure

Brimacombe and Webber, (1964) reported that the chitosan consists of repeating units of beta (1-4) 2-amino-2-deoxy-D-glucopyranon (D-glucosamine) (Fig.1). The primary unit of chitin is 2-acetamido-2-deoxy-D-glucose, while that of chitosan is 2-amino-2-deoxy-D-glucose with beta, 1-4 glucosidic linkages (Muzarelli, 1985). Chitosan is a natural polysaccharide derived from chitin by its deacetylation (Dang and Leong, 2006). Chitosan shares a number of chemical and structural similarities with collagen. These similarities form the basis for the development of HAp/chitosan nano-composites for use in bone tissue engineering (Wilson et al., 2007

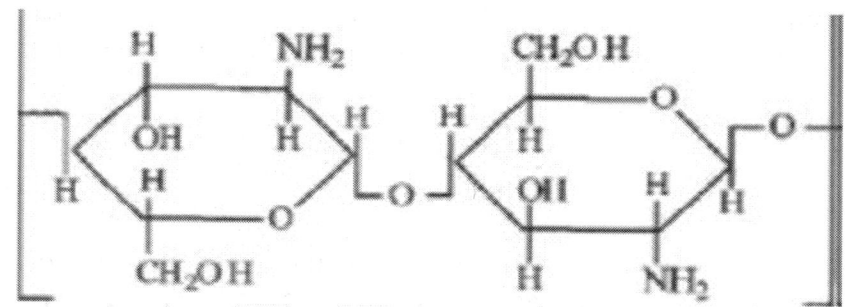

Figure 1. Structure of chitosan

3.2. Origin

The history of chitosan dates back to the 19th century, when Rouget, 1859 discussed the deacetylated form of chitosan. Chitin, the source material for chitosan, is one of the most abundant organic materials, being second only to cellulose in the amount produced annually by biosynthesis. It is an important constituent of the exoskeleton in animals, especially in crustacean, molluscs and insects. It is also the principal fibrillar polymer in the cell wall of certain fungi (Eugene and Lee, 2003). Chitosan is one of the most abundant naturally occurring polysaccharides, primarily obtained as a sub-product of seafood, containing amino and hydroxyl groups (Muzarelli, 1985 and Manjubala et al., 2006).

3.3. Properties

3.3.1. Physical and Chemical Properties

The size of chitosan molecule plays an important role in drug delivery system where the release rate was reported to be inversely proportional to the molecular weight of the reservoir. Chitosan is insoluble in water, alkali and many organic solvents but soluble in many dilute aqueous solutions of organic acids, of which most commonly used are formic acid and acetic acid. Tomihata and Ikada, (1997)concluded that molecular weight of chitosan affect the crystal size and morphological character of its cast film. Gorbunoff, (1984) concluded that the chemical interaction between NH^{3+} and HPO_4^{2-} control the adsorption of chitosan on octacalcium phosphate (OCP) crystal surface which is similar to the binding of protein to the surface of apatite crystals. Chitosan permeability increased in the amorphous region of the membrane (Brine, 1984). Chitosan forms a polycation in aqueous acid solutions like acetic acid and hydrochloric acid via protoration of amine functions (Muzarelli, 1985).

Blair et al., (1987) reported that chitosan prepared from crab shell chitin has a lower molecular weight than that prepared from prawn shells and tensile strength versus elongation decrease at the break with prolonged treatment in alkali solutions and both increase proportional to higher molecular weight. Also, crystallinity of membrane increased with decreasing molecular weight of chitosan and it is hydrophilic i.e it contains a large number of OH groups which are easy to combine with another group and form new bond (Ogawa et al., 1992). The larger molecular weight of chitosan used, the higher tensile strength and higher tensile elongation of membrane were obtained (Chen and Hwa, 1995). The degree of deacetylation affects the physical properties of chitosan membrane and it does not change during ultrasonic degradation on chitosan. The molecular weight of the prepared chitosan depends on the severity of the deacetylation process.

Chitosan is flexible and has a high resistance upon heating due to the intra-molecular hydrogen bonds formed between hydroxyl and amino groups (Lee et al., 1999). The degradation rate is inversely related to the degree of crystallinity which is controlled mainly by the degree of deacetylation (DD). Highly deacetylated forms (85 %) exhibit relatively a low degradation rate and may take several months in vivo, whereas, the forms with lower DD degrade more rapidly. The degradation rates also inherently affect both the mechanical and solubility properties (Kamiyama et al., 1999).

The cationic nature of chitosan also allows for pH-dependent electrostatic interactions with anionic glycosaminoglycans (GAG) and proteoglycans distributed widely throughout the body and other negatively charged species. This property is one of the important elements for tissue engineering applications because numbers of cytokines/growth factors are known to be bound and modulated by GAG including heparin and heparin sulfate (Nishikawa et al., 2000). Chitosan is easy to handle for its resistive nature to heating due to intra-molecular hydrogen bonds between hydroxyl and amino groups (Itoh et al., 2003). Chitosan is a very versatile biopolymer with film, fiber, and micro/nanoparticle forming properties. It is biocompatible, biodegradable and non-toxic (Yilmaz, 2004). Deacetylation of chitin with a degree of deacetylation more than 50 % gives chitosan, which is soluble in organic acids such as acetic or formic acid, and has been more widely used than chitin as films, membranes, fibres and particles (Kang et al., 2006). Chitosan has three types of reactive functional groups, an amino group as well as both primary and secondary hydroxyl groups at the C(2), C(3), and C(6) positions, respectively. These groups allow modification of chitosan like graft copolymerization for specific applications, which can produce various useful scaffolds for tissue engineering applications. The chemical nature of chitosan in turn provides many possibilities for covalent and ionic modifications that allow extensive adjustment of mechanical and biological properties (Kim et al., 2008). When the temperature is 30°C, the crystallinity of chitosan membrane is relatively high and its crystal particles are much small, which makes the mechanical properties poor. When the temperature is as high as 90°C, the temperature will cause the change of chitosan properties and the chitosan membrane is nearly not a crystal structure at this moment, resulting in the fall of mechanical properties (Xianmiao et al., 2009).

3.3.2. Biological Properties

Chitosan potentiates the differentiation of osteoprogenitor cells and may facilitate the formation of bone. Rao and Sharma, (1997) concluded that chitosan is an ideal non-toxic biopolymer and the cell binding and cell activating properties of chitosan play a crucial role in its potential action. Chitosan degrades in the body to non-harmful and non-toxic compounds and has been used in various fields such as nutrition, metal recovery and biomaterials (Muzzarelli et al., 2001). It has gained much attention as a biomaterial in diverse tissue engineering applications due to its low cost, large-scale availability, anti-microbial activity, and biocompatibility (Khora and Limb, 2003). Chitosan was suggested as an alternative polymer for use in orthopedic applications to provide temporary mechanical support to the regeneration of bone cell in-growth due to its good biocompatible (Khora and Limb, 2003), non-toxic, biodegradable and inherent wound healing characteristics (Eugene and Lee 2003). Chitosan had been used in various forms such as zero dimension microsphers, two-dimension membrane, three-dimension pin or rod (Hu et al., 2003). Therefore, much attention has been paid to chitosan-based biomedical materials, for instance, as a drug delivery carrier or a wound-healing agent. Chitosan is structurally similar to glycosaminoglycan (GAG) and has many desirable properties as tissue engineering scaffolds (Kuma, et al., 2004). It was reported as being neither antigenic in mamalian test system nor thrombogenic and chitosan reported to improve hemostasis, decreased fibroplasias with enhanced tissue organization as well as normal bone formation (Mohamed, 2004). Chitosan marginally supports biological activity of diverse cell types (Sarasam and Madihally, 2005).

A number of natural and synthetic polymers have been studied for overcoming the weak points as bone substitutes. Chitosan has been found in a broad spectrum of applications along with unique biological properties including biocompatibility, biodegradability to harmless products, non-toxicity, physiological inertness, remarkable affinity to proteins, antibacterial, haemostatic, fungi-static, anti-tumoral and anti-cholesteremic properties (Kim et al., 2008). Chitosan has been shown to degrade in vivo, which is mainly by enzymatic hydrolysis. The degradability of a scaffold plays a crucial role on the long-term performance of tissue-engineered cell/material construct because it affect many cellular process, including cell growth, tissue regeneration, and host response. If a scaffold is used for tissue engineering of skeletal system, degradation of the scaffold biomaterial should be relatively slow, as it has to maintain the mechanical strength until tissue regeneration is almost completed. Lysozyme is the primary enzyme responsible for in vivo degradation of chitosan, which appears to target acetylated residues (Kim et al., 2008).

3.4. Applications

Chitosan of biopolymer have been used as blood coagulant, in artificial kidney membrane, digestive sutures, hypercholesterolemic agents, media for the slow release of drugs and hemostatic agent (Mohamed, 2004). It is a good candidate for biomedical applications such as for wound healing, vaccine delivery, as well as tissue regeneration (Yilmaz, 2004). It has been extensively investigated in biotechnological, biomedical, and environmental fields (Dang and Leong, 2006). Chitosan polymer is used in dentistry, because it prevents the formation of plaque and tooth decay. Since chitosan can regenerate the connective tissue that covers the teeth near the gums, it offers possibilities for treating periodontal diseases such as gingivitis and periodontitis (Elizalde-Pen et al., 2007).

3.4.1. Membrane

Aiba et al., (1986) used chitosan in membrane separation, chemical engineering, medicine and biotechnology areas. It was also found that the water adsorption and the mechanical properties of fibroin membrane were improved by blending chitosan (Chen et al., 1998).

3.4.2. Skin

Muzzarelli et al., (1988) used chitosan as artificial skin substitute and they reported that no adverse effect after implantation in tissue. In general, these materials have been found to evoke a minimal foreign body reaction, with little or no fibrous encapsulation. It observed the typical course of healing with formation of normal granulation tissue, often with accelerated angiogenesis (Suh and Matthew, 2000). Also, chitosan has many advantages for wound healing such as hemostasis, accelerating the tissue regeneration and stimulating the fibroblast synthesis of collagen (Mi et al., 2001). Chitosan possesses the properties favorable for promoting rapid dermal regeneration and accelerate wound healing suitable for applications extending from simple wound coverings to sophisticated artificial skin matrices (Kim et al., 2008).

3.4.3. Bone Substitutes

Sapelli et al., (1986) used chitosan powder to promote healing of periodontal pockets, palatal wounds and extraction sites. Malette et al., (1986) proved enhanced leg bone regeneration in dogs using chitosan. It was reported to accelerate wound healing and was applied for bone wound repair in dogs (Borah et al., 1992) as well as bone growth in critical size metacarpal fibular defects. Klokkevold et al., (1992) concluded that chitosan solution may enhance the formation of bone. Later, it was reported to improve osseous healing of defects in femoral coundyl of sheep and stimulated cell proliferation and organized the hystoarchitectural tissue structure (Muzzarelli et al., 1994). Chitosan has been also extensively used in bone tissue engineering since after exploring its capacity to promote growth and mineral rich matrix deposition by osteoblasts in culture. Also, chitosan is biocompatible (additional minimizes local inflammation), biodegradable, and can be molded into porous structures (allows osteoconduction) (Martino et al., 2005 and Kim et al., 2008).

3.4.4. Drug Delivery System (DDS)

Chitosan as an inert and hydrophilic material, its gel is suitable for application as matrices for enzyme/cell immobilization (Roberts, 1992) and for separation processes (Li et al., 1992). Felt et al., (1998) discussed the use of chitosan to manufacture sustained release systems deliverable by other routes such as nasal, ophthalmic, transdermal and implantable devices. Tarsi et al., (1998) suggested that low molecular weight chitosan may be very interesting as potential antidental caries agents. Chitosan has been reported to enhance drug delivery across the nasal or mucosal layer without damage (van der Lubben et al., 2001). The selectivity of membrane is a critical parameter in membrane separation and several factors are affecting its selectivity such as pore size, thus it is very suitable for the use as DDS and in artificial kidney (Mohamed, 2004). The cationic properties of chitosan offer valuable properties for drug delivery systems, gene delivery systems, and tissue engineering, that is, the formation of ion complexes between chitosan and anionic drugs or DNA can be used as a delivery vehicle (Kim et al., 2007).

3.4.5. Anti-Bacterial

The experiments of antibacterial activity of chitosan-graft-polyethylene terephthalate (PET) against S. aureus showed a high growth inhibition in the range of 75–86% and still maintained a 48–58% bacterial growth inhibition after laundering (Hu et al., 2003). Chitosan and chito-oligosaccharides grafted membranes showed antibacterial activity against Escherichia coli, Pseudomonas aeruginosa, methicilin-resistant Staphylococcus aureus (MRSA), and S. aureus (Hu et al., 2003). Chitosan is a biomaterial with antiseptic property and the influence of the release or positive migration of protonated glucosamine fractions from the biopolymer into the microbial culture is the responsible event for the antimicrobial performance of the biopolymer (Beherei et al., 2009).

3.4.6. In-Vitro Application

Chitosan and collagen have intrinsic properties that support growth and differentiation of osteoblasts. Collagen was combined with chitosan and cross-linked to improve the biological stability and strength of chitosan–collagen composite sponges to reach the demand of an application in bone tissue engineering. The incorporation of chitosan into a collagen scaffold increases the mechanical strength of the scaffold and reduces the biodegradation rate against collagenase (Arpornmaeklong et al., 2007).

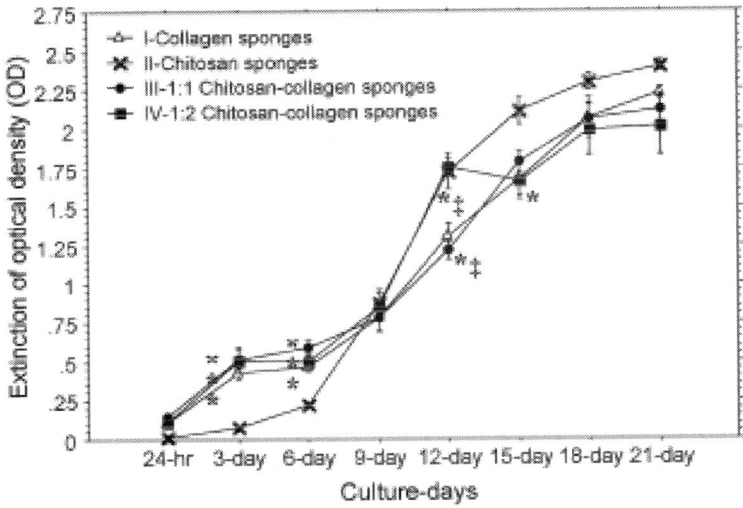

Figure 2. Cell viability of osteoblasts cultured for 21 days on the sponges (Arpornmaeklong et al., 2007).

The collagen, chitosan and chitosan–collagen sponges were biocompatible. All sponges supported growth of cells on three-dimensional structures in a similar manner. Chitosan sponges had a tendency to promote growth of cells to a greater extent than the other groups (Fig.2). It is postulated that strong attraction between positive charges on the chitosan surface and negative charges on the cell surface enhanced the metabolic activity of cells on chitosan sponges (Mi et al., 2001). In vivo, chitosan enhances angiogenesis and wound healing, and supports growth and differentiation of osteoblasts (Lee et al., 2004).

3.4.7. Other Applications

a) Blood vessels:The chitosan fibers offer the potential of being fabricated into blood vessels and their blood compatibility results demonstrated for applications where hemocompatibility is required (Khora and Limb, 2003). Kim et al., (2008) reported that the effort has made to overcome both incomplete endothelialization and smooth muscle cell hyperplasia, which are two of the problems contributing to the poor performance of existing small-diameter (4 mm) vascular grafts, through complexation of GAGs with porous chitosan scaffolds. GAG-based material should promise because of their growth inhibitory effects on vascular smooth muscle cells and their anti-coagulant activity. However, few data regarding chitosan as a scaffold of tissue engineered blood vessels have been reported. Chitosan itself was documented to promote migration of endothelial cells and fibroblasts so as to accelerate wound healing. b) Nerve: Chitosan has been studied as a candidate material for nerve regeneration due to its properties such as antitumor, antibacterial activity, biodegradability and biocompatibility. Neurons that were cultured on the chitosan membrane can grow well and that chitosan tube can greatly promote the repair of the peripheral nervous system, also the chitosan fibers supported the adhesion, migration and proliferation of Schwann cells (SCs), which provide a similar guide for regenerating axons to Büngner bands in the nervous system (Yuan et al., 2004). c) Liver: Chitosan as a promising biomaterial can be applied in liver tissue engineering due to its various properties such as its structure that is similar to glycosamineglycans (GAGs), which are components of the liver extracellular matrix (ECM) (Li et al., 2003). d) Cartilages: Chitosan is one of the most abundant polysaccharides and thus shares some bioactivities with various glycosaminoglycans and hyaluronic acid present in articular cartilage (Suh and Matthew, 2000). Lu et al., 1999 has demonstrated that the chitosan solution injected into the knee articular cavity of rats lead to a significant increase in the density of chondrocytes in the knee articular cartilage, indicating that chitosan could be potentially beneficial to the wound healing of articular cartilage (Lee et al., 2004).

4. BONE STRUCTURE

Bone is a specialized tissue comprising mineral substances, organic tissue, and water (Otto et al., 1997). Cortical bone is largely a composite of collagen, fiber and biological apatite (Fricain et al., 1998). The inorganic component of bone (bone mineral) is calcium phosphate that contains up to 8 wt% carbonate. Substitution of carbonate or other ions HA can occur in two distinct atomic sites in the lattice (Suchanek et al., 2002). These ions can partially substituted in the lattice for hydroxyl ions (OH), known as the A site, and/or for phosphate ions, known as the B site (Gibson and Bonfield, 2002). Skeletal bone is of two types: cortical bone and trabecular bone. Cortical bone is the outermost mineralized cortex. It is compact, strong, and densely packed as an intricate calcium matrix. Cortical bone comprises 85% of the skeleton, specifically 75% in the femoral neck, 75% in the distal radius, and 95% in the midradius. Cortical bone has no contact with marrow. Trabecular bone is the inner spongy structure composed of the sturdy collagen matrix. It comprises 15% of the skeletal mass and has structural rigidity and elasticity to withstand mechanical stress. Trabecular bone contains hematopoietic tissue in its central cavity. The bone of diaphysis consists of cancellous bone covered with a shell of cortical bone (Fig.3). The flat bones of the skull have a middle layer of cancellous bone sandwiched between two relatively thick layers of cortical bone (Liu et al., 2009).

Bone remodeling occurs continuously throughout the lifetime, although the process slows with age. The balance of osteoclastic and osteoblastic activity results in breakdown and reconstruction, which ensures skeletal integrity and maintains mineral homeostasis. Osteoclasts and osteoblasts are interconnected and influence the activity of each other. Osteoclasts are large multinuclear cells that develop from monocyte-macrophage precursors. They are imbedded in the bone matrix at or near the site of bone resorption and dissolve first the calcium and then the organic matrix of the bone. Osteoblasts arise from mesenchymal cells and are found layered over the bone. They deposit calcium into the matrix that is building up cortical bone and produce collagen and other proteins to synthesize the bone matrix. Osteoclasts that are in close proximity can increase sensitivity of osteoblasts to growth factors. When a bone is broken there is usually bleeding into the space between the bone and the periosteum. This produces a hematoma, swelling due to blood. The osteoblasts, bone-producing cells, near the hematoma invade it along with small blood vessels from the bone. In a short time, the hematoma is replaced by bone tissue produced by the osteoblasts. In general, this bone is cancellous, but the added bone makes the broken junction much thicker than it was. The swollen bone is called the callus. A remarkable process follows. Where there is strain, the newly formed cancellous bone condenses into compact bone. Where strain is absent, the cancellous bone disappears through the activity of osteoclasts, bone destroying cells. This process seems to determine the development of the skeleton in normal growth (Roodman, 2004).

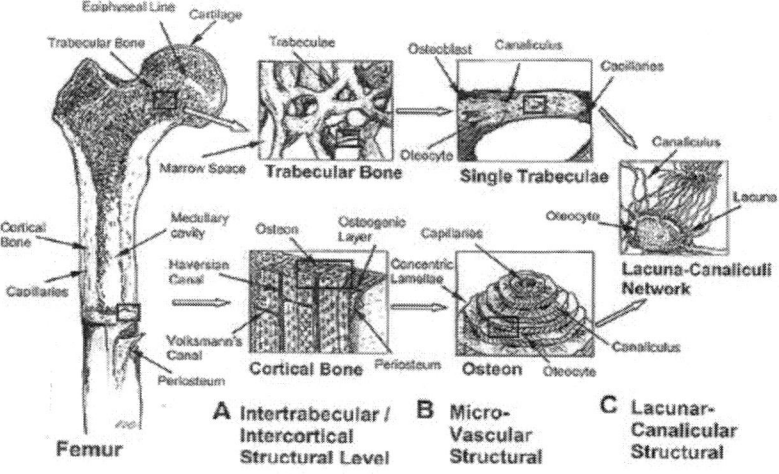

Figure 3. Bone structure

5. HYDROXYAPATITE/CHITOSAN BIOCOMPOSITES: INTRODUCTION

For the increase of bioactivity and mechanical property, some composites of polymer and bioactive ceramics have been developed for bone tissue engineering. These composites fulfill the mechanical properties required for their function as skeleton, teeth and cells of organisms. Among these composites, HAp/polymer composites have attracted much attention since such composites may have osteconductivity due to the presence of HAp, which has a similar chemical composition and structure to the mineral phase of human bones and hard tissues. Thus, HA/polymer composite scaffolds are of interest for biomedical applications (Jin et al., 2008). Kikuchi et al., (2004) prepared a self organized HA-collagen nano-composite by a biomimetic co-precipitation method. It was reported that the composite had similar microstructure to native bone and showed osteoclastic resorption and good osteoconductivity. However, a major concern over HA-collagen composite is the high cost of collagen, which limits its clinical application in healing bone defect to its insouciant formability and flexibility. Polymers such as chitosan have a higher degradation rate than bioceramics. Incorporation of HA into a chitosan polymer matrix has been shown to increase osteoconductivity and biodegradability with significant enhancement of mechanical strength (Yamaguchi et al., 2001). Chitosan's primary attractive features including its biocompatibility, biodegradability, flexibility, adhesiveness and anti-infectivity, make it as a feasible wound healing agent and an ideal polymeric matrix for HA ceramic (Rusu et al., 2005).

5.1. Structure

The free amino groups of chitosan (C-NH₂) was protonated to C-NH₃⁺, when chitosan was dissolved in acetic acid (HAc) solution, which was shown as follows:

$$C - NH_2 + HAc \longrightarrow C - NH_3^+ + Ac \qquad pH = 4.2$$

The presence of calcium and phosphate ions in chitosan solutions leads to formation of HA\chitosan composites through electrostatic interactions between C-NH₃⁺ and Ca²⁺ and\or PO₄³⁻ ions to form C-Ca and C-PO₄ complexes. There is also an interaction between OH of chitosan and OH of HA via hydrogen bond (Hu et al., 2004).

5.2. Properties

Chitosan can be utilized in combination with other bioactive inorganic ceramics, especially HA to further enhance tissue regenerative efficacy and osteoconductivity. Incorporation of HA with chitosan, the mineral component of bone, could improve the bioactivity and the bone bonding ability of the chitosan/HA composites (Wang et al., 2002). Chitosan just plays in a role of adhesive to dissolve the problem of difficulty of HA specific shape and migration of HA powder when implanted. HA/chitosan nano-composites are prepared by the precipitation. It is proposed that the nano-structure of HA/chitosan composite will have the best biomedical properties in the biomaterials applications (Chen et al., 2002). Also, it has been demonstrated that chitosan-hydroxyapatite composite induces osteoconductivity in osseous defects and could act as drug vehicle. It is important to be able to load these composites with short-time life and controlled action anti-inflammatory to reduce or eliminate undesirable inflammatory processes (Larena et al., 2004).

It is desirable to develop a composite material with favorable properties of chitosan and HA. The designed composites are expected to have an optimal mechanical performance and a controllable degradation rate as well as eminent bioactivity and this will be of great importance for bone remodeling and growth (Zhang et al., 2005). It must be emphasized at this point that the successful design of a bone substitute material requires an appreciation of the structure of bone. Thus, the use of a hybrid composite that makes up of chitosan and calcium phosphate resembles the morphology and properties of natural bone. This may be one-way to solve the problem of calcium phosphates brittleness, besides possessing good biocompatibility, high bioactivity and great bone-bonding properties (Ding, 2007). The mechanical strength of the chitosan/calcium phosphate composite fiber with core-shell structure increased with an increased concentration of chitosan solution (Matsuda et al., 2004). Among the composites studied, the 30/70 chitosan/n HA exhibits the maximum value of compressive strength, about 120 MPa, which is strong enough to be used in load-bearing sites of bone tissue. In contrast the compressive strength of pure HA compact prepared by the similar method has been reported as 6.5 MPa about one twentieth of the maximum value of the composite. In general, the proper stress transfer occurring between the reinforcement and the matrix governs the mechanical characteristics of filled polymers. The chemical and mechanical interlocking between n-HA and chitosan are accounts for the efficient stress-transfer in the composite system. Besides, the interactions such as hydrogen bonding and chelation between the two phases, also contribute to the good mechanical properties of chitosan/n-HA composite (Zhang et al., 2005). The biodegradable composites based on chitosan and calcium phosphate have been prepared using a simple mixing and heating method. The detrimental effects of the simulated physiological environment on mechanical properties of the hybrid composites resulted in the significantly decrease in strength and modulus. The chitosan/calcium phosphate composites containing 10 wt/v % might an optimal material in terms of initial strength and degradation behavior. Although susceptibility to solution attack, this type of chitosan/calcium phosphate composites with high initial strength might be acceptable for use in bone tissue repair (Ding, 2007).

Three-dimensional biodegradable chitosan/nHA composite scaffolds were characterized by superior mechanical, physicochemical, and biological properties compared to pure chitosan scaffolds for bone tissue engineering. The nanocomposite scaffolds were characterized by a highly porous structure and the pore size was similar for scaffolds with varying n-HA content. The nano-composite scaffolds exhibited greater compression modulus, slower degradation rate and reduced water uptake, but the water retention ability was similar to that of pure chitosan scaffolds. Favorable biological response of pre-osteoblast on nanocomposite scaffolds included improved cell adhesion, higher proliferation, and well spreading morphology in relation to pure chitosan scaffold (Thein-Han and Misra, 2009).

Mechanical properties of biocomposites: hydroxyapatite/chitosan (HA/CS) nano-composite rods were reinforced via a covalently cross-linking method. The bending strength and bending modulus of the cross-linked HA/CS (5/100, wt/wt) rods could arrive at 178 MPa and 5.2 GPa, respectively, increased by 107% and 52.9% compared with uncross-linked HA/CS (5/100, wt/wt) rods (Takagi et al., 2003). The presence of HA-DBM filler into the grafted chitosan copolymer matrix resulted in compressive strength properties are quite close to those of cancellous bone (2–12 MPa). This result is due to effect of the presence of demineralized bone matrix (DBM) powder and pMMA having bone cement formation within this composite (Mohamed et al., 2008). The E-modulus and compressive strength for the three composites HA/, 90%HA-10Ti/, and 70%HA-30%Ti/grafted chitosan copolymer composites recorded comparable values compared to the cancellous bone. Therefore, the presence of HA filler or HA filler containing titania content up to 30% into the copolymer resulted in compressive strength properties that are quite close to those of cancellous bone (2–12 MPa) (Mohamed et al., 2007).

Collagen/apatite composite membranes exhibited significantly improved mechanical properties compared with their pure collagen equivalent; their mechanical properties were still lower than those of natural bone (Teng et al., 2009). It was notified that hardness of calcium pyrophosphate (CPP)/ chitosan composite (66.80) was increased compared to chitosan copolymer (60.88) and the hardness of CPP/chitosan-grafted composite (68.23) was also increased compared to chitosan–gelatin copolymer (84.12) proving the polymer/filler interaction and adhesion. Also, the compressive strength of CPP/chitosan composite (6.53 MPa) was increased compared to chitosan copolymer (5.11 MPa). These values of compressive strength were comparable to those of human cancellous bone (Kokubo et al., 2003). As a result, CPP filler powder into the chitosan copolymer matrix containing chitosan or chitosan–gelatin polymer resulted in more effective reinforcement of the composite, then, stiffer composite (El-Kady et al., 2009). The CPC–chitosan composites were more stable in water than conventional calcium phosphate cement (CPC). They did not disintegrate even when placed in water immediately after mixing. The CPC–chitosan paste hardened within 10 min in all cases. The authors demonstrated that CPC–chitosan composites are stable in a wet environment and have acceptable mechanical strengths for clinical applications (Wang et al., 2010).

5.3. Preparation

Although powder ceramics remain the form of choice for filling small irregular defects, the therapeutic effect of the filling implant was lost by migration of particles from the defect site. Furthermore, it was difficult to be handled and kept in place compactly for convenient fabrication and operation of block-type ceramics (Lin et al., 1998). Thus, it is necessary to mix a suitable binder with the granular material to overcome these problems. Presently, the approaches to obtain chitosan/hydroxyapatite (HAp) composite materials are based either on mixing or co-precipitation methods. Yamaguchi et al., (2001) have developed one of the co-precipitation methods that lead to a type of chitosan/HAp composites. In this approach, the composite was co-precipitated in one step, by dropping a chitosan solution containing phosphoric acid into a calcium hydroxide suspension. Other approaches employ either the biomineralization of chitosan in a solid form (especially as membranes) in simulated body fluids (SBF) (Beppu and Santana, 2001) or by mixing of a chitosan solution with different calcium phosphate fillers followed by their precipitation as hydrogel composite. The use of these approaches leads to incorporation of inorganic fillers into the structure of composites (Schwarz and Epple, 1998), either as nano-sized or micro-sized particles.

Different preparation methods of HAp/chitosan composites have been reported, such as mechanical mixing of HAp powder in a chitosan solution, coating of HAp particles onto a chitosan sheet, coating of HAp crystals onto a tendon chitosan (Yamaguchi et al., 2003) and a co-precipitation method. The chitosan/HAp hybrid fiber has also been reported by Chung and Korean, (2002). However, these materials were shown in the macroscopically homogeneous. The conventional method to fabricate chitosan/HA composite is that HA powder was mixed with chitosan, dissolved in 2% acetic acid solution, then the mixture was impressed into mold, finally was freezing-dried to make sponge composite. Surface modification and polyblend methods can be used to change the physicochemical properties of chitosan–gelatin membranes or scaffolds by incorporating hyaluronic acid. Adding hyaluronic acid can improve the mechanical, biological and anti-degradation properties of the membranes or the scaffolds (Mao et al., 2003).

Hydroxyapatite (HAp) was prepared by precipitation method, while the biphasic hydroxyapatite/tricalcium phosphate (HA/B-TCP) was prepared by heating the prepared HAp at 900°C for 5 hours in air. To improve bioactivity both HAp and HAp/TCP fillers were loaded onto chitosan grafted with two monomers, hydroxyethylmethacrylate (HEMA) and methylmethacrylate (MMA) during copolymerization process (Hashem and Mohamed, 2007). Also, biocomposites containing HA-DBM mixture powder loaded onto the copolymer matrix containing the grafted chitosan with poly methylmethacrylate (pMMA) and its derivative during copolymerization were fabricated (Mohamed et al, 2007).

The chitosan mineralization in the case of using a stepwise co-precipitation approach involves the following stages: First, chitosan chains change their conformation as a function of environmental parameters, such as pH. Starting from extended conformations such as worm-like, as seen in Fig.4 (1), adopted at low pH (until 3–3.5), chitosan turns to the more compact conformations such as extended random coil Fig.4(2) and even to more compacted random coil conformations, at higher pH values (from 3.5 until 6). Between 5.5 (the pH at which brushite is precipitated) and 6.5–6.7 (the pH at which chitosan is precipitated) is the pH range that leads to formation of an interconnected three-dimensional net work between chitosan and brushite that can be approximated as a dendritic-like structure, as shown in Fig.4(3). This structural model is characterized by highly dense irregular shaped cores which are linked by the chitosan bridging segments. In the dendritic core, the chitosan chains are randomly packed as amorphous regions while in the bridging regions more extended conformations are found, in some parts parallel oriented chains domains are presented. Since the dendritic core is in the range between 200 and 600nm hence it is formed by the many compact random coil chitosan units that are approaching one another chitosan chain segments can interact with the CaPs phases (i.e. seeds of HAp already formed at pH 5.5 and identified by its XRD pattern, brushite and some other ACP). In this way, they achieve of so called "anchoring regions" by different specific interactions such as ion-dipole or/and through the complexation of Ca with chitosan. Considering that a dendritic-like structure is formed in the earlier stage of composite formation, one can explain the complex bimodal distribution of HAp nano-crystallites in the chitosan matrix. Inside the dendritic core, the chitosan chains density is much greater than outside of them, shown as inter-connection blob regions. We assume that the probability to find a certain number of HAp crystallite seeds per a chitosan chain unit is the same, inside and outside the dendritic core regions. Since the chitosan chain density is much greater inside the cores, this leads to the conclusion that the higher density of HAp crystallite seeds is located inside the dendritic cores. Consequently inside the core favorite the formation of the "small HAp nanocrystallites (their growth is spatially limited) whereas outside the cores, along the interconnection regions where the space constraint is not that much limited, "large" HAp nanocrystallites are favorized to be formed. Finally, the cluster-like and scattered-like size domains are generated in this way, as seen in Fig.4(4) This theoretical model is supported by the experimental data, since we could demonstrate that the amount of chitosan in the composite can be used to control the HAp nano-crystallite size, otherwise no influence should be observed (Rusu et al., 2005).

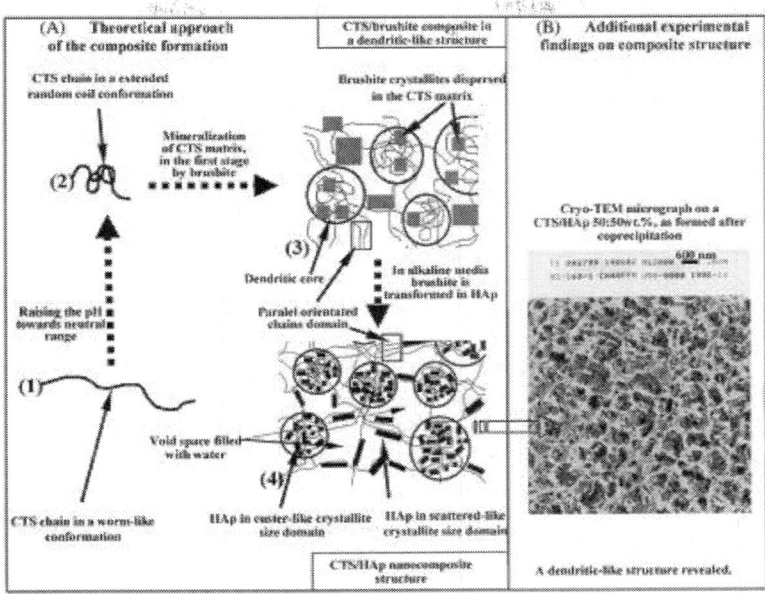

Figure 4. Sketch of chitosan (CTS) mineralization through nanosized HAp. The main stages in the formation of the composite structure are outlined by the theoretical approach, as shown on the left side (A). On the right side (B), we present a TEM micrograph a CTS/HAp 50:50wt% composite, which emphasizes the dendritic-like structure of sample at this stage of formation (in gel-like form). Furthermore, when the excess water is released a solid, rigid composite is obtained.

An interesting approach was reported by Hu et al., (2004) in which the chitosan hydrogel is mineralized via in situ hybridization by the ionic diffusion processes in a controlled manner. It should be mentioned that, each approach of those cited above leads to a particular type of chitosan/HAp composite materials with respect to their structure and properties. In order to prepare such types of composites, we have reported a stepwise co-precipitation method in which the pH of the chitosan solution is gradually increased in a stepwise fashion (Ng et al., 2008). Biodegradable hydroxyapatite/chitosan-gelatin polymeric biocomposites were fabricated by using HA powder and HA filler containing titania powder (10 and 30%) with a chitosan and gelatin grafted co-polymeric matrix during copolymerization process (Mohamed and Mostafa, 2008). Preparation of a model HAp/CTS (30:70 in mass ratio) nanocomposite nanofibers using a two-step method, which involves firstly preparing HAp/CTS nanocomposites by a co-precipitation synthesis approach and then fabricating the resultant HAp/CTS nanocomposites, aided with a fiber-forming additive–

ultrahigh molecular weight poly(ethylene oxide), into nanofibers via the electrospinning process (Zhang et al., 2008).

5.4. Characterization

The FT-IR spectra show that the two characteristic bands of amide I (1655 cm^{-1}) and amide II (1599 cm^{-1}) for chitosan shift to lower wavenumber after being compounded, which suggests that interaction must take place between chitosan and n-HA, including hydrogen bonds between -NH$_2$ and -OH of n-HA as well as the chelation between -NH$_2$ and Ca^{++}. The more shift of these bands to lower wave number, the stronger the hydrogen bonds between these groups and also the stronger the interaction between these molecules (Zhang et al., 2005). The XRD pattern of precipitated HA shows un-differentiated broad peaks with poor crystallinity around the characteristic region. However, the crystallographic structure of precipitated HA nano-crystals is more identical with natural bone mineral (biological apatite). Hence, the prepared HA nano-crystallites in this investigation have more similarity with natural bone mineral in terms of degree of crystallinity and structural morphology. The calcined HA exhibited all the characteristic diffracted peaks of stoichiometric HA with higher degree of crystallinity. Rising in the calcination temperature shapes the diffracted peaks more sharper, which is a good sign for the improvement of crystallinity of precipitated HA. The obtained results did not show any peaks corresponding to calcium carbonate and calcium oxide and hence suggesting that the ingredients were reacted completely and produced a homogeneous HA. The XRD analysis of HA/chitosan composites proved that, a broad peak assigned to chitosan at 20° becomes wider and weaker with increase of n-HA. It suggests that the addition of n-HA obviously affects the crystallinity of chitosan. The characteristic peaks at 25.8° (002) and 39.6° (310) are used to calculate the n-HA crystal sizes (Xianmiao et al., 2009). The TGA mass loss of HA\C composite increased from 3.3–6.5 mass% as the chitosan concentration increased from 0–2.5 mass%. The amount of chitosan that adsorbed on HA was 2.8–3.1 mass% based on Carbon–Hydrogen–Nitrogen (CHN) analysis. The specific surface area of HA increased after aging in chitosan acetate gel solutions and attained a high value of 160 m^2/g in comparison to 85 m^2/g for untreated HA (Wilson and Hull, 2008).

The SEM micrograph of precipitated HA (Fig.5a) exhibited nano-sized crystals with almost uniform particles size. The HA particles prepared were not only stoichiometric but also mono-dispersive and roughly particles were not fused together with other crystals. It can be inferred that majority of the particles were of single crystals, regular shape and cleaner contours with no agglomeration which are highly beneficial for coating of nano HA onto biomedical implants. On the other hand, composites bone paste (Figs.5b and c) showed heterogeneous phases with complete fusion of HA crystallites into chitosan matrix. Major changes in the crystal size of composites were monitored as compared to single phase HA due to the presence of chitosan macromolecules. The physical appearances of the composites are quite different from the starting material and also apparent that ultra-fine particles of HA are found to aggregate into large clusters and precipitate in the chitosan matrix. One of the possible reason may be due to some of the HA nano-particles might have partially dissolved in the acidic chitosan solution that permitting the HA particles more easily to penetrate into the chitosan matrix. The particles of composites showed a high tendency to agglomerate and hence it can have capability to prevent the particle mobilization after post-implantation. Both the SEM pictures (b and c) of composites exhibited porous surfaces, but the pores were not uniform. The average pore size was found to 105 and 80mm for 5 and 10 wt chitosan composites, respectively. The composites containing porous structure on their surfaces will be more beneficial for tissue in-growth (Murugan and Ramakrishna, 2004). Also, the TEM micrographs of HA\chitosan =100\5 (wt\wt) proved that the HA particle size was 100 nm in length and 20-50 nm in width which dispersed well in chitosan matrix homogenously (Hu et al., 2004).

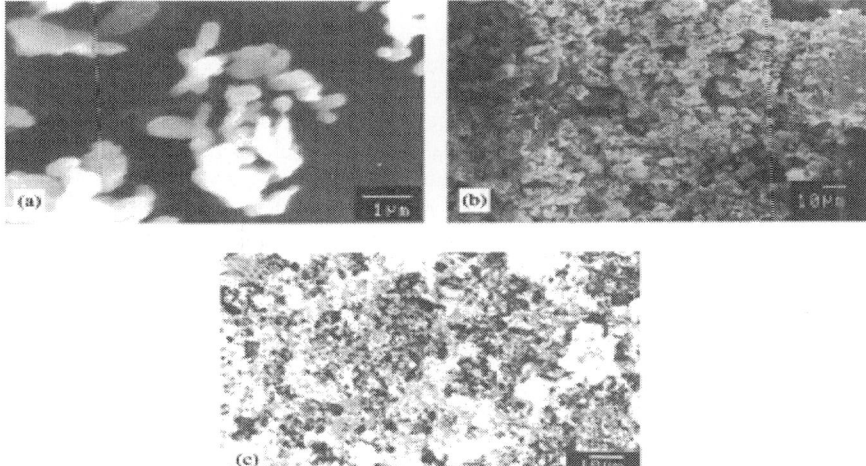

Figure 5. SEM photographs of (a) precipitated nano HA; (b) composite with 5% chitosan sol; and (c) composite with 10% chitosan sol.

5.5. Applications

One of the present trends in implantable applications requires materials that are derived from nature. The impetus is twofold. First, such "natural" materials have been shown to better promote healing at a faster rate and are expected to exhibit greater compatibility with humans. Second, new concepts in implantable medical devices especially tissue engineering derived from a combination of biomaterial onto which cells are seeded, require "temporal" features dictating the biomaterial to matrix was progressively resorbed. Therefore, HA or other calcium containing materials incorporated into chitosan has been a primary research area where orthopedic or bone substitution and periodontal applications were the focus (Khor and Lim, 2003)
.

5.5.1. Bone Substitutes

Maruyama and Ito, 1996 reported that the strength of chitosan-HA hardened composite was comparable to that the cancellous bone derived from tibial eminentia. Pal et al., (1997) prepared different varieties of HA in conjunction with chitosan, as binder, to know its unique biological behavior in bone bonding. These hybrid materials displayed good blood compatibility (Chen et al., 1998). Apatite cement (AC) was almost completely surrounded by mature bone at eight weeks. No promotion or production of osteoconductivity was observed by chitosan even though it is considered to promote bone formation. Then, they concluded that there is enhancement of bone formation (Takechi et al., 2001).

Chitosan and silk fobroin (SF) together as a complex organic matrix for HA granules were employed attempting to obtain a novel composite HA/chitosan–silk fobroin (HA/CTS–SF) with good osteoconductivity, enhanced mechanical strength and sufficient formability and flexibility. Additionally, chitosan and SF are easily derived from naturally abundant chitin and silk cocoon, respectively, which offers a great promise for the potential use of HA/CTS–SF composite as bone scaffold material. HA/CTS–SF composite was obtained via a simple co-precipitation method at room temperature with chitosan and SF serving as a complex organic matrix. The inorganic component in the composite is identified as mono-phase poorly crystalline HA containing carbonate ions. The chemical interactions between the inorganic and organic constituents in the composite, probably take place via the chemical bonding between Ca^{2+} and the amino group of chitosan or the amide bands of SF. The involvement of chitosan and SF endows the composite with higher compressive strengths compared to pure HA. These findings suggest that HA/ CTS–SF composite may be a promising biomaterial for bone in-growth and implant fixation (Wang and Li, 2007).

A novel bone repair material can be obtained by incorporating carboxymethyl cellulose (CMC) into n-HA/CS system. Not only did it compound uniformity by chemical interactions and resembled natural bone apatite in composition morphology and size, but also it improved the compressive strength compared with n-HA/CS composite and had controllable degradation rate via adjusting the CS/CMC weight ratio (Liuyun et al., 2008). HA can promote the formation of bone-like apatite on its surface. Polymers combined with HA are capable of promoting osteoblast adhesion, migration, differentiation and proliferation, especially useful for potential applications in bone repair and regeneration. HA particles have been incorporated into chitosan matrices to enhance the bioactivity of tissue engineering scaffolds for hard tissue regeneration. Therefore, composite membrane of HA and chitosan is expected to be a good degradable barrier membrane for guided bone regeneration (GBR) technique (Xianmiao et al., 2009).

5.5.2. Bone Tissue Engineering

The scaffold is a key component of tissue engineering (Langer and Vacanti, 1993). The study of inorganic crystal assembly in or on an organic polymer matrix is an important focus of bio-mineralization to produce nano composites, which can mimic natural bone. The 3D macro porous scaffolds play an important role in the formation of new tissues and provide a temporary scaffold to guide new tissue in-growth and regeneration (Nikalson and Langer, 1997). Chitosan has been proposed to serve as a non-protein matrix for 3D tissue growth. Chitosan could provide the biological primer for cell-tissue proliferation and reconstruction. One of the most promising features of chitosan is its excellent ability to be processed into porous structures for use in cell transplantation and tissue regeneration. In tissue engineering, the porous structure of chitosan provides a scaffold for bone cells to grow in and seed new bone regeneration. For rapid cell growth, the scaffold must have optimal micro architecture such as pore size, shape and specific surface area (Madihally and Matthew, 1999).

In bone tissue engineering, the biodegradable substitutes act as a temporary skeleton inserted into the defective sites of skeleton or lost bone sites, in order to support and stimulate bone tissue regeneration while they gradually degrade and are replaced by new bone tissue (Service, 2000). Chitosan-based scaffolds possess some special properties for use in tissue engineering. The major goal in fabricating scaffolds for bone tissue engineering is to accurately control pore size and porosity. Porous chitosan structures can be formed by freezing and lyophilizing chitosan–acetic acid solutions in suitable moulds (Chow and Khor, 2000). Bone regeneration research needed to deal with various clinical bone diseases such as bone infections, bone tumors and bone loss by trauma (Braddock et al., 2001). To combine the osteoconductivity of calcium phosphate and good biodegradability of polymers, composites have been developed for bone tissue engineering either by directly mixing the components or by a biomimetic approach (Wei and Ma, 2004). Polymer-ceramic composite scaffolds are expected to mimic natural bone, in the way that natural bone is also a composite of inorganic compounds (calcium phosphates especially substituted carbonated hydroxyapatite) and organic compounds (collagen, protein matrix, etc.). The hydroxyapatite–chitosan–alginate porous network has been reported and demonstrated to be suitable for bone tissue engineering applications using osteoblast cells (Zhao et al., 2003).

Research advances in bone regeneration in tissue engineering have focused on the development of three dimensional (3D) porous scaffolds that can serve as a support, reinforce and in some cases organize the tissue regeneration or replacement in a natural way (Sachlos et al., 2003). Several studies have been focused on chitosan–calcium phosphates (CP) composites for this purpose in bone tissue engineering. Beta-tricalcium phosphate (β-TCP) and hydroxyapatite (HA) of CP bioceramics are excellent candidates for bone repair and regeneration because of their similarity in chemical composition with inorganic components of bone (Zhang et al., 2003). Tissue engineering is regarded as an ultimately ideal medical treatment for diseases that have been too difficult to be cured by existing methods. This biomedical engineering is designed to repair injured body parts and restore their functions by using laboratory-grown tissues, materials and artificial implants. For regeneration of failed tissues, this biomedical engineering utilizes three fundamental tools: living cell, signal molecules, and scaffold. The choice of chitosan as a tissue support material is governed among others by multiple ways by which its biological, physical and chemical properties can be controlled and engineered under mild conditions (Krajewska, 2005). The 3D macro porous scaffolds play an important role in the formation of new tissues and provide a temporary scaffold to guide new tissue in-growth and regeneration. The fabrication of biodegradable and osteoconductive scaffolds with a 3D interconnected porous network has been a formidable challenge. The feasibility of producing cost effective organic–inorganic scaffolds for tissue engineering to mimic bone by the diffusion method was performed. The porous structure of chitosan scaffold was homogeneously mineralized using this technique of apatite formation at room temperature. The mineralized scaffolds were found to be non-cytotoxic and better for cell proliferation and growth, as indicated by the enzyme activity and protein levels, than un-mineralized scaffold. This suggested that it could be used for further osteoconductivity studies. A biodegradable matrix with sufficient mechanical strength, optimized architecture and suitable degradation rate, which could finally be replaced by newly formed bone, is most desirable (Manjubala et al., 2006). Although the chitosan based composite biomaterials need to improve their mechanical properties for bone tissue engineering, no doubt that chitosan is a promising candidate scaffold material in clinical practice due to the worthiest ability to bind anionic molecules such as growth factors, GAG and DNA. Especially, the ability to link chitosan to DNA may render this material a good potential as a substrate for gene activated matrices in gene therapy application in orthopedics (Kim et al., 2008).

Shen et al., (2007) performed that with the increase of pH after the addition of ammonia, carboxyl groups of citric acid may begin to act as nucleation center for calcium phosphate formation. These negatively charged carboxyls in the reaction system can bond Ca^{2+} strongly and thus forms a large scale of local super saturation microenvironment, and strong electric field resulted from high concentration of negatively charged carboxyls are favor of the interaction that with the most positively charged crystalline plane, so there are many nucleation sites in the network of hydrogel template, each point of nucleation can result in microcrystal. Here, to our attention, biocompatible citric acid took the place of acetic acid in this work because three carboxyl of citric acid could provide more nucleation sites which were appropriate to formation of ultra fine nano-sized carbonate apatite. And it has been conjectured that appropriate increase of citrate ions can benefit the bone resorption and ossification through the formation of dissociated calcium citrate complexes in the surrounding body fluid (Rhee and Tanaka, 1999). Each citric acid molecule can provide three negatively charged carboxyls which act as nucleation center for calcium phosphate formation. Increase of nucleation center can be appropriate to fine crystallites. Furthermore, cross-linking chitosan hydrogel was provided with three dimensional network microstructures, its compartment effect limited the growth of inorganic mineral particles, so the inorganic nano-particles were limited to aggregate in the compartment of the chitosan hydrogel template according to orientation of preferential growth of crystal plane. This multiple-order template effect based on multiple-point nucleation of citric acid and compartment of hydrogel network had a very obvious mediation in the formation process of homogeneous composites (Shen et al., 2007) (Fig.6). In bone tissue engineering, the biodegradable scaffold is a temporary template introduced at the defective site or lost bone to initiate bone tissue regeneration, while it gradually degrades and is replaced by newly formed bone tissue. Finally, an ideal scaffold is characterized by excellent biocompatibility, controllable biodegradability, cytocompatibility, suitable microstructure (pore size and porosity) and mechanical properties. Additionally, it must be capable of promoting cell adhesion and retaining the metabolic functions of attached cells (Thein-Han and Misra, 2009).

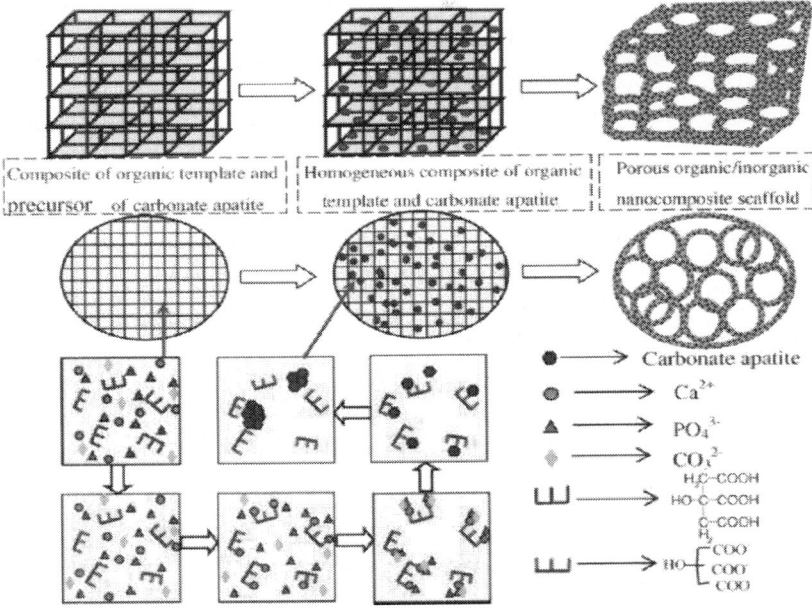

Figure 6. The scheme of formation of homogeneous chitosan/carbonate apatite composite and 3D nanocomposite scaffold (Shen et al., 2007).

5.5.3. In Vitro Applications

The process of apatite formation on the bioactive materials in living body could be reproduced in simulated body fluid (SBF), which means that in vivo bone bioactivity of a material can be predicted by assessing apatite formation on its surface in SBF. They confirmed that there are two types of material which inserted into living body. One of them was able to have apatite form on its surface in SBF, and consequently has apatite produced on its surface in the living body, and bonds to living bone through this apatite layer. The other type was directly bond to living bone without the formation of apatite on their surfaces, so examination of apatite formation on the surface of a material in SBF is useful for predicting the in vivo bone bioactivity of the material, not only qualitatively but also quantitatively (Kokubo and Takadama, 2006).

5.5.3.1. In simulated body fluid (SBF)

From SEM photos (Fig.7), it can be known that chitosan in the chitosan /nHA composite gradually degraded during the soaking in SBF solution, which resulted in plenty of macro-and micropores on the surface of and inside the specimens. At the same time, a lot of tiny apatite crystals deposited on the surface of the specimens, and till the 8th week, a thin layer of bone-like apatite, being highly bioactive was formed. At the first 4 weeks, the degradation rate of chitosan was higher than the deposition rate of apatite on the surface of specimens, which corresponding to a continuous increase of the rate of weight loss. After that, the deposition of apatite is prior to the degradation of chitosan, so the rate of weight loss decreased. This was also confirmed by the rate of water adsorption with the degradation of chitosan during the specimen's soaking in SBF solution, a more sponge-like structure was formed, which can hold more water. However, with more apatite crystals deposition, some of these pores were filled or covered, so water adsorption decreased (Zhang et al., 2005). Kong et al., (2006) reported that chitosan/nano-hydroxyapatite composite scaffolds analysis showed that after incubation in simulated body fluid on both of the scaffolds (the apatite-coated composite scaffolds and apatite-coated chitosan scaffolds), carbonate hydroxyapatite was formed. With increasing nano-hydroxyapatite content in the composite, the quantity of the apatite formed on the scaffolds increased. Compared with pure chitosan, the composite with nano-hydroxyapatite could form apatite more readily during the biomimetic process, which suggests that the composite possessed better mineralization activity (Kong et al., 2006).

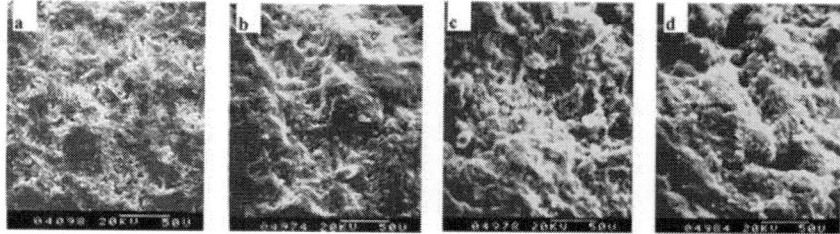

Figure 7. The SEM images of chitosan/nHA composites after soaking in SBF solutions for (a) 0 week, (b) 1week, (c) 4 weeks and (d) 8 weeks, Magn. × 400.

The swelling properties and degradation behavior proved the stability of hydroxyapatite-titania/chitosan-gelatin polymeric biocomposites into the media. In-vitro test behavior confirmed that the prepared composite enhanced the deposition of Ca^{++} and P ions onto the surface that is in the favor of the formation of apatite layer. FT-IR and SEM-EDAX of copolymer and three composites post-immersion verified the formation of spherical apatite particles onto the copolymer surface; therefore, it was expected to enhance the apatite nucleation onto the filler composite surface especially hydroxyapatite-titania/chitosan-gelatin (AK1) composite containing 10% content of titania (Mohamed and Mostafa, 2008).

5.5.3.2. Bone Tissue engineering (Scaffold)

Anti-washout scaffold paste could be directly applied to fit complex shapes of bone defects, without involving machining as in the case of sintered hydroxyapatite. The synergistic use of a reinforcing agent (e.g., chitosan) and a pore-forming agent (e.g., mannitol) in a bone graft may be applicable to other tissue engineering materials. In developing strong and macro-porous calcium phosphate cement (CPC) scaffolds by incorporating chitosan and water-soluble mannitol. The new CPC–chitosan formulation was biocompatible and supported the adhesion, spreading, proliferation and viability of osteoblast cells. The cells were observed to infiltrate into the pores of the scaffold and establish cell–cell interactions. The increased strength and macroporosity of the new apatite scaffold may help facilitate bone ingrowth, implant fixation, and more rapid new bone formation (Fig.8) (Hockin et al., 2005).

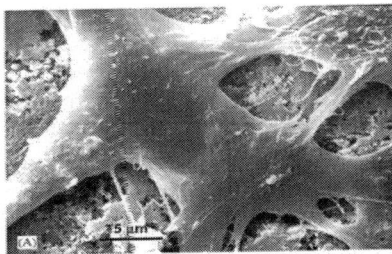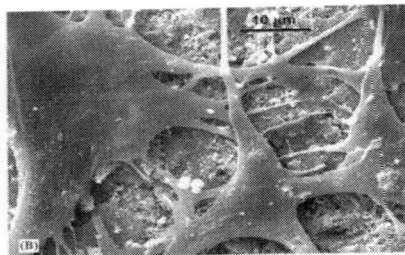

Figure 8. The SEM of cell attachment on (A) CPC control and (B) CPC chitosan composite. The cells developed cytoplasmic processes with lengths ranging from approximately 20 to 50 mm, and the materials exhibited similar cell attachment and cytoplasmic processes development (Hockin et al., 2005).

Kong et al., (2006) reported that pre-osteoblast cells cultured on the apatite-coated scaffolds showed different behavior. On the apatite-coated chitosan/nano-hydroxyapatite composite scaffolds cells presented better proliferation than on apatite-coated chitosan scaffolds. The cells on composite scaffolds showed a higher alkaline phosphatase activity which suggested a higher differentiation level. The results indicated that the addition of nano-hydroxyapatite improved the bioactivity of chitosan/nano-hydroxyapatite composite scaffolds. MSCs do not appear to be rejected by the immune system, allowing for large-scale production, appropriate characterization and testing, and the subsequent ready availability of allogeneic tissue repair enhancing cellular therapeutics. Overall it can be said that, for now, MSCs present more advantages than other cells and have already been widely used in bone tissue engineering. All the superiorities of MSCs encourage us to introduce MSCs into n-HA composite scaffolds for tissue engineering application (Wang et al., 2007).

The morphology and behavior of bone marrow stem cell (BMSCs) cultured in-vitro with the n-HA/chitosan (CS) composite membranes are observed under phase-contrast microscope. Fig.9 shows representative phase-contrast micrographs of cell attachment on the membrane with a n-HA/CS ratio of 4:6 after culture for 1 day, 7 days and 11 days. At the first day, only a few BMSCs are present with the elongated fusing form shape. At 7 days, a large amount of cells proliferate and form cell colony. At 11 days, the population of cells increases manifestly and cells fully attach to the membrane. Obviously, the n-HA/CS composite membrane has no negative effect on the cell morphology, viability and proliferation (Xianmiao et al., 2009).

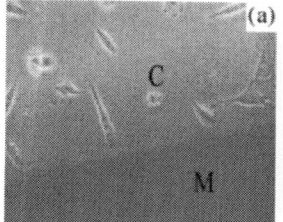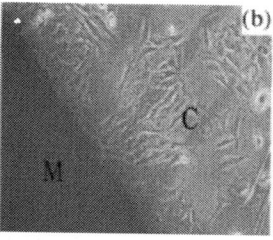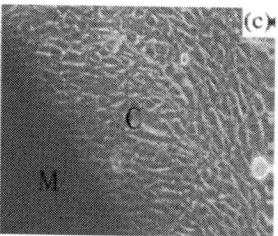

Figure 9. Phase-contrast micrographs of the BMSCs (denoted as C) attached to n-HA/CS (4:6) membrane (denoted as M) after in vitro culture for 1d (a), 7d (b) and 11 d (c).

In chitosan–nHA scaffolds, the presence of extensive filopodia, flat morphology, and excellent spreading in and around the interconnected porous structure, indicated strong cellular adhesion and growth (Fig.10e–h). Furthermore, cell density, cell–cell contact, sheet-like structure, and formation of extracellular matrix and cytoplasmic extensions were more pronounced on the chitosan–nHA surface than on pure chitosan. They believe based on Fig.10e–h that the steps involved in the development of sheet-like morphology involves clustering of cells and bridge formation between the pore walls with consequent formation of a multilayer structure. These steps occurred during early stages in the nano-composite constructs in relation to chitosan scaffolds, suggesting that pre-osteoblasts have high affinity to the surface of chitosan–nHA composite, which is attributed to its increase surface area and composition. Chitosan–nanocrystalline calcium phosphate scaffolds characterized by a relatively rough surface and approximately 20 times greater area/unit mass than chitosan scaffold indicated increase adsorption of fibronectin and improved cell attachment (Thein-Han and Misra, 2009).

5.5.4. In-Vivo Application

The extracellular matrix (ECM) is a powerful regulator of cell adhesion and indeed cells respond to the ECM by means of integrins, which couple the component of the ECM with the actin cytoskeleton. This structure thus mediates adhesion to the ECM and therefore to the implant material. In this respect, the possibility of bonding osteoinductive polymers such as modified chitosan to the ceramic substrate could enhance cell proliferation and consequently anchorage to the implant (Mattiolibelmonte et al., 1998). Sections from chitosan coated HA implants exhibited an evident mesenchymal reaction between bone and implant with several features of osteoinduction. Bone trabeculae penetrating the HA implant were also observed.

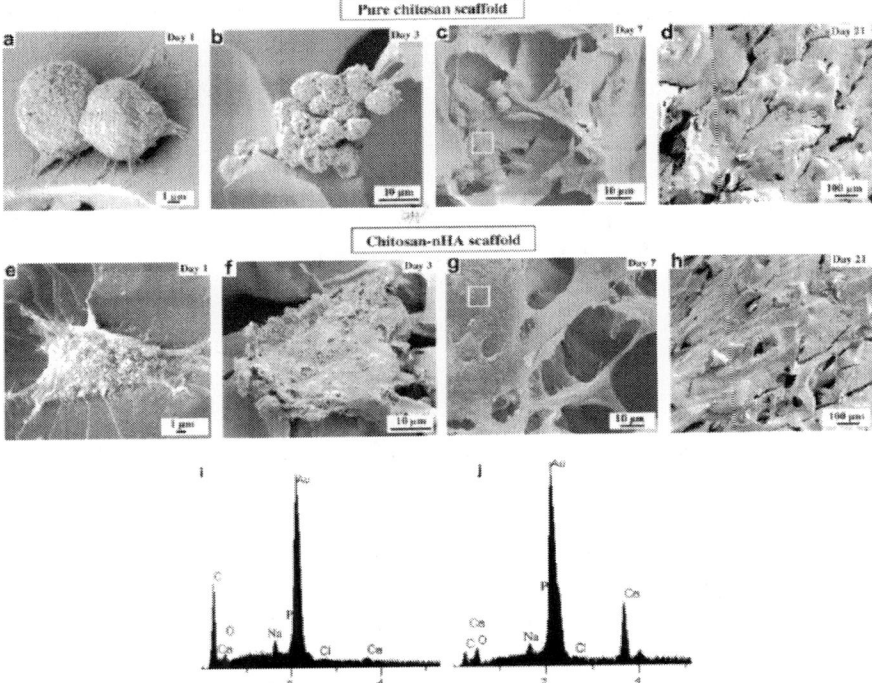

Figure 10. Scanning electron micrographs illustrating morphology of pre-osteoblasts seeded on high-MW chitosan (CH) and chitosan–nHA (CH1) scaffolds (Saggital section). Pre-osteoblasts on chitosan surface after (a) day 1, (b) day 3, (c) day 7 and (d) day 21 of cell culture; and on chitosan–nHA surface after (e) day 1, (f) day 3, (g) day 7 and (h) day 21 of cell culture. EDS spectra for the boxed region in (c) and (g) are presented in (i) and (j), respectively, showing the presence of Ca and P. The P peak is merged with the Au peak, which is due to conductive gold coating on the sample.

Sunny et al., (2002) have reported the preparation of HA-chitosan microspheres as potential bone and periodontal filling materials. HA powder was mixed with chitosan solution followed by paraffin oil, hexane and a surfactant and the microsphere production process commenced. Subsequently, glutaraldehyde was added to crosslink chitosan to give spherical particles ranging from 125 to 1000 mm. When the chitosan/n-HA composite implanted in body using as tissue scaffold, the degradation of chitosan makes room for the growth of new bone and then is substituted by new bone completely. It has been reported that chitosan can promote nucleation and growth of apatite and calcite crystals as well. Moreover, the surface of chitosan is hydrophilic, which can facilitate cell adhesion, proliferation and differentiation (Fig. 11). So, the chitosan/n-HA composite, used as bone substitutes, are hopeful to activate the regeneration and remodeling of bone tissue (Zhang et al., 2005).

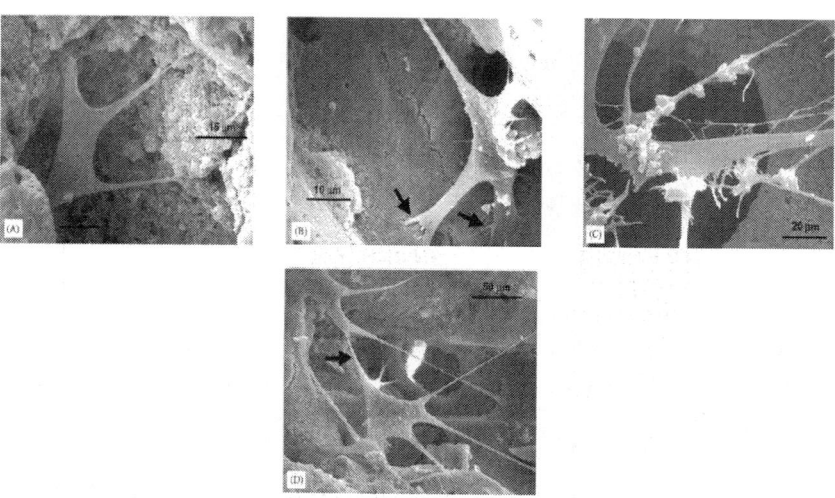

Figure 11. A) SEM of cell infiltration into a macropore. (B) Cell attachment (arrows) to the bottom of a pore. (C) Cells inside a large pore near an opening at the bottom of the pore. (D) Cell-cell interactions inside a pore (arrow indicates a cell-cell junction).

The development of suitable three-dimensional scaffold for the maintenance of cellular viability and differentiation is critical for applications in periodontal tissue engineering. The different ratios of porous nanohydroxyapatite/chitosan (HA/chitosan) scaffolds were prepared through a freeze-drying process. The results indicated that the porosity and pore diameter of the HA/chitosan scaffolds were lower than those of pure chitosan scaffold. The HA/chitosan scaffold containing 1% HA exhibited better cytocompatibility than the pure chitosan scaffold. These scaffolds are evaluated in vitro by the analysis of microscopic structure, porosity, and cytocompatibility. The expression of type I collagen and alkaline phosphatase (ALP) activity are detected with real-time polymerase chain reaction (RT-PCR). Human periodontal ligament cells (HPLCs) transfected with enhanced green fluorescence protein (EGFP) are seeded onto the scaffolds, and then these scaffolds are implanted subcutaneously into athymic mice after implanted in vivo, EGFP transfected with HPLCs) not only proliferate but also recruit surrounding tissue to grow in the scaffold. The degradation of the scaffold significantly decreased in the presence of HA. This study demonstrated the potential of HA/ chitosan scaffold as a good substrate candidate in periodontal tissue engineering (Zhang et al., 2007).

REFERENCES

1. I. Aiba, S. Izume, M. N. Minoura, Y. Fujiwara, In chitin in Nature and Technology Ed. R.A.A. Muzzarelli, C. Jeuniaux and G.M. Goodey, Plenum Press, New York, 1986 396-398.

2. P. Arpornmaeklong, N. Suwatwirote, P. Pripatnanont, K. Oungbho, Growth and differentiation of mouse osteoblasts on chitosan-collagen sponges. Int. J. Oral Maxillofac. Surg. 36 2007 2007 328337.

3. H. H. Beherei, K. R. Mohamed, A. I. Mahmoud, J. Egyptian, Chemistry. of, xxx, 2009 xxx-xxx. Under press.

4. Beppu M.M. and Santana C.C. In vitro biomineralization of chitosan. Key Eng Mater 2001 192-195:31-4.

5. H. S. Blair, J. Guthrie, T. Law, p. Turkington, Chitosan and modified membranes. App. Polymer. Sci., 33 1987 641- 656.

6. G. Borah, G. Scott, K. Wortham, Bone induction by chitosan in endochondral bones of the extremities" Brine, C., Sanford P.A., Zikakis 5 5th Int. Conf. Chitin and Chitosan, 1991, Princeton, N.J, London : Elsevier Applied Science, (1992) 47- 53.

7. M. Braddock, P. Houston, C. Campbell, P. Aschroft, again. Born, tissue. bone, for. engineering, repair. bone, News Physiol 16 16 2001 20813.

8. Brine C.J. Introduction: Chitin: Accomplishments and Perspectives. In Zikakis JP, ed chitin, chitosan and related enzymes. London: Academic Press 1984

9. H. Chen, X. Tian, H. Zou, Preparation and blood compatibility of new silica-chitosan hybrid biomaterials. Artif. Cells Blood Substit. Immobil. Biotechnol., 264 (1998) 431-6.

10. F. Chen, Z. Wang, C. Lin, Preparation and characterization of nano-sized hydroxyapatite particles and hydroxyapatite/chitosan nano-composite for use in biomedical materials Materials Letters 57 2002 2002 858861.

11. Chen R.H. and Hwa, H.D. Effect of molecular weight of chitosan with the same degree of deacetylation on the thermal, mechanical, and permeability properties of the prepared membrane. J. Carbohydrate Polymers, 29 1995 1995 353358.

12. K. S. Chow, E. Khor, Novel fabrication of open-pore chitin matrixes. Biomacromolecules 1 12000 2000 617.

13. Y. S. Chung, J. Korean, Soc. Fiber, 395 (5) (2002) 532- 536.

14. J. M. Dang, K. W. Leong, Drug. Adv, Rev.. Delivery, 2006 487-499.

15. S. Ding, Biodegradation behavior of chitosan/calcium phosphate composites. J. Non-Crystalline Solids 353 2007 2007 23672373.

16. E. A. Elizalde-Pen, N. Flores-Ramirez, G. Luna-Barcenas, S. R. Va´squez-Garcı, G. Ara´mbula-Villa, B. Garcı´a-Gaita, J. G. Rutiaga-Quinones, J. Gonza´lez-Herna´ndeze, Synthesis and characterization of chitosan-g-glycidyl methacrylate with methyl methacrylate. European Polymer Journal 43 2007 2007 39633969.

17. K. Eugene, Y. L. Lee, Implantable applications of chitin and chitosan. Biomaterials 24 2003 2003 33949.

18. El-Kady A.M., Mohamed K.R., El-Bassyouni G.T, Fabrication, characterization and bioactivity evaluation of calcium pyrophosphate/ polymeric biocomposites. Ceramics International 35 2009 2009 29332942.

19. Felt O., Buri P. and Gurny R. Chitosan: a unique polysaccharide for drug delivery. J. Drug Dev. Ind. Pharm, 24 (11) (1998) 979- 93.

20. J. C. Fricain, R. Bareille, F. Ulysse, B. Dupuy, J. J. Amedee, Mater. Biomed, . Res, 1998 96-102.

21. T. Furukawa, Y. Matsusue, T. Yasunaga, Y. Shikinami, M. Okuno, T. Nakamura, . Biomaterials, 2000 889.

22. I. R. Gibson, W. J. Bonfield, Mater. Biomed, . Res, 2002 697-708.

23. J. M. Gomez-Vega, E. Saiz, et al.. Biomaterials, 2000 105.

24. J. M. J. Gorbunoff, Interaction of proteins with hydroxyapatite. Anal. Biochem., 136 1984 1984 425445.

25. C. J. Grande, F. G. Torres, C. M. Gomez, M. C. Bano, Nanocomposites of bacterial cellulose/hydroxyapatite3 for biomedical application. Acta Biomaterialia xxx 2009 xxx-xxx.

26. W. J. E. M. Habraken, J. G. C. Wolke, J. A. Jansen, Ceramic composites as matrices and scaffolds for drug delivery in tissue engineering, Advanced Drug Delivery Reviews 59 2007 2007 234248.

27. Hashem A.H and Mohamed K.R. Chitosan graft copolymer-HA/DBM biocomposites: Preparation, characterization and in-vitro evaluation Chitosan graft copolymer-HA/DBM biocomposites: Preparation, characterization and in-vitro evaluation. Egyptian J. of Chemistry, 505 (5) (2007) 625-644.

28. H. Hockin, K. Xua, G. Carl, Jrb. Simon, Fast setting calcium phosphate-chitosan scaffold: Mechanical properties, biocompatibility and Biomaterials 26 2005 2005 13371348.

29. H. K. Hockin, C. G. Xua, J. Simon, Fast setting calcium phosphate-chitosan scaffold: mechanical properties and biocompatibility. Biomaterials 26 2005 13371348.

30. Q. Hu, B. Li, M. Wang, J. Shen, Preparation and characterization of biodegradable chitosan/hydroxyapatite nanocomposite rods via in situ hybridization: a potential material as internal fixation of bone fracture. Biomaterials 255 (2004) 779-85

31. Q. L. Hu, X. Z. Qian, B. Q. Li, J. C. Shen, Study on chitosan rods prepared by in situ precipitation method, Chinese. J Chem Univ 243 (2003) 528-31.

32. S. G. Hu, C. H. Jou, M. C. J. Yang, Applied Polymer Science 8812 (12) (2003) 2797-2803.

33. J. Huang, S. M. Best, W. Bonfieu, R. A. Brooks, et al. In vitro assessment of the biological response to nano-size Hydroxyapatite. J Mater Sci-Mater 15 15 2004 4415.

34. S. Itoh, I. Yamaguchi, M. Suzuki, S. Ichinose, K. Takakuda, H. Kobayashi, K. Shinomiyag, J. Tanaka, Hydroxyapatite-coated tendon chitosan tubes with adsorbed laminin peptides facilitate nerve regeneration In vivo Brain Research 993 2003 2003 111123.

35. K. Kamiyama, H. Onishi, Y. Machida, Biodisposition characteristics of N-succinyl-chitosan and glycol-chitosan in normal and tumor bearing mice. Biol Pharm Bull 222 (1999)179-86.

36. H. Kang, Y. Cai, P. Liu, characterization. Synthesis, sensitivity. thermal, chitosan-based. of, copolymers. graft, Carbohydrate Research 341 2006 2006 28512857.

37. T. Kasuga, Y. Ota, et al.. Biomaterials, 2001 19.

38. E. Khora, L. Y. Limb, Implantable applications of chitin and chitosan Biomaterials 24 2003 2003 23392349.

39. M. Kikuchi, H. N. Matsumoto, T. Yamada, Y. Koyama, K. Takakuda, J. Tanaka, Glutaraldhyde cross-linked hydroxyapatite/collagen self-organization nanocomposites. Biomaterials, 25 2004 2004 6369.

40. D. G. Kim, Y. I. Jeong, J. W. J. Nah, Polym. Appl, 1. Sci, 2007 3246-3254.

41. S. B. Kim, Y. J. Kim, Y. J. , T. L. Yoon, S. A. Park, I. H. Cho, E. J. Kim, I. A. Kim, J. W. Shin, The characteristics of a hydroxyapatite-chitosan-PMMA bone cement Biomaterials 25 2004 2004 57155723.

42. Y. Kim, S. Seo, H. Moon, M. Yoo, I. Park, B. Kim, C. Cho, Chitosan and its derivatives for tissue engineering applications. Biotechnology Advances 26 2008 2008 121

43. T. Kokubo, H. Takadama, How useful is SBF in predicting in vivo bone bioactivity? Biomater 2715 (2006) 2907-15.

44. Kokubo T., Kim H., Kawashita M., Novel bioactive materials with different mechanical properties, Biomaterials 24 2003 2003 21612175.

45. L. Kong, Y. Gao, G. Lu, Y. Gong, N. Zhao, X. A. Zhang, on. study, bioactivity. the, chitosan/nanohydroxyapatite. of, scaffolds. composite, bone. for, engineering. tissue, European Polymer Journal 42 2006 2006 31713179.

46. B. Krajewska, Membrane-based processes performed with use of chitin/chitosan materials. Sep Purif Technol 41 2005 2005 30512.

47. R. Langer, J. P. Vacanti, engineering. Tissue, 2. Science, 1993 920-6.

48. A. Larena, D. A. Caceres, C. Vicario, A. Fuentes, Release of a chitosan-hydroxyapatite composite loaded with ibuprofen and acetyl-salicylic acid submitted to different sterilization treatments. Applied Surface Science 238 2004 2004 518522

49. S. B. Lee, Y. H. Kim, M. S. Chong, Y. M. Lee, Preparation and characteristics of hybrid scaffolds composed of beta-chitin and collagen. Biomaterials 25 2004 23092317.

50. Y. L. Lee, E. Khor, C. E. Ling, J. Biomed, Res.. Mater, 1999 111.

51. G. Li, E. T. Dunn, E. W. Grandmaison, M. F. A. Goosen, Applications and properties of chitosan. J. Bioact. Compat. Polyss, 7 71992 1992 370397.

52. J. Li, J. Pan, L. Zhang, X. Guo, Y. Yu, Culture of primary rat hepatocytes within porous chitosan scaffolds. J Biomed Mater Res 67 67 (2003a) 93843

53. X. Li, Y. Tsushima, M. Morimoto, H. Saimoto, Y. Okamoto, S. Minami, et al. Biological activity of chitosan-sugar hybrids: specific interaction with lectin. Polym Adv Technol 11 2000 2000 1769.

54. F. H. Lin, C. H. Yao, J. S. Sun, H. C. Liu, C. W. Huang, Biological effects and cytotoxicity of the composite composed by tricalcium phosphate and glutaraldehyde cross-linked gelatin. Biomaterials 19 1998 1998 90517.

55. Z. Liu, J. Han, T. Czernuszka, Gradient collagen/nano-hydroxyapatite composite scaffold: Development and characterization. Acta Biomaterialia 5 52009 2009 661669.

56. J. Liuyun, L. Yubao, Z. Li, L. Jianguo, Preparation and properties of a novel bone repair composite: nano-hydroxyapatite/chitosan/carboxymethyl cellulose. J Mater Sci: Mater Med 2008 19:981-987.

57. J. X. Lu, F. Prudhommeaux, A. Meunier, L. Sedel, G. Guillemin, Effects of chitosan on rat knee cartilages. Biomaterials 20 1999 1999 193744.

58. Madihally S.V. and Matthew H.W.T. Porous chitosan scaffolds for tissue engineering. Biomaterials 20 1999 113342.

59. W. G. Malette, H. J. Quigley, E. D. Adickes, Chitin in nature and Technology Muzzarelli R, Jeuniauxc, Gooday, G.W, eds: Chitosan effect in vascular surgery Tissue culture and Tissue regenerations, New York: Plenum press, 1988 435- 442.

60. I. Manjubala, S. Scheler, J. Bossert, K. D. Jandt, Mineralization of chitosan scaffolds with nano-apatite formation by double diffusion technique. Acta Biomaterialia 2 22006 2006 7584.

61. J. S. Mao, H. F. liu, Y. J. Yin, K. D. Yao, The properties of chitosan-gelatin membranes and scaffolds modified with hyaluronic acid by different methods. J. Biomaterials 24 2003 2003 16211629.

62. Martino A.D., Sittinger M., Risbud M.V. Chitosan: a versatile biopolymer for orthopaedic tissue-engineering. Biomaterials 26 2005 2005 598390.

63. A. Matsuda, T. Ikoma, H. Kobayashi, J. Tanaka, Sci. Mater, Eng, 24 24 (2004) 723.

64. M. Mattiolibelmonte, A. De Benedittis, R. A. A. Muzzarelli, P. Mengucci, et al

65. Bioactivity modulation of bioactive materials in view of their application in osteoporotic patients. J Mat. Sci: Mat. In Med. 9 91998 1998 485492.

66. F. L. Mi, S. S. Shyu, Y. B. Wu, S. T. Lee, J. Y. Shyong, R. N. Huang, Biomaterials, 222 (2) (2001) 165.

67. Y. Miyamato, K. I. Shikawa, Basic properties of calcium phosphate cement containing atelocollagen in its liquid or powder phases. Biomaterials 19 1998 707-15).

68. Mohamed K.R. Preparation of ceramic/ceramic and/or ceramic/biopolymer composites in the system of 23 Al2O3- CaO- 2O5 and their characterization as bioceramic" Ph.D, Biophysics Dept., Faculty of Science, Cairo University.

69. Mohamed K.R and Mostafa A.A. Preparation and bioactivity evaluation of hydroxyapatite-titania/chitosan-gelatin polymeric biocomposites J. Materials Science and Engineering 28 28 2008 10871099.

70. Mohamed K.R., El Bassyouni G.E., Beheri H.H. Chitosan graft copolymer-HA/DBM biocomposites: Preparation, characterization and in-vitro evaluation. J. Applied Polymers Scienc, 105 2007 2007 25532563.

71. J. L. Moreau, H. H. K. Xu, Mesenchymal stem cell proliferation and differentiation on an injectable calcium phosphate-chitosan composite scaffold. Biomaterials, xxx 2009 1-8.

72. R. Murgan, S. Ramakrishna, Crystallographic study of hydroxyapatite bioceramics derived from various sources: Cryst Growth 5 5 (2005)111-2).

73. R. Murugan, S. Ramakrishna, Bioresorbable composite bone paste using polysaccharide based nano hydroxyapatite. Biomaterials 25 2004 2004 38293835

74. Muzarelli R.A. In: Aspinall GOA, editor. The polysaccharides. New York, NY: Academic Press; 1985 417-25.

75. R. Muzzarelli, V. Baldassarre, F. Conti, P. Ferrara, G. Biagini, G. Gazzanelli, V. Vasi, Biological activity of chitosan : Ultrastructural study. Biomaterials, 93 (3) (1988) 247-52.

76. Muzzarelli R.A., Mattioli-Belmonte M., Tietz C., et al., Stimulatory effect on bone formation exerted by a modified chitosan. Biomat. 15 15 (13) (1994) 1075-1081.

77. R. A. A. Muzzarelli, G. Biagini, et al.Polym.. Carbohydr, 2001 35.

78. C. H. Ng, V. M. Rusu, M. G. Peter, Formation of chitosan hydroxyapatite composites in the presence of different organic acids. Adv Chitin Sci;7, 2008

79. Nikalson L.E. and Langer R.S. Advances in tissue engineering of blood vessels and other tissues. Trans Immunol 5 51997 1997 3036.

80. H. Nishikawa, A. Ueno, S. Nishikawa, J. Kido, M. Ohishi, H. Inoue, et al. Sulfated glycosaminoglycan synthesis and its regulation by transforming growth factor-beta in rat clonal dental pulp cells. J Endod 26 2000 1697.

81. K. Ogawa, T. Yui, M. J. Miu, Biotech. Bioscince, Biochem, 566 (6) (1992) 858.

82. A. W. Otto, J. W. Klau, J. M. Johnson, S. M. George, Thin solid films 292 12 (1997) 135-144.

83. Rao S.B. and Sharma, C.P. Use of chitosan as a biomaterial: Studies on its safety and hemostatic potential" Biomed Mat. Res., 34 1) (1997) 21-28, Jan.

84. Rehman I., Smith R., Hench, L.L., Bonfield W. J. Biomed. Mater. Res. 29 (1995) 1287-1294.

85. S. H. Rhee, J. Tanaka, Effect of citric acid on the nucleation of hydroxyapatite in a simulated body fluid. Biomaterials 20 1999 1999 215560.

86. G. A. F. Roberts, Chemistry. Chitin, macmillan 1992

87. Roodman G.D. Mechanisms of bone metastasis. NEJM. 350 2004 16551664.

88. Rouget, 1859 : Book of chitin, Edited by Muzzarelli, R.A.A., (1977), Ancona, Italy, 60100, Pergamon Press.

89. V. M. Rusu, C. H. Ng, M. Wilke, B. Tiersch, P. Fratzl, M. G. Peter, Size-controlled hydroxyapatite nanoparticles as self-organized organic-inorganic composite materials. Biomaterials, 26 2005 2005 54145426.

90. E. Sachlos, N. Reis, C. Amsley, B. Derby, J. T. Czernuska, Novel collagen scaffolds with predefined internal morphology made by solid free form fabrication. Biomaterials 4 2003 148797.

91. P. L. Sapelli, V. Baldassare, R. A. A. Muzzarelli, M. Emanuelli, in. Chitosan, dentistry, Chitin in Nature and Technology, 1986 507-512.

92. A. Sarasam, S. V. Madihally, Characterization of chitosan-polycaprolactone blends for tissue engineering applications. Biomaterials 26 2005 2005 55008.

93. K. Schwarz, M. Epple, Biomimetic crystallisation of apatite in a porous polymer matrix. Chem Eur 410 4(10) (1998) 1898-903.

94. Service R.F. Tissue engineers build new bone. Science 289 2000 1498500.

95. X. Shen, H. Tong, T. Jiang, Z. Zhu, P. Wan, J. Hu, Science. Composites, . Technology, 2007 2238-2245

96. W. L. Suchanek, P. Shuk, K. Byrappa, R. E. Riman, Huisen. K. S. Ten, V. F. Janas, . Biomaterials, 2002 699-710.

97. W. Suchanek, M. Yoshimura, Processing, of. properties, biomaterials. hydroxyapatite-based, use. for, tissue. hard, implants. replacement, J Mater 13 13 1998 94117.

98. Suh J.K.F and Matthew H.W.T., Application of chitosan-based polysaccharide biomaterials in cartilage tissue engineering: a review. Biomaterials 21(24) (2000) 2589-98.

99. Sunny M.C., Ramesh P.,Varma H.K. Microstructured microspheres of hydroxyapatite ceramic. J Mater Sci Mater 13 13 2002 62332.

100. Y. Takahashi, M. Yamamoto, Y. Tabata, Enhanced osteoinduction by controlled release of bone morphogenetic protein-2 from biodegradable sponge composed of gelatin and β-tricalcium phosphate. Biomaterials 26 2005 2005 48565.

101. Takagi S., Chow L., Hirayama S., Eichmiller F., Properties of elastomeric calcium phosphate cement-chitosan composites, Dental Materials, 198 (8) (2003) 797-804.

102. Takechi M., Ishikawa K., Miyamoto Y., Nagayama M., Suzuki K. Tissue responses to anti-washout apatite cement using chitosan when implanted in the rat tibia JMat.Sci. In Medicine 12, (2001)597-602

103. R. Tarsi, B. Corbin, C. Pruzzo, R. A. Muzzarelli, Effect of low molecular weight chitosan on the adhesive properties of oral streptococci. Oral Microbial Immunol., 134 (4) (1998) 217-24.

104. S. Teng, E. Lee, B. Yoon, D. Shin, H. Kim, J. Oh, Chitosan/nano-hydroxyapatite composite membranes via dynamic filtration for guided bone regeneration. J. Biomed. Mater. Res. Part A, 883 (2009) 569-580.

105. Thein-Han W.W and Misra R.D.K. Biomimetic chitosan-nanohydroxyapatite composite scaffolds for bone tissue engineering. Acta Biomaterialia, 54 (2009) 1182-1197.

106. K. Tomihata, Y. Ikada, In-vitro and in-vivo degradation of films of chitin and its deacetylated derivatives. J. Biomater., 187 (7) (1997) 567- 75.

107. I. M. Van Der Lubben, J. C. Verhoef, G. Borchard, H. E. Junginger, Drug. Adv, Rev. Delivery, 52 2001 136144.

108. S. Viala, M. Freche, J. L. Lacout, Chim. Ann, Mater.. Sci, 1998 69.

109. H. Wang, Y. Li, Y. Zuo, J. Li, S. , L. Cheng, Biocompatibility and osteogenesis of biomimetic nano-hydroxyapatite/polyamide composite scaffolds for bone tissue engineering. Biomaterials 28 2007 2007 33383348.

110. L. Wang, C. Li, Preparation, properties. physicochemical, a. of, hydroxyapatite/chitosan-silk. novel, composite. fobroin, Polymers.. Carbohydrate, 2007 740-745.

111. M. Wang, W. Bonfield, Biomaterials, 22 (2001)1311.

112. X. Wang, J. , Y. Wang, B. He, Bone repair in radii and tibias of rabbits with phosphorylated chitosan reinforced calcium phosphate cements. Biomaterials 23 2002 2002 416776.

113. Z. Wang, Q. Hu, Preparation, of. properties, hydroxyapatite/chitosan. three-dimensional, rods. nano-composite, Biomed. Mater. 2010 5 045007 doi:10.1088/17486041 /5/4/045007.

114. G. Wei, P. X. , Structure and properties of nano-hydroxyapatite/ polymer composite scaffolds for bone tissue engineering. Biomaterials 25 2004 2004 474957.

115. O. C. Wilson, J. R. Hull, et al. 2008 "Surface modification of nanophase hydroxyapatite with chitosan" Materials Science and Engineering C 28 434437.

116. Xianmiao 29., Yubao L., Yi Z., Li Z., Jidong L., and Huanan W., "Properties and in vitro biological evaluation of nano-

hydroxyapatite/chitosan membranes for bone guided regeneration" Materials Science and Engineering C 29, (2009) 29-35.

117. I. Yamaguchi, S. Iizuka, A. Osaka, H. Monma, J. Tanaka, The effect of citric acid addition on chitosan/hydroxyapatite composites. Colloids and Surfaces A: Physicochem. Eng. Aspects 214 2003 2003 111118.

118. I. Yamaguchi, S. Itoh, M. Suzuki, A. Osaka, J. Tanaka, . Biomaterials, 2003 3285- 3292.

119. I. Yamaguchi, K. Tokuchi, H. Fukuzaki, Y. Koyama, K. Takakuda, H. Monma, et al. Preparation and microstructure analysis of chitosan/hydroxyapatite nanocomposites. J of Biomedical Materials Research, 55 2001 2001 2027.

120. E. Yilmaz, Exp. Adv, 5. Biol, 2004 59.

121. Y. Yuan, P. Zhang, Y. Yang, X. Wang, X. Gu, The interaction of Schwann cells with chitosan membranes and fibers in vitro. Biomaterial.25 (8) 2004 4273.

122. Y. Zhang, M. Ni, M. Zhang, B. Ratner, Calcium phosphate chitosan composite scaffolds for bone tissue engineering. Tissue 9 9 2005 33745.

123. Y. F. Zhang, X. R. Cheng, Y. Chen, B. Shi, X. Chen-H, D. Xu-X, J. Ke, Three dimensional Nanohydroxyapatite/chitosan scaffold as potential tissue engineered periodontal tissue J of Biomaterials Applications, 21 21 4 (2007) 333-349

124. Y. Zhang, J. R. Venugopal, Ramakrishna. S. Adel-Turki El, Su. Bo, C. T. Lim, Electrospun biomimetic nano-composite nano-fibers of hydroxyapatite/chitosan for bone tissue engineering. Biomaterials 29 2008 2008 43144322.

125. A. F. Zhao, W. L. Graysona, T. Maa, B. Bunnellb, W. W. Luc, Effects of hydroxyapatite 3 3-D chitosan-gelatin polymer network on human mesenchymal stem cell construct development Biomaterials 27 2006 18591867.

126. F. Zhao, Y. Yin, W. W. Lu, C. Leong, W. Zhang, J. Zhang, et al. Preparation and histological evaluation of biomimetic three-dimensional hydroxyapatite/chitosan-gelatin network composite scaffolds. Biomaterials 23 2003 2003 322734.

CHAPTER 6

New Composite Materials in the Technology for Drinking Water Purification from Ionic and Colloidal Pollutants

Marjan S. Ranđelović[1], Aleksandra R. Zarubica[1] and Milovan M. Purenović[1]

[1] University of Niš, Faculty of Science and Mathematics, Department of Chemistry, Niš,, Serbia

1. INTRODUCTION

Composite materials (composites) are inherently heterogeneous and represent a defined combination of chemically and structurally different constituent materials, ensuring the required properties such as mechanical strength, stiffness, low density, or other specific characteristics depending on their purpose. Therefore, composite material is a system composed of two or more physically distinct phases whose combination produces a synergistic effect and aggregate properties that are different from those of its constituents. Favorable characteristics of composite materials were known to the people even in the period BC (before Christ-Century) and were used in order to improve the quality of human daily life. For example, it is known that in the ancient period, people made bricks that were reinforced with straw, and thus secured greater longevity and durability of their buildings. The incorporation of the straw improves the strength, toughness and thermal insulation properties of these composites. In principle, the degree of reinforcement (volume fraction of straw) and the level of alignment of the straw stalks (and their lengths) may be adjusted so that not only the properties but their anisotropy may be optimised differently in various parts of the structure [1]. Significant development and application of composites began in the second half of the 20th century, wherein their diversity and areas of application are constantly increasing. Development of composite materials is resulted mainly from the increasing need for materials with better mechanical characteristics that would be used as components in various constructions. For this purpose, such composites should have an adequate strength, stiffness, good oxidation resistance and low weight. Intensive study of composite materials and their processing methods has caused that these materials replace metals and alloys and become indispensable in the manufacture of parts for automobiles, spacecrafts, sports equipment etc. In terms of exploiting modern engineering composites this remains a central principle. Modern composites can be said to have "designed micro- and nanostructures" which means that the constituents of composites have much more finely divided structures and tend to have sizes in the micrometre or nanometre range. Basic factors affecting properties of composites are as follows:

- Properties of phases;
- Amount of phases;
- Bonding and the interface between the phases;
- Size, distribution and shape (particles, flakes, fibers, laminates) of the dispersed phase - reinforcement;
- Orientation of the dispersed phase - reinforcement (random or preferred).

Good bonding (adhesion) between matrix and dispersed phase provides a high level of mechanical properties of the composite via the interface. In addition, interfaces are responsible for numerous processes of electron transfers and play crucial role in redox processes, heterogeneous catalysis, adsorption etc. Usually, there are three forms of interface between the two phases within the composite:

1. Direct bonding with no intermediate layer. In this case adhesion ("wetting") is provided by either covalent bonding or van der Waals force;
2. Intermediate layer in form of solid solution of the matrix and dispersed phases constituents;
3. Intermediate layer (interphase) in form of a third bonding phase (adhesive).

Current challenges in the field of composite materials are associated with the extension of their application area from structural composites to functional and multifunctional composites. In this respect, a great improvement of composite materials through processing has been made enabling the development of composite materials for electrical, thermal and other functional applications that are relevant to current technological needs. Examples of functions are joining, repair, sensing, actuation, deicing (as needed for aircraft and bridges), energy conversion (as needed to generate clean energy), electrochemical electrodes, electrical connection, thermal contact improvement and heat dissipation (*i.e.*, cooling, as needed for microelectronics and aircrafts) [2]. Modern processing includes the use of additives (which may be introduced as liquids or solids), the combined use of fillers at the micrometer and nanometer scales, the formation of hybrids, the modification of the interfaces in a composite and control over the microstructure. Therefore, it can be said that the development of composite materials for current technological needs must be application driven and process oriented. The conventional composites engineering approach, which is focused on mechanics and purely structural applications, is in contrast to mentioned modern practice.

On the contemporary level of science development it is known that materials of certain characteristics can be obtained only by strictly defined procedures of processing and depend on their chemical composition and structure. Since composites are heterogeneous systems, as already has been noted, the matrix is of great importance whose structure and chemical composition determine the most dominant features of the composite as a unit. However, it should be noted here that the composite does not possess properties of a single component but exhibits qualitatively new features, because of which it is considered as a new material. In addition to the dominant use of composites as structural elements, important application of composite materials is in the water purification technologies. In this field of application, composites usually have the role of adsorbent, electrochemically active materials, catalysts, photocatalysts etc. Bearing in mind that the material efficiency in the removal of harmful substances from water is higher if greater is its surface area, there are tends of scientists to develop these materials with required and defined nanostructures. In addition to the specific surface area increasing, nanostructured materials exhibit a qualitatively new properties compared to the related structure at the micro or macro scale. In this manner, it is developed specific procedure for certain metal hydroxides and natural organic matter layering onto alumosilicate matrix as well as procedures of microalloying which both lead to significant changes of the surface acido-basic and electrical properties of the alumosilicate matrix. The nano-scale composites provide an opportunity to study the phase boundaries and phenomena occurring at the surface, interface boundaries and within intergranular area during composites synthesis or during their interaction with aqueous solutions.

2. AN OVERVIEW AND TRENDS IN USE OF COMPOSITES IN INDUSTRIAL PLANTS

Nanocomposites based on polymers represent an area of significant scientific interest and developing industrial practice. Despite the proven benefits of polymer based nano-composites in the scope of their mechanical properties, and some distinctive combination/synergism of improved structural features, the real application remains still relatively isolated and not well discussed.

An insight in the historical (re)view on polymer nano-composites showed on the first type used based on the combination of natural fillers and polymers in the 90s [3-6] up to estimated 145 million USD spent at huge market of polymer based nano-composites in 2013 [7].

3. THE CONCEPTS OF INTERPHASE BOUNDARIES MODIFICATION, MICROALLOYING AND COATING/LAYERING IN THE COMPOSITE SYNTHESIS

Methods and techniques for managing properties of composite materials include the selection and modification of constituent materials as well as changing the interface boundaries within the composite. Some composites are most commonly fabricated by impregnation (infiltration) of the matrix or matrix precursor in the liquid state into the appropriate filler preform. The connection between the constituents depends on the microstructure and chemistry of the interface boundary. The matrix and filler are connected by chemical bonds, interdiffusion, van der Waals forces and mechanical interlocking [2]. The first three interactions require very close filler-matrix contact that can be achieved if the matrix or matrix precursor wetting the surface of filler during the infiltration of matrix or matrix precursors in the filler preform. Effective wetting means that the liquid is evenly distributed over the surface of filler, while a poor wetting means that the liquid drops formed on the surface. Wettability can be increased by applying the coatings, adding wetting agents or by chemical surface functionalization (the introduction of functional groups on the surface that increase wettability) thereby changing the surface energy. If the filler is carbon fiber, surface treatments involve oxidation treatments and the use of coupling agents, wetting agents, and/or coatings. Often, metals or ceramics are used as coatings for carbon fillers. Metallic coatings are usually formed by coating carbon fiber reinforcements with metals *i.e.* Ni, Cu and Ag. Examples of ceramic coatings are TiC, SiC, B_4C, TiB_2, TiN which are distributed by using Chemical Vapor Deposition (CVD) technique or by solution coating methods starting from organometalic compounds. Therefore, these are examples of application of coatings on carbon materials to illustrate the method of modification of surface properties.

In the case of metal-ceramic composites, certain liquid metals react with ceramic preform during infiltration. For instances, composites based on the Al–Al_2O_3 system can be obtained by Reactive Metal Penetration (RMP) method which is based on infiltration of ceramic preforms by a liquid metal, generally aluminium or aluminium alloys [8,9]. During the process, a liquid metal simultaneously reacts and penetrates the ceramic preform, usually silica or a silicate, resulting in a metal-ceramic composite characterized by two phases that are interpenetrated. Another example is the reaction between SiC and Al during the infiltration of molten aluminum in a preheated preform:

$$4Al + 3SiC \rightarrow Al_4C_3 + 3Si$$

From the equation it can be seen that Si is generated during the reaction which is then dissolved in molten aluminum, while Al_4C_3 occurs at the SiC-Al interfacial boundary. The degree of reaction increases with increasing temperature. On the contrary, there are metals that in liquid state difficult wet the surface of the ceramic resulting in metal infiltration hindering. The difficulty of wetting and bonding of liquid metals to ceramic surfaces is related to atomic bonding in the ceramic lattice and can be improved by application of coatings. Coated particles (composite particles) are composed of solid phase covered with thinner or thicker layer of another material [10.11]. These coatings - layers on the surface are important for several reasons. In such way, the surface characteristics of the initial solid phase are modified and sintering conditions as well as molten metal infiltration can be better controlled.

As can be seen from examples, the processing of composite materials often involves high temperature and pressure to cause the joining of constituent materials forming a cohesive material. Generally, the matrix dictates the required temperature, pressure and processing time during composite synthesis. Sintering is an important factor in achieving the desired microstructure of ceramic based composites and includes very complex processes. In addition to surface coatings, an important influence on sintering has been exhibited by an addition of microalloying components, which significantly determine a microstructure and properties of ceramics [12]. The presence of small amounts of impurities in the starting material can vastly influence their mechanical, optical, electrical, color, diffusivity, electrical conductivity, and dielectric properties of matrix. Microalloying, as a known modern procedure for changing the intrinsic semiconductor properties, by authors' original works (Purenovic et al.), get more and more important role in the control of some structurally sensitive properties of metals, alloys, ceramics, composites and other materials. It is known that the nature of matter is determined by its composition and structure. There are many structurally sensitive properties of materials, but among the most sensitive are the conductivity, electrode potential, magnetic, catalytic and mechanical properties. Microalloying means adding certain elements in small (ppm) quantities, thereby modified structure results in a significant change in the value of conductivity and the electrode potential. Conducted own investigation and the results obtained showed an excellent rational electrochemical behavior of composites such as microalloyed aluminum, microalloyed magnesium, as well as composite ceramics and quartz sand microalloyed with aluminum and magnesium, in contact with aqueous solutions of electrolytes or water which contain harmful ingredients in ionic, molecular and colloidal state. Microalloyed and structurally modified composite ceramics have high porosity (30%), with the macro-, meso-, micro- and submicropores. There is direct relationship between porosity and structure of these composite materials, especially when it comes to nanostructured fragmented crystals. It is worth to emphasize the domination of amorphous phases with crystalline substructure, which is impossible to be removed, and it would be inappropriate to be removed, because the contact of crystals with amorphous layer is responsible for numerous processes of electrons exchange. By certain processes and reactions in the solid phase, the amorphous microalloyed aluminum, microalloyed amorphous magnesium, amorphous-crystalline structure of composite microalloyed ceramics and amorphous-crystalline structure of microalloyed quartz sand could be obtained. Many metals, alloys and composite electrode materials manifested significant differences in the reversible thermodynamic potential and the steady corrosion potential.

The manufacturing processes used to make composite ceramics can cause the development of liquid phases during sintering, and their retention as remnant glass at triple junctions and along grain boundaries and interphase boundaries after cooling to room temperature. Formed thin intergranular films are relevant to creep behavior at high temperatures, and also responsible for the strength of the bonding at interfaces. However, the heat treatment at elevated temperatures which is used for joining constituent materials and establishing the cohesive forces shows a disadvantage because cooling can lead to disturbance of established bonds between phases. Namely, during the cooling, differences in coefficients of thermal expansion could result in unequal contraction by which established bonds are broken. This problem is particularly evident in metal-ceramic composites, where high temperatures are usually applied during synthesis.

4. PREPARATION OF MODERN NANO-COMPOSITES

Processing of nanocomposites based on layered silicates is rather challenging activity to achieve the full technical and engineering potential, which is the field with the largest growth forecast [13-16]. The modification of silicates by use of organic components is needed to allow intercalation, and also in order to improve compatibility/nano-distribution some additional ingredients have to be applied. The thermal treatment as step in processing sequence helps proper stabilisation of nanocomposites that has to take into consideration the oxidative stability of the polymer substrate, the influence of the nano-filler and the impact of modifiers and compatibilisers.

Montmorillonite of natural origin is among the most used nano-fillers. Traditional nano-fillers contain metal ions and other contaminants that may influence the thermooxidative stability and features of the nanocomposites. Organic modification of the (natural/traditional) clay is usually realized by cation exchange with a long-chain amines or quaternary ammonium salts. Content of such involved organic material content within the clay may be up to 40 mas.%. Therefore, the total thermal resistance of the composite material highly depends on the thermal stability of the organic ingredient. The thermal stability of the ammonium salts is limited at the processing temperatures applied (ex. extrusion, injection molding, etc.). Namely, thermal degradation of ammonium salts starts at 180°C and may be even tentatively reduced by catalytically active sites on the alumosilicate layer [17].

The compatibiliser applied as organically modified filler is often polypropylene-g-maleic anhydride in amount from 5 to 25% in the final composite formulation. The inferior stability of such low molecular weight filler comparing to the parent polymer affects the total stability of the final polymer based nanocomposites.

5. AN IMPROVEMENT OF COMPOSITES STABILITY

Nanocomposites may show higher stability due to increased barrier to oxygen, or lower stability because of undergone to hydrolysis through entrapped water [18,19]. In conventional practice stabilizer systems based on phenolic antioxidants and phosphites are applied, and in recent investigations new found components of filler degradation deactivators has been tested [20].

A traditional state-of-art polypropylene (PP) nanocomposite consisting of maleated PP and nano-clay is traditionally stabilized by a proven combination of phenolic antioxidant and phosphites. The polymer degradation may be completely prevented even after 5 extrusion cycles by using the patented stabilizer system AO-2 (based on oxazoline, oxazolone, oxirane, oxazine and isocyanate groups) [20], additionally improving mechanical properties of the resulting nano-composites and discoloration during processing and application.

The underlined thermal instability of the usual ammonium organic modifiers can be diminished by using the phosphonium, imidazolium, pyridinium, tropylium ions [21]. An alternative way to produce thermally stable nano-composites is the use of unmodified clays in combination with selected copolymers playing role of dispersants, intercalants, exfoliants and compatibilisers for PP nano-composites. In current processing of nano-composites different structures are identified such as polyethyleneoxide based nonionic surfactants [22] and amphiphilic copolymers based on long-chain acrylates [23]. Recently, more specifically poly(octadecylacrylate-co-maleic anhydride) and poly(octadecylacrylate-co-N-vinylpyrrolidone) in the form of gradient copolymers are applied with unmodified montmorillonite for processing PP nano-composites. Such obtained nano-composites show partial exfoliation, the final visual appearance is similar to the classical ammonium modified systems, however better thermal and thermo-oxidative stability is proven [23]. The most important improvement is achieved in the mechanical vales comparing to the conventional polymer system.

6. NANOCOMPOSITES USE IN A COMPETITIVE ENVIRONMENT OF THE MATERIALS

Nanocomposites materials are very attractive from the scientific and practical point of view, although some other materials are also interesting, such as plastics, fillers, blends, and different additives fulfilling the specified product profile. In such competence, the lowest cost solution comprising acceptable material structure and properties/resistances would dominate. Even more, competitive (nano)composite materials would benefit from nanocomposites developments and keep their application fields with improved features. Most of nanocomposites materials applications are intended for long-term and outdoor use. This is important aspect on the need for relevant nanocomposites stability. Namely, it is known that inorganic fillers often show a negative effect on the oxidative stability to a varying extent. The interactions of the filler and the stabilizers over adsorption/desorption mechanisms are mainly responsible for the impact. The specific surface area of the filler and pore volumes, surface functionality, hydrophilicity, thermal and photo-sensation properties of the filler and transition metal content (ex. manganese, titanium, iron) have been found to be potential factors/elements of the interaction [24].

Polypropylene/montmorillonite nanocomposites, additionally stabilized with antioxidant, degrade much faster under photo-oxidative conditions than pure polypropylene [25,26].This phenomenon is attributed to active species/sites in the clay generated by photolysis or photo-oxidation, and by consequence interaction between antioxidant, montmorillonite and maleic anhydride modified polypropylene. In natural clay present iron may additionally play an active role in the dramatic modification of material oxidation conditions [27], and nanoparticles also catalyze the decomposition process [28]. The use of so-called filler deactivators or coupling agents is potential solution for diminishing the negative influence of fillers on the (photo)oxidative stability by blocking active sites on the filler surface. Amphiphilic modifiers with reactive chemical groups in the form of polymers, olygomers or low molecular weight molecules such as bisstearylamide or dodecenylsuccinic anhydride have been proposed [29].Thus, stabilizer systems containing filler deactivators should have an affirmative effect in nano-composites for long-term stability.

7. NANO-COMPOSITES MATERIALS FOR WATER TREATMENTS: STATE-OF-THE-ART AND PERSPECTIVES

Clean drinking water is essential to human health, and also so-called technical water is a critical feedstock in a variety of key industries including electronics, pharmaceuticals and food processing industries. Taking into consideration that available supplies of fresh water are limited (due to population growth, extended deficiency, stringent health regulations, and competing demands from a variety of users/consumers) the world is facing with challenges to satisfy demands on high water quality standards and quantities (volumes). Benefits and trends in nano-scale science, chemistry and engineering impose that many of the current problems regarding green chemistry may be resolved using nano-sorbents, nano-catalysts, nanoparticles and nanostructured catalytic membranes. Nano-materials are characterized by a number of key physicochemical properties being particularly attractive for water purification treatments. Nanomaterials have much large specific surface area than bulk respect particles (mass to volume ratio), also they can be functionalized with reactive chemical groups specific in affinity to a given model compound. These materials may possess redox features and take part in shape- and structural-dependent catalyzed reactions of water purification. In aqueous solutions, they can serve as sorbents/catalysts for toxic metal ions, radionuclides, organic and inorganic solutes/anions [30]. Moreover, nano-materials can be used in selective targeting of biochemically constituents of aquatic bacteria and viruses. The nano-materials seems to be key components in future environmental friendly and cost-effective functional materials to desalinate public and polluted waters world-wide, for purification of water contaminated by pesticides, pharmaceuticals, phenol and other aromatics. The presence of heavy metals in water exhibits a variety of harmful effects on the living organisms in polluted ecosystems. The removal of heavy metals from water includes the following procedures: chemical precipitation, coagulation/flocculation, membrane processes, ion exchange, adsorption, electrochemical precipitation, etc. [31,32]. However, the application of composite materials in the controlling of pollutants in the environment and drinking water is significant [33,34], as described in further text.

The use of zeolites, natural or synthetic ones in waste water treatments is highly limited due to low adsorption capacity in the case of former and relatively small grain size in latter. Modification of natural or synthetic zeolites toward composite material which would satisfy both essential properties is a challenging task. Tailoring synthetic zeolite resulted in a composite porous host supporting microcrystalline active phase of vermiculite matrix [35]. The vermiculite-based composite showed the same hydraulic properties as natural clinoptilolite with similar grain size (2-5 mm), while the rate of adsorption and maximal adsorption capacity was improved four times. In other words, cation exchange capacity is increased when compared to natural zeolite with a comparative grain size, ion-exchange kinetics are substantially improved in comparison to natural zeolite, and hydraulic conductivity is considerably higher that synthetic powdered zeolite [35].

The development of new composite material based on use of inorganic polymeric flocculants as a combination of anionic and cationic poly-aluminium chloride (PACl) in one unique polyelectrolyte is proposed [36]. The incorporation of the anionic polyelectrolyte into PACl structure noticeably affects its initial properties (*i.e.* turbidity, Al species distribution, pH and conductivity). Interactions are taking place between Al species and polyelectrolytes molecules over hydrogen bonding (amino/amidic groups of the polyelectrolyte, and the –OH and –H groups of Al species are involved) and electrostatic forces/interactions. This resulted in new composite material. The main advantage of composite coagulants is lower residual aluminium concentration that remains in the treated sample, and more efficient treatments of waters (organic matter removal) can be realized [36]. Additional benefit is in cost effective process in the absence of specific equipment for handling the polyelectrolyte (ex. pumping system, etc.). Taking into account faster flocculation, increased efficiency and cost effectiveness, such new composite material seems to be promising one.

Porous ceramic composites can be prepared by silver nanoparticles-decoration using a silver nanoparticle colloidal solution and an aminosilane coupling agent [37]. The interaction between the nanoparticles and the ceramics comprises the coordination bonds between the –NH$_2$ group and the silver atoms on the surface of the nanoparticles. The composite can be stored for long periods without losing of nanoparticles, also being highly resistance to ultrasonic irradiation and washing. Such composite has shown high sterilization property as an antibacterial water filter [37]. This low cost composite, bearing in mind commonly available synthesis, simple preparation, the use of cheap and non-toxic reagents in the procedure, may be imposed as a potential solution for widespread use in water treatments.

Ultrafine AgO particles-decorated porous ceramic composites are prepared based on the main ingredient, cristoballite. The results on composite structure show that silver(II)oxide decorated diatomite-based porous ceramic composites possess crystal structure, and are composed of tetragonal cristoballite, monoclinic silver(II)oxide and cubic silver(I)oxide [38]. Such AgO-decorated porous ceramic composites show a strong antimicrobial activity and an algal-inhibition capacity. As the extension time is longer, the antibacterial effects are enhanced up to 99.9% [38].

Actual nanostructured composite materials based on multi-walled carbon nanotubes (MWCNT) and titania exhibited strong interphase structure between MWCNT and titania. This contact and interaction facilitated a homogeneous deposition/coverage of titania over MWCNT [39]. The photo-catalytic activity of the prepared composite materials was tested in the conversion of phenol from model watery solution under UV or visible light. The results showed higher photo-catalytic activity of the composite MWCNT and titania than over mechanical mixture proving an assumption on the existence of the interphase structure effect [39].

Nanocomposite membranes based on silica/titania nanotubes over porous alumina supports membranes were prepared [40]. An inserting of amorphous silica into nanophase titania caused the surpressed of phase transformation from anatase to rutile, and decreased the titania particle size. Good photo-catalytic activity of organic contaminants degradation, and wettability of composite membrane under UV-irradiation, helped to obtain high permeate flux across the composite membrane [40].

8. NEW ALUMOSILICATE BASED COMPOSITES CHEMICALLY MODIFIED BY COATINGS/THIN LAYERS – TESTED IN THE REMOVAL OF COLLOIDAL AND IONIC FORMS OF HARMFUL HEAVY METALS FROM WATER

Without new materials, there are no new technologies. Having in mind this fact, electrochemically active and structurally modified composites were obtained through microalloying and certain metals hydroxides layering, starting from bentonite as alumosilicate precursor. The composites have prognosed electrochemical, ion-exchanging and adsorption properties, as very sensitive structural and surface properties of materials. After the series of experiments, including composites interaction with synthetic waters, the obtained results are presented, analyzed and then systematized in the form of appropriate models of interactions.

8.1. Alumosilicate Composite Ceramic Microalloyed by Sn For The Removal of Ionic and Colloidal Forms of Mn

Usually, manganese does not present a health hazard in the household water supply. However, it can affect the flavor and color of water because it typically causes brownish-black staining of laundry, dishes and glassware [32]. Although manganese is one of the elements that are at least toxic, concentrations of manganese much higher than the maximum allowed concentration during long-term exposure can cause health damage. A number of known procedures for the manganese removal are not suitable for an elimination of its all chemical species due to reversible release of manganese into water systems. Therefore, some of these used procedures are at the edge of techno-economical viability. In order to remove ionic and colloidal forms of manganese, a new aluminosilicate-based ceramic composite with defined electrochemical activity was synthesized [41]. Synthesis procedure of the composite material consists of two phases. Firstly, composite particles were synthesized by applying Al/Sn oxide coating on the bentonite particles in an aqueous suspension. In the second phase, aluminium powder was added to the previously obtained plastic mass and after shaping in the form of spheres 1 cm in diameter and drying, sintering was performed at 900°C. Fig. 1 a), b) and c) presents the microstructure of composite by using different magnifications.

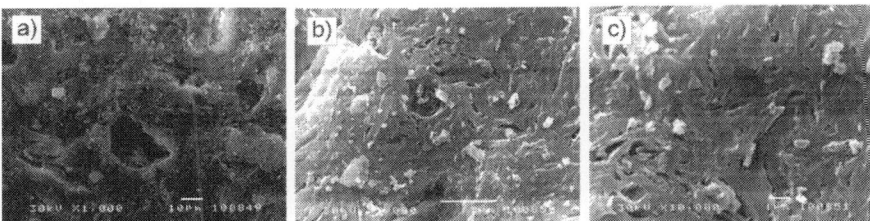

Figure 1. SEM images of the composite recorded at: a) low, b) medium and b) high magnifications.

During sintering a microalloying of composite by Sn occurred causing crystal grain surface layer amorphization and a creation of non-stoichiometric phases of Al_2O_3 with a metal excess [42,43]. In this way, microalloying causes electrochemical activity, which manifests itself in contact with the aqueous solutions of electrolytes and harmful substances in water. Therefore, the ceramics is unstable in contact with water and susceptible to corrosion because surface electrochemical processes taking place. The composite influence redox properties of water and electrochemically interacts with ionic and colloidal forms of manganese in synthetic water systems.

Alumosilicate matrix, whose particles are coated with Al/Sn oxides, was filled with a metal phase which is mostly aluminum with a small quantity of tin as a microalloying component. During the thermal treatment, liquid aluminium simultaneously reacts and penetrates the ceramics preform, resulting in metal/ceramic composite, where the all phases are interpenetrated forming a porous structure. In fact, the reduction of tin(II) occurred according to the following reaction:

$$2Al + 3SnO \rightarrow Al_2O_3 + 3Sn$$

The first reaction step is the reduction of Sn(II) to elemental Sn and its dispersion from the ceramics into the melt. Therefore, during the reaction, Sn is liberated into the liquid metal and diffuses towards the Al source. Moreover, oxygen partial pressure within the composite, at the Al-Al$_2$O$_3$ interface, can be estimated on the basis of thermodynamic parameters and calculated using the following equation [44]:

$$\Delta G^\circ = RT\ln PO_2$$

The standard free energy of the reaction:

$$4/3Al + O_2 \rightarrow 2/3Al_2O_3$$

at 900°C, given by Ellingham diagram [44] is -869 KJ/mol, and corresponding oxygen partial pressure: $PO_2 = 2.02 \cdot 10^{-39}$ Pa. Therefore, this low oxygen partial pressure during sintering provides reducing environment and the formation of nonstoichiometric oxide phases, with the metal excess, or with vacancies in oxygen sublattice. Nevertheless, Al$_2$O$_3$ belongs to the oxides of stoichiometric composition or with a negligible deviation from stoichiometry, it can occur as an amorphous and nonstoichiometric oxide with a metal excess during oxidation of aluminium. Common nonstoichiometric reactions occur at low oxygen partial pressures when one of the components (oxygen in this case) leaves the crystal [45,42]. A corresponding defect reaction is [45]:
(oxygen in this case) leaves the crystal [45,42]. A corresponding defect reaction is [45]:

$$O_O^x \rightleftarrows \frac{1}{2}O_2(g) + V_O^{\bullet\bullet} + 2e^{'}$$

As the oxygen atom escapes, an oxygen vacancy $(V_O^{\bullet\bullet})$ is created. Taking in mind that the oxygen is to be presented in neutral form, two resulting electrons would be easily excited into the conduction band.

Al–Sn alloys show a great activity compared to the thermodynamic Al^{3+}/Al potential of −1.66V vs. NHE, which stands for a pure aluminium. The activation is manifested by a shifting of the pitting potential in the negative direction and significant reducing of the passive potential region [43,46]. The addition of microalloying Sn to aluminium produced a considerable shift of the open circuit potential (OCP) in the negative direction [46].

During the process of composite ceramics sintering, significant changes in the structure of alumosilicate matrix were occurred. Namely, the polycrystaline alumosilicate matrix with amorphised grain and sub-grain boundary were obtained, where a main role possesses metallic aluminum itself, then a microalloyed tin and nonstoichiometric excess of these elements in ceramics, creating macro-, meso- and micro- pores with the reduced mobility of grain boundaries and termination of grain growth [47]. Aluminum and tin in conjunction with other admixtures present in composite ceramics cause drastic changes in the structure-sensitive properties and electrochemical activity. An active composite ceramics in contact with synthetic water containing manganese reduce and deposit the manganese in the macro-, meso- and micro- pores (eq. 6). Electrochemical activity is provided by electrochemical potential of Al atoms and free electrons that participate in redox processes.

$$2Al + 3Mn^{2+} \rightarrow 2Al^{3+} + 3Mn$$

The deposited manganese on microcathode parts of the structure can further form separate clusters and the adsorption layer [48,49]. Reduction processes take place until the Al^{3+} ions continue to solvate themselves in water. A part of Al^{3+} ions reacts with OH- ions giving insoluble $Al(OH)_3$.

8.1.1. Interaction of Composite Material with Ionic and Colloidal Forms of Mn in Synthetic Water

Interaction of the composite material with water manifests itself as decreasing in the redox potential of water, as shown in Fig. 2. This confirms the fact that the composite is electrochemically active in contact with water. During the interaction with water, aluminium from the composite is electrochemically dissolved into water providing electrons which can participate in the number of redox reactions of water yielding reduced species (molecules, ions and radicals) such as H_2, OH•, etc. [47].

TDS value of distilled water immediately after contact with ceramics increases. It seems that increasing the TDS value is due to dissolution of Al^{3+}, Mg^{2+}, Na^+, SiO_3^{2-} from the bentonite based composite. Al^{3+} and SiO_3^{2-} ions are subjected to hydrolysis and polymerization reactions which are followed by spontaneous coagulation-flocculation processes and appearance of sludge after a prolonged period of time.

A reduction of manganese concentration in synthetic waters is shown in Fig. 3.

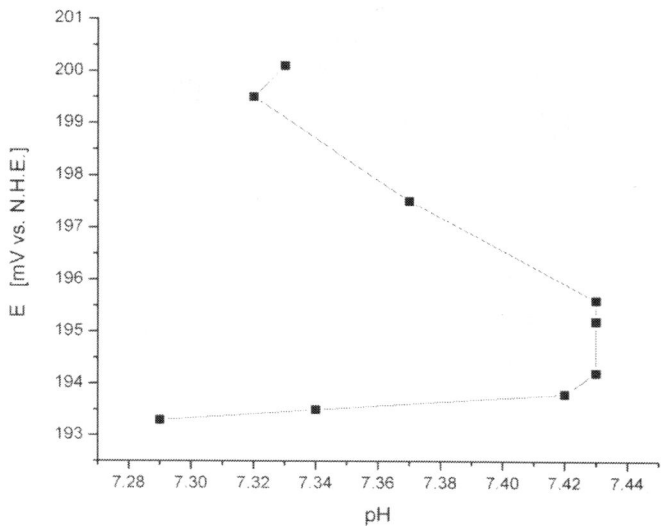

Figure 2. Redox potential of water dependence on pH during interaction of the composite with distilled water.

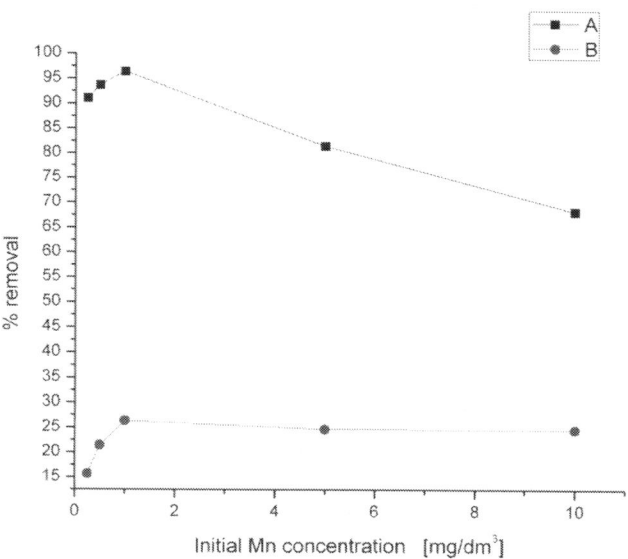

Figure 3. Percentage removal of Mn^{2+} (A) and colloidal MnO_2 (B) from synthetic waters (the composite dosage, 2 g/dm^3; contacting time, 20 min; initial Mn concentrations in range $0.25 - 10$ mg/dm^3; initial pH 5.75 ± 0.1; temperature, $20 \pm 0.5°C$).

Average initial pH of the synthetic waters was 5.75. After 20 min of contact with the composite material average pH was 6.70.

During the interactions of composite with synthetic waters, the colloidal MnO_2 was removed to a lesser degree than Mn^{2+}. The authors imposed that colloidal manganese possesses the following structure of micelles:

$$\{m[MnO_2]nSO_4{}^{2-}\ 2(n-x)K^+\}2xK^+$$

Potential-determining ions in the structure of micelles are $SO_4{}^{2-}$. They are primarily adsorbed on MnO_2 and responsible for the stability of colloids. Therefore, it is clear that the reduction of manganese is more difficult and there is an electrostatic repulsion between colloidal particles and a composite with dominantly negatively charged surface sites. Thus, the removal efficiency of colloidal manganese is significantly lower compared with the ionic form of Mn^{+2}. During the electrochemical interactions of synthetic water containing Mn^{2+} and colloidal MnO_2 with the composite material, transferring of Al^{3+} ions in a solution increases the TDS value, as shown in Table 1.

Table 1. The results of synthetic waters analysis before and after treatment with composite material.

Co(Mn) mg/dm³	TDS (mg/dm³)	pH	C(Mn) mg/dm³	TDS (mg/dm³)	pH
Before Mn²⁺ synthetic water treatment			After Mn²⁺ synthetic water treatment		
0.25	3	5.75	0.0223	17	6.65
0.50	7	5.73	0.0318	21	6.71
1.0	10	5.71	0.0363	25	6.72
5.0	14	5.70	0.9271	29	6.70
10.0	28	5.76	3.9773	39	6.58
Before colloidal MnO₂ synthetic water treatment			After colloidal MnO₂ synthetic water treatment		
0.25	3	5.82	0.2108	18	6.73
0.50	6	5.75	0.3928	22	6.71
1.0	11	5.71	0.7366	25	6.72
5.0	14	5.72	3.768	29	6.75
10.0	28	5.75	7.549	39	6.67

The initial dissolution of the Al based alloys introduces both aluminium and alloying ions into the solution, and then the reposition of microalloying tin onto active sites at surface occurs [46], so it was not detected by ICP-OES analysis.

Aluminium ions generated during electrochemical processes of manganese removal may form monomeric species such as $Al(OH)^{2+}$, $Al(OH)_2{}^+$ and $Al(OH)_4{}^-$. During the time, these monomers have tendency to polymerize in the pH range 4–7 which results in oversaturation and formation of amorphous hydroxide precipitate according to complex precipitation kinetics. Many polymeric species such as $Al_6(OH)_{15}{}^{+3}$, $Al_7(OH)_{17}{}^{+4}$, $Al_8(OH)_{20}{}^{+4}$, $Al_{13}O_4(OH)_{24}{}^{+7}$, $Al_{13}(OH)_{34}{}^{+5}$ have been reported [50]. Average concentration of aluminium, immediately after 20 min of composite interaction with Mn^{2+} synthetic waters, was 0.2131 mg/dm³ and included all mentioned monomeric and polymeric species which were not coagulated. After a prolonged period of time concentration of aluminum has a tendency to decrease reaching values that are below 0.1 mg/dm³, due to precipitation of $Al(OH)_3$ sludge.

The increase in the pH during the experiments can be explained in terms of the electrochemical and the chemical reactions that take place in the system composite-synthetic water. Water reduction at cathodic parts of composite (eq. 8), the electrochemical dissolution of aluminum (eq. 9) and protolytic reactions (eq. 10-14) increase the pH value [51].

$$H_2O + e^- \leftrightarrows 1/2H_2 + OH^-$$

$$2Al + 6H_2O \leftrightarrows 2Al^{3+} + 3H_2 + 6\,OH^-$$

$$Al(OH)_4^- + H^+ \leftrightarrows Al(OH)_3 + H_2O$$

$$Al(OH)_3 + H^+ \leftrightarrows Al(OH)_2^+ + H_2O$$

$$Al(OH)_2^+ + H^+ \leftrightarrows Al(OH)^{2+} + H_2O$$

$$Al(OH)^{2+} + H^+ \leftrightarrows Al^{3+} + H_2O$$

$$Al(OH)_3(s) \leftrightarrows Al^{3+} + 3OH^-$$

8.2. Bentonite Modified By Mixed Fe, Mg (Hydr)Oxides Coatings For The Removal of Ionic and Colloidal Forms of Pb(Ii)

Lead (Pb) is heavy metal which presents one of the major environmental pollutants due to its hazardous nature. It diffuses into water and the environment through effluents from lead smelters as well as from battery, paper, pulp and ammunition industries. Scientists established that lead is nonessential for plants and animals, while for humans it is a cumulative poison which can cause damage to the brain, red blood cells and kidneys [52].

8. New alumosilicate based composites chemically modified by
coatings/thin layers – Tested in the removal of colloidal and ionic

137

In this subchapter, a cheap and effective composite material as a potentially attractive adsorbent for the treatment of Pb(II) contaminated water sources has been described. The procedure for obtaining a bentonite based composite involves the application of mixed Fe and Mg hydroxides coatings onto bentonite particles (0.375 mmol Fe and 0.125 mmol Mg per gram of bentonite) in aqueous suspension and subsequent thermal treatment of the solid phase at 498 K [53]. Bearing in mind layered structure of montmorillonite, the quite limited extent of isomorphous substitution of Mg for Fe in iron (hidr)oxides and significant differences in acid-base surface properties between these two (hydr)oxides, formation of heterogeneous coatings onto bentonite and specific structure of obtained composite have been achieved [54]. Different adsorption sites on such heterogeneous surface provide efficient removal of numerous chemical species of Pb(II) over a wide pH range.

The structural changes of montmorillonite during composite synthesis are mainly reflected in the reduction of d_{001} diffraction peak intensity in X-ray diffractograms and its shifting towards the higher values of 2θ. Moreover, it can be observed that the peak is broadened suggesting that the distance between the layers is non-uniform with disordered and partially delaminated structure. The crystallographic spacing d_{001} of montmorillonite in the native bentonite and the composite, computed by using Bragg's equation ($n\lambda = 2d \sin \theta$), is 1.54 nm and 1.28 nm, respectively. These changes in the structure took place because the d-spacing is very sensitive to the type of interlamellar cations, and the degree of their hydration [55].

The XRD patterns of the composite and starting (native) bentonite are presented in Fig. 4a and b,respectively.

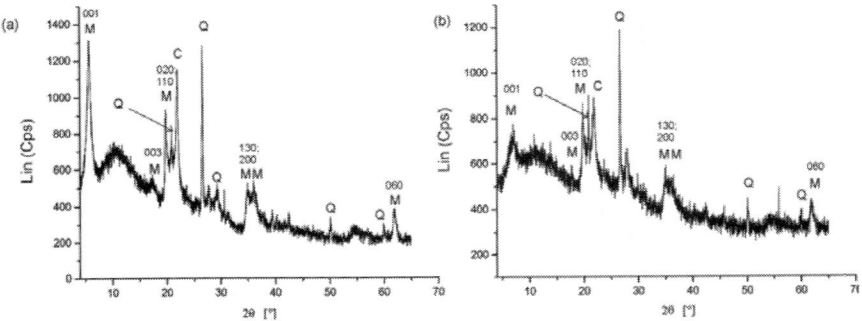

Figure 4. X-ray diffractograms of (a) composite and (b) native bentonite

SEM micrographs (Fig. 5 a, b and c) show that bentonite and composite are composed of laminar particles arranged in layered manner, forming the aggregates with diameters up to 50 μm.

Figure 5. a) SEM of synthesized composite, (b) SEM of composite after interaction with Pb(II) solution and (c) surface morphology of the native bentonite

No significant changes in the microstructure of composite occurred during the interaction with the aqueous solution of Pb(II).

Despite a thorough washing process, a large amount of NO_3^- is retained in the composite. A vibration mode at ca. 1389 cm^{-1} in FTIR spectrum confirms the NO_3^- stretching which indicates that some positive charged sites exist on the surface of composite and that they are counterbalanced by the NO_3^-which can be exchanged by other anions [53]. In addition, the formation of poorly crystallized magnesium hydroxonitrate in pH range 9-11 [56,57], where Fe/Mg coprecipitation was performed over bentonite particles, is very likely.

8.2.1. Specific Surface Area Determined by N₂ Adsorption/Desorption Using Bet Equation

The Fig. 6. shows the comparative nitrogen adsorption-desorption isotherms of native bentonite and composite.

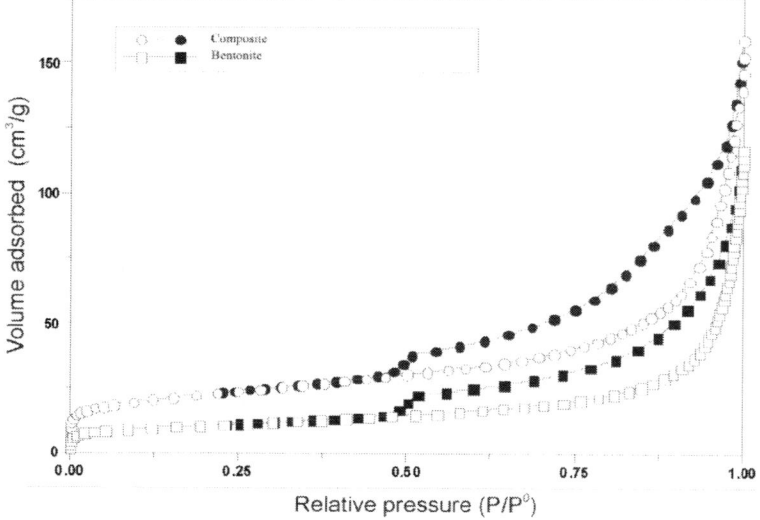

Figure 6. Nitrogen adsorption-desorption isotherms of native bentonite and composite.

The isotherms can be assigned to Type II isotherms, corresponding to non-porous or macroporous adsorbents. The hysteresis loops of Type H3 in the IUPAC classification occur at $p/p^0 > 0.5$, which is not inside the typical BET range. Furthermore, hysteresis loops of these isotherms indicate that they were given by either slit-shaped pores or, as in the present case, assemblages of platy particles of montmorillonite. Porous structure parameters are summarized in Table 2.

Table 2. Specific surface area and porosity of native bentonite and composite, determined by applying BET, BJH and D-R equation to N_2 adsorption at 77 K

Sample	S_{BET} (m²/g)	Median mesopore diameter (nm)	Cumulative mesopore area (m²/g)	Cumulative mesopore volume (cm³/g)	Micropore volume (cm³/g)
Bentonite	37.865	13.629	53.329	0.1202	0.0153
Composite	80.385	11.021	82.675	0.1716	0.0316

Compared to native bentonite, during the composite synthesis additional meso- and micropores were generated. Pore volumes (Gurvich) at p/p^0 0.999 for bentonite and composite are 0.180 cm³/g and 0.243 cm³/g, respectively. It was found that isotherms gave linear BET plots from p/p^0 0.03 to 0.21 for bentonite and from 0.03 to 0.19 for composite.

The composite has the specific surface area that is twice the size compared to the surface area of the native bentonite. This can be explained by the structural changes that occurred during the chemical and thermal modification of the native bentonite. The structural changes include delamination as well as the decrease of the distance between the layers of montmorillonite particles, because the interlayer water was lost under heating. The higher surface area of composite mainly results from the interparticle spaces generated by the three-dimensional co-aggregation of magnesium polyoxocations, iron oxide clusters and plate particles of montmorillonite. Macro- and mesopores arose from particle-to-particle interactions, while micropores were generated in the interlayer spaces of clay minerals due to irregular stacking of layers of different lateral dimensions [58].It is apparent that the changes of montmorillonite structure are responsible for the creation of new pore structure in the composite, which is then stabilized by the thermal treatment with the removal of H_2O molecules. The changes that involve partial dehydroxylation and cationic dehydration are brought about by thermal activation and they lead to various forms of cross-linking between oxides and smectite framework. As a result, composite does not swell and can be easily separated from water by filtration or centrifugation. There is a wide pore size distribution which supports disordered structure consisting of the delaminated parts with mesoporosity and the layered parts with microporosity.

The pH of the Pb(II) solution plays an important role in the adsorption process, influencing not only the surface charge of the adsorbent and the dissociation of functional groups on the active sites of the adsorbent but also the solution Pb(II) chemistry. The adsorption of Pb(II) on the composite decreased when pH decreased as shown in Fig. 7.

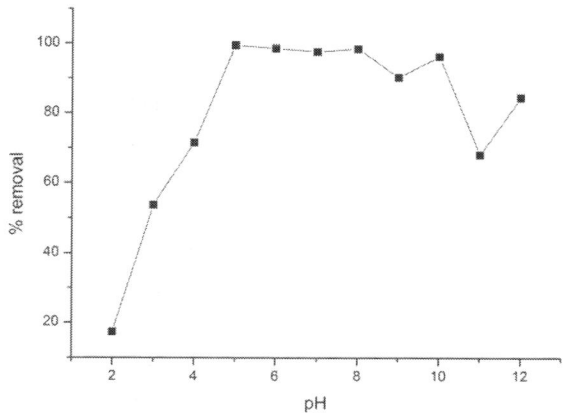

Figure 7. Effect of pH on adsorption of Pb(II) onto composite

The adsorptive decrease at pH below 5 was caused by the competition between H^+ and Pb^{2+} for the negatively charged surface sites. Maximum retention is in the pH range 5-10. The main Pb(II) species in the pH range 6.5-10 are $Pb(OH)^+$ and $Pb(OH)_2$ which can easily form colloidal micelles characterized with the following imposed structure:

$$\{m[Pb(OH)_2]nPb(OH)^+\cdot(n-x)NO_3^-\}xNO_3^-$$

The potential – determining ion is $Pb(OH)^+$ and that is the reason for the positive ZP of colloidal Pb(II) at the pH below 10 [59,60]. Therefore, colloidal micelles were easily attracted by the negatively charged composite surface. Particle size of colloidal Pb(II) at pH 7±0.1 was determined to be 268.7 ± 16.7 nm. At the pH range of 10-12 the predominant Pb(II) species are $Pb(OH)_2$ and $Pb(OH)_3^-$ which give rise to the formation of negatively charged colloidal micelles with the following structure:

$$\{m[Pb(OH)_2]nPb(OH)_3^-\cdot(n-x)Na^+\}xNa^+$$

ZP values for Pb(II) colloidal solutions at pH 11.8 were - 50.7±3.6 mV with particle size of 252.7±28.2 nm. Having in mind surface heterogeneity of the composite and high point of zero charge value of $Mg(OH)_2$ (between pH 12 and pH 13) [61], negative ions and particles can be adsorbed on the positively charged surface sites at pH 10-12. Removal efficiency of $Pb(OH)_3^-$ was higher than negatively charged colloids, probably because the ionic species were involved in the process of ion exchange and chemisorption, while colloidal micelles could be bound to the surface dominantly by electrostatic forces.

8. New alumosilicate based composites chemically modified by
coatings/thin layers – Tested in the removal of colloidal and ionic

141

8.3. Bentonite Based Composite Coated With Immobilized Thin Layer of Organic Matter

Synthesis of bentonite based composite material, described in this section, was carried out by applying thin coatings of natural organic matter, obtained by alkaline extraction from peat, mostly comprised of humic acids [62]. Humic acids have high complexing ability with various heavy metal ions, but it is difficult to use them as the sorbent because of their high solubility in water. However, they form stabile complexes with the inorganic ingredients of bentonite (montmorillonite, quartz, oxides, etc.) and can be additionally insolubilized and immobilized by heating at 350°C. After immobilization, humic acids represent an important sorbent for heavy metals, pesticides and other harmful ingredients from water. Humic acid are insolubilized by condensation of carboxylic and phenolic hydroxyl groups. Therefore, the aim was to remove manganese from aqueous solutions by treating it with synthesized composite as well as to study and explain the mechanism of composite interaction with manganese aqueous solutions. The composite does not release significant quantity of organic matter in water because it is tightly bonded to bentonite surface [63-65]. The degree of manganese removal was more than 94% at a range of initial manganese concentrations from 0.250 to 10 mg/l.

The result of conductometric titration is given in Fig. 8. Equivalence point was located at the intercept of the first and second linear part of the titration curve. The value of the total acidic group content is calculated to be 215.18 μmol/g.

The experimental data of manganese adsorption onto composite are very well fitted by the Freundlich isotherm model (Fig. 9.) with a very high correlation coefficient value of 0.9948. The good agreement of experimental data with the Freundlich model indicates that there are several types of adsorption sites on the surface of the composite. The amount of adsorbed Mn(II) increases rapidly in the first region of adsorption isotherm and then the slope of isotherm gradually decreases in the second region. The adsorption capacity of composite is 11.86 mg/g, at an equilibrium manganese concentration of 16.28 mg/l.

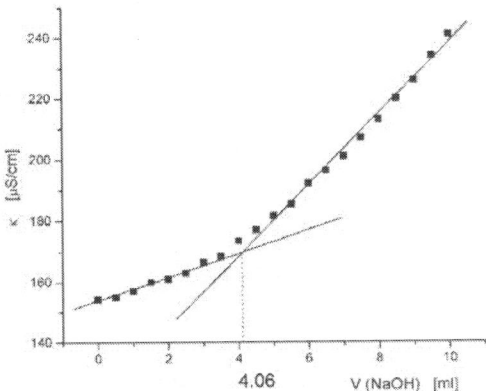

Figure 8. The conductometric titration of composite suspension (1 g in 250 ml of 1mM NaCl solution as background electrolyte) with 0.053 M NaOH.

After the treatment of model water with composite for the period of 20 min, the following results were obtained (Table 3)

Table 3. The results of water analysis before and after treatment with composite

Before water treatment			After water treatment			
$C_0(Mn)$m g/l	pH	Conductivity µS/cm	pH	Conductivity µS/cm	C(Mn) mg/l	%Mn Adsorption
0	6.43	8.01	6.67	11.43	0	0
0.250	6.37	9.57	7.11	13.76	0.0030	98.8
0.490	6.32	10.67	7.15	15.31	0.0039	99.2
1.0	6.30	14.67	7.12	31.10	0.0090	99.1
2.5	6.20	20.70	6.96	37.20	0.0187	99.25
5.0	6.19	32.80	6.83	49.40	0.0646	98.71
10.0	6.16	55.30	6.70	68.90	0.5314	94.69

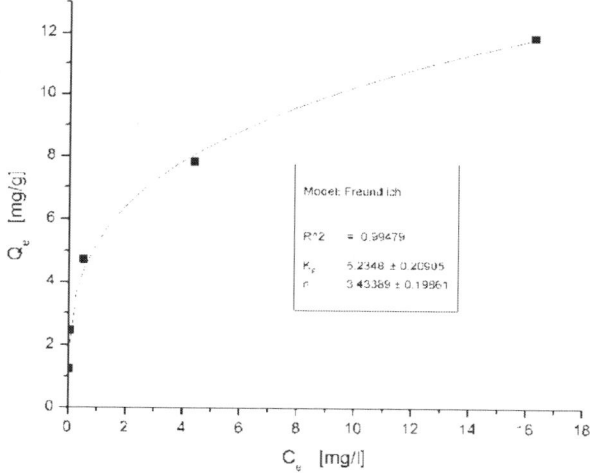

Figure 9. Freundlich adsorption isotherm for manganese adsorption onto composite.

During the thermal treatment in nitrogen atmosphere at 350 °C, the condensation of carboxyl and adjacent alcohol and phenol groups occurs. In this way the solubility of organic matter immobilized on bentonite matrix surface decreases [65]. Moreover, a part of carboxyl groups is decomposed by decarboxylation reaction, releasing CO_2 and CO. However, despite of this, a part of oxygen functional groups remains on the surface, and these groups act as sites that bind bivalent manganese forming inner-sphere complexes.

Besides organic functional groups, there are also Si-OH and Al-OH groups on the sites of crystal grain breaks, as well as permanent negative charge due to isomorphic substitution in clay minerals. They all contribute to the reduction of manganese concentration in the aqueous solution. Manganese retention by the formation of outer-sphere complexes, including ion exchange, can be showed by an Eq. (17) [66].

$$(\equiv S\text{-}O^-)_2...C^{n+}{}_{3\text{-}n} + Mn^{2+} \rightleftharpoons (\equiv S\text{-}O^-)_2...Mn^{2+} + (3\text{-}n)\ C^{n+}$$

in which C represents the cation that is exchanged.

The formation of inner-sphere complexes is represented by the Eqs. (18) and (19) and involves the release of hydrogen ions and the change of solution pH.

$$\equiv S\text{-}OH + Mn^{2+} \rightleftharpoons \equiv S\text{-}O\text{-}Mn^+ + H^+$$

$$\equiv 2S\text{-}OH + Mn^{2+} \rightleftharpoons (\equiv S\text{-}O)_2\text{-}Mn + 2H^+$$

According to these equations, it can be concluded that the pH value of the solutions decrease after the treatment. However, an opposite phenomenon can be experimentally observed (Table 3). The explanation for it is that hydrogen ions which are released during manganese retention participate in the protonation of surface groups:

$$\equiv S\text{-}OH + H^+ \rightleftharpoons \equiv S\text{-}OH_2{}^+$$

$$\equiv S\text{-}O^- + H^+ \rightleftharpoons \equiv S\text{-}OH$$

Therefore, the pH value of the Mn^{2+} aqueous solutions after treatment with composite had a higher value than the initial pH. This indicates that more hydrogen ions are bound to the surface than released by manganese binding. Namely, the composite exhibits amphoteric character due to the surface sites that act either as proton acceptors or as proton donors.

Organic matter decreases the PZC value of bentonite and neutralizes positive electric charge that comes from interlaminated cations, thus increasing composite affinity to manganese, even at lower pH values (67). Fig. 10. presents the pH dependence of residual Mn concentration, for the initial Mn concentration of 5 mg/l. The residual concentration of Mn decreases gradually with pH increasing in the range of 3.5-7 and then increases in the range of 7-10, with the apparent minimum at pH 7.

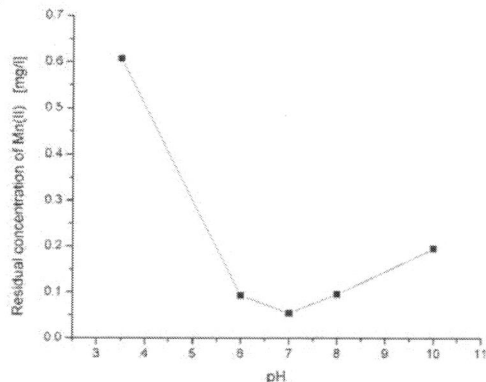

Figure 10. Residual concentration of Mn(II) as a function of model water pH.

The increase of pH value has dual effect on the removal of manganese. The increase of the pH value favours manganese removal due to increase of the number of deprotonated sites that are available for the binding of manganese. However, there is an increase in the solubility of organic matter which has been applied on the bentonite particles. The dissolved organic matter (humic acids) reacts with manganese forming complexes which bear a negative charge and have a weaker binding affinity for the composite surface than Mn^{2+}. Fig 10. indicates two opposite effects of the pH on manganese removal. The pH dependence of released organic matter (expressed as permanganate number) and turbidity (NTU) of solutions are shown in Fig. 11.

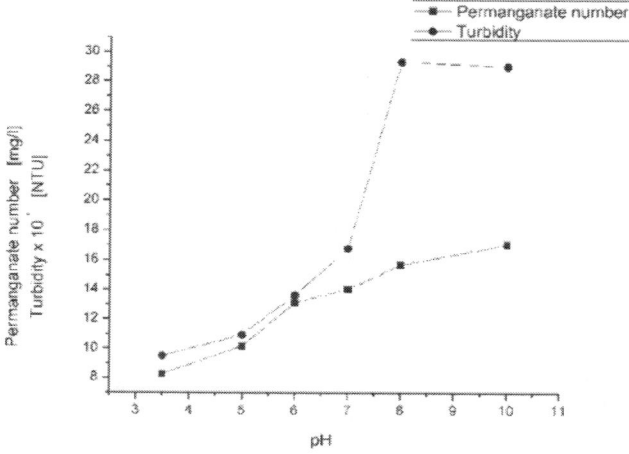

Figure 11. Premanganate number and turbidity of filtrate as function of pH (0.2 g of composite and 100 ml of 1mM Na_2SO_4 as background electrolyte).

The released organic matter contributes to the increased turbidity at higher pH values.

9. SUMMARY

The widespread industrial areas where nanocomposites can be applied are primary and conversion industry, modern coating technologies, constructional regions, and environmental (water, air) purification. In addition to the dominant use of composites as structural elements, important application of composite materials is in the water purification technologies. In this field of application, composites usually have the role of adsorbent, electrochemically active materials, catalysts, photocatalysts etc.

Bentonite is a natural and colloidal alumosilicate with particle size less than 10 μm, which is effectively used as sorbent for heavy metals and other inorganic and organic pollutants from water. Due to its positive textural properties and high specific surface area it can be used as low-cost matrix for synthesis of adsorbents or electrochemically active composite materials for the removal of pollutants in ionic and colloidal form from water. In this respect, three new/modified bentonite based composite materials have been synthetised and characterized.

Coated or composite particles are composed of solid phase covered with thinner or thicker layer of another material. These coatings - layers covering the surface of matrix are important for several reasons. In such way, the surface and textural characteristics of the initial solid phase are modified and sintering conditions can be better controlled. An important factor in achieving the desired microstructure of ceramics is sintering procedure that includes rather complex processes. A considerable influence on sintering has been exhibited by an addition of microalloying components, which significantly determined a microstructure and resulted properties of ceramics. The presence of small amounts of impurities in the starting material can vastly influence their mechanical, optical, electrical, color, diffusive, and dielectric properties of alumosilicate matrix. In summary, the process of diffusion mass transport in ceramic crystal regions are affected by temperature, oxygen partial pressure and concentration of impurities. A procedure for the removal of manganese in ionic (Mn^{2+}) and colloidal (MnO_2) forms from synthetic waters, by reduction and adsorption processes on electrochemically active alumosilicate ceramics based composite material has been described. Synthesis procedure of the composite material consists of two phases. Firstly, composite particles were synthesized by applying Al/Sn oxide coating onto the bentonite particles in an aqueous suspension. In the second phase, aluminium powder is added to the previously obtained plastic mass and after shaping in the form of spheres 1 cm in diameter and drying, sintering was performed at 900°C. Elemental tin, resulting from the reduction of Sn^{2+}-ion, comes into contact with liquid aluminum in the pores of the matrix performing aluminum microalloying and activation. Moreover, due to a low partial pressure of oxygen, nonstoichiometric oxides with metal excess are obtained, and they play an important role in the electrochemical activity of the composite material. In accordance with this, a redox potential of water is changed in contact with composite.

Another effective composite material as a potentially attractive adsorbent for the treatment of Pb(II) contaminated water sources has been synthesized by coating of bentonite with mixed iron and magnesium (hydr)oxides. The procedure for obtaining a bentonite based composite involves the application of mixed Fe and Mg hydroxides coatings onto bentonite particles in aqueous suspension and subsequent thermal treatment of the solid phase at 225°C. Formation of heterogeneous coatings on bentonite results in changes of bentonite acid-based properties, high specific surface area and positive adsorption characteristics. Different adsorption sites on such heterogeneous surface provide an efficient removal of numerous chemical species of Pb(II) (ionic and colloidal) over a wide pH range.

Third bentonite based composite material was obtained by applying thin coatings of natural organic matter, extracted from a peat, mostly based on humic acids. Humic acids are known due to high complexing ability to various heavy metal ions, but it is difficult to use them directly as the sorbent because of their high solubility in water. However, they form stabile complexes with the inorganic ingredients of bentonite (montmorillonite, quartz, oxides, etc.) and can be successfully insolubilized and immobilized by heating at 350°C. After immobilization, humic acids represent an important sorbent for heavy metals, pesticides and other harmful ingredients from water. Humic acid are insolubilized by condensation of carboxylic and phenolic hydroxyl groups. The composite such obtained can be effectively used as the sorbent for heavy metals.

ACKNOWLEDGEMENT

The authors acknowledge financial support from the Ministry of Education and Science of the Republic of Serbia.

REFERENCES

1. B. Ralph, H. C. Yuen, W. B. Lee, The processing of metal matrix composites- an overview. Journal of Materials Processing Technology 19976 63339

2. D. Chung, Materials. Composite, Science. . Springer, Media. B. Business, B.2010 2010

3. Usuki A, Kawasumi M, Kojima Y, Fukushima Y, Okada A, Kurauchi T. Swelling behavior of montmorillonite cation exchanged for ω-amino acids by ε-caprolactam. Journal of Materials Research 1993; 8 1179-1184.

4. Y. Kojima, A. Usuki, M. Kawasumi, Y. Fukushima, A. Okada, T. Kurauchi, properties. Mechanical, of, 6 clay hybrid. Journal of Materials Research 1993; 81185 .

5. E. Giannelis, Silicate. Polymer-layered, Advanced. Nanocomposites, Materials, 1996 829 .

6. R. Pfaendner, Industrial. Nanocomposites, of. opportunity, Polymer. challenge?, Stability. Degradatio, 20109 95369 .
7. Nanocomposites and nanotubes conference,1112 th March, 2008 www.nanosconference.com/home.asp;. accessed 30.04.2012.
8. 2010D. Manfredi, M. Pavese, S. Biamino, A. Antonini, P. Fino, C. Badini, Microstructure and mechanical properties of co-continuous metal/ceramic composites obtained from Reactive Metal Penetration of commercial aluminium alloys into cordierite. Composites: Part A 2010; 41639 .
9. 1999E. Saiz, S. Foppiano, Chan. W. Moberly, A. P. Tomsia, Synthesis and processing of ceramic-metal composites by reactive metal penetration. Composites: Part A 1999; 30399 .
10. Haq, E. Matijevic, Preparation and properties of uniform coated inorganic colloidal particles. 12 Tin and its compounds on hematite. Progress in Colloid and Polymer Science 1998; 109185 .
11. F. Bergaya, B. K. G. Theng, G. Lagaly, Handbook of clay science. Elsevier; 2006
12. L. Olmos, C. L. Martin, D. Bouvard, Sintering of mixtures of powders: Experiments and modelling. Powder Technology 200919 190134 .
13. Michael, P. Dubois, silicate. Polymer-layered, preparation. nanocomposites, properties, of. a. uses, class. new, materials. of, Materials Science and Engineering 20002 281 .
14. S. S. Ray, M. Okamoto, Silicate. Polymer/layered, A. Nanocomposite, from. Review, to. Preparation, Processing, Progress in Polymer Science 20032 281539 .
15. F. Hussain, M. Hojjati, M. Okamoto, R. E. Gorga, article. Review, nanocomposites. polymer-matrix, manufacturing. processing, an. application, Journal. overview, Composite. of, Materials, 20084 401511 .
16. J. Moszo, B. Pukanszky, micro. Polymer, Structure. nanocomposites, properties. interactions, Journal of Industrial and Engineering Chemistry 20081 14535 .
17. W. Xie, Z. Gao, W. P. Pan, D. Hunter, A. Singh, R. Vaia, Thermal degradation chemistry of alkyl quaternary ammonium montmorillonite. Chemistry of Materials 20011 132979 .
18. Leszczynska, J. Njuguna, K. Pielochowski, J. R. Banerjee, (montmorillonite. Polymer/clay, with. nanocomposites, thermal. improved, Part. I. I. A. properties, study. review, thermal. on, of. stability, nanocomposites. montmorillonite, on. based, polymeric. different, matrixes, Thermochimica Acta 2007 454 1 122 .
19. Pandey J.K, Reddy K.R, Kumar A.P, Singh R.P, An overview on the degradability of polymer nanocomposites. Polymer Degradation and Stability 20058 88234 .
20. H. Wermter, R. Pfaender, patent. E. P. European, 1592741 assigned to Ciba Holding Inc., 2009.
21. Leszcynska, J. Njuguna, K. Pielichowski, J. R. Banerjee, Thermal stability of polymer/montmorillonite nanocomposites: Part I: Factors influencing thermal stability and mechanisms of thermal stability improvement. Thermochimica Acta 200745 45375 .

22. G. Moad, K. Dean, L. Edmond, N. Kukaleva, G. Li, R. T. A. Mayadunne, Poly(ethylene. Non-Ionic, Surfactants. oxide)-based, Dispersants. Intercalants, for. Exfoliants, Nanocomposites. Polypropylene-Clay, Macromolecular Materials and Engineering 200629 29137 .

23. G. Moad, K. Dean, L. Edmond, N. Kukaleva, G. Li, R. T. A. Mayadaunne, R. Pfaendner, A. Schneider, G. Simon, Wermter, 2006 2006. Novel Copolymers as Dispersants/Intercalants/Exfoliants for Polypropylene-Clay Nanocomposites. Macromolecular Symposia 2006; 233170 .

24. N. S. Allen, M. Edge, T. Corrales, A. Childs, C. M. Liauw, F. Catalina, C. Peinado, A. Minihan, D. Aldcroft, Ageing and stabilisation of filled polymers: an overview. Polymer Degradation and Stability 19986 61183 .

25. H. Qin, C. Zhao, C. Zhang, G. Chen, M. Yang, Photo-oxidative degradation of polyethylene/montmorillonite nanocomposite. Polymer Degradation and Stability 20038 81497 .

26. S. Morlat-Therias, B. Mailhot, D. Gonzalez, J. L. Gardette, Photooxidation of Polypropylene/Montmorillonite nanocomposites Part II : Interactions with antioxidants. Chemistry of Materials 20051 171072 .

27. S. Morlat, B. Mailhot, D. Gonzalez, J. L. Gardette, of. Photooxidation, Montmorillonite. Polypropylene, Part. I. . nanocomposites, of. Influence, nanoclay, agent. compatibilising, of. Chemistry, Materials, 2004 16 3 377383 .

28. S. Chmela, A. Kleinova, A. Fiedlorova, E. Borsig, D. Kaempfer, R. Thormann, Photo-oxidation of sPP/organoclay nanocomposites. Applied Chemistry 2005 42 7 821829 .

29. B. Rotzinger, P. P. A. Talc-filled, concept. new, maintain. to, term. long, stability. heat, Polymer Degradation and Stability 20069 912884 .

30. N. Savage, M. S. Diallo, Nano materials and water purification: opportunities and challenges. Journal of Nanoparticle Reserch 2005 7331 .

31. Da Fonseca M.G, De Oliveira M.M, Arakaki L.N.H. Removal of cadmium, zinc, manganese and chromium cations from aqueous solution by a clay mineral. Journal of Hazardous Materials 2006 B137288 .

32. Dimirkou, M. K. Doula, Use of cl2optilolite and an Fe-overexchanged clinoptilolite in Zn2+ and Mn2+ removal from drinking water. Desalination 2008; 224280 .

33. Bladergroen B.J, Linkov V.M. Electrosorption ceramic based membranes for water treatment. Separation and Purification Technology 20012 25347 .

34. T. Kanki, S. Hamasaki, N. Sano, A. Toyoda, K. Hirano, Water purification in a fluidized bed photocatalytic reactor using TiO2 -coated ceramic particles. Chemical Engineering Journal 200510 108 155-160.

35. C. D. Johnson, F. Worrall, Novel granular materials with microcrystalline active surfaces-Waste water treatment applications of zeolite/vermiculite composites. Water Research 20074 412229 .

36. Tzoupanos N.D, Zouboulis A.I. Preparation, characterisation and application of novel composite coagulants for surface water treatment Water Research 20114 453614 .

37. Y. Lv, H. Liu, Z. Wang, S. Liu, L. Hao, Y. Sang, D. Liu, J. Wang, R. I. Boughton, Silver nanoparticle-decorated porous ceramic composite for water treatment. Journal of Membrane Science 200933 33150 .

38. W. Shen, L. Feng, H. Feng, Z. Kong, M. Guo, silver. I. I. Ultrafine, particles. oxide, porous. decorated, composites. ceramic, water. for, treatment, Chemical Engineering Journal 201117 175592 .

39. W. Wang, P. Serp, P. Kalck, Silva. C. Gomes, Faria. J. Luıs, Preparation and characterization of nanostructured MWCNT-TiO2 composite materials for photocatalytic water treatment applications. Materials Research Bulletin 20084 43958 .

40. H. Zhang, X. Quan, S. Chen, H. Zhao, Fabrication and characterization of silica/titania nanotubes composite membrane with photocatalytic capability. Environmental Science and Technology 20064 406104 .

41. M. Ranđelović, M. Purenović, A. Zarubica, J. Purenović, I. Mladenović, G. Nikolić, Alumosilicate ceramics based composite microalloyed by Sn: An interaction with ionic and colloidal forms of Mn in synthetic water. Desalination 2011 279(1-3) 353-358.

42. Jeurgens L.P.H, Sloof W.G, Tichelaar F.D, Mittemeijer E.J. Composition and chemical state of the ions of aluminium-oxide films formed by thermal oxidation of aluminium. Surface Science 200250 506313 .

43. Purenovic M.M. Influence of Some Alloying Elements and Admixtures on Electrochemical Behaviour of the System Aluminium- Oxide Layer-Electrolyte. PhD Thesis. Faculty of Technology and Metallurgy, University of Belgrade; 1978

44. R. W. Chan, P. Haasen, Metallurgy. Physical, edition. fourth, Science. B. V. Elsevier, Amsterdam, Netherlands; 1996

45. Rahaman M.N. Ceramic processing and sintering, second edition. Marcel Dekker, Inc. New York; 2003

46. S. Gudic, I. Smoljko, M. Kliskic, Electrochemical behaviour of aluminium alloys containing indium and tin in NaCl solution. Materials Chemistry and Physics 201012 121561 .

47. Cvetković V.S, Purenović J.M, Jovićević J.N. Change of water redox potential, pH and rH in contact with magnesium enriched kaolinite-bentonite ceramics. Applid Clay Science 20083 38268 .

48. Q. Wei, X. Ren, J. Du, S. Wei, S. Hu, Study of the electrodeposition conditions of metallic manganese in an electrolytic membrane reactor. Minerals Engineering 20102 23578 .

49. Cvetković V.S, Purenović J.M, Purenović M.M, Jovićević J.N. Interaction of Mg-enriched kaolinite-bentonite ceramics with arsenic aqueous solutions. Desalination 200924 249582 .

50. O. T. Can, M. Bayramoglu, M. Kobya, Decolorization of reactive dye solutions by electrocoagulation using aluminum electrodes. Industrial and Engineering Chemistry Research 20034 423391 .

51. P. Canizares, F. Martinez, M. Carmona, J. Lobato, M. A. Rodrigo, Continuous Electrocoagulation of Synthetic Colloid-Polluted Wastes. Industrial and Engineering Chemistry Research 20054 44 (8171-8177.

52. Lead in Drinking-water, Background document for development of WHO Guidelines for Drinking-water Quality, World Health Organization; 2003.

53. M. Ranđelović, M. Purenović, A. Zarubica, J. Purenović, B. Matović, M. Momčilović, Synthesis of composite by application of mixed Fe, Mg

(hydr)oxides coatings onto bentonite- a use for the removal of Pb(II) from water. Journal of Hazardous Materials 2012 199-200 367-374.

54. R. M. Cornell, U. Schwertmann, Iron. The, Structure. Oxides, Reactions. Properties, Occurences, second. Uses, ed, W. I. L. E. Y. -V, C. H. Verlag, H. Gmb, K. Co, A. Ga, Weinheim; 2003

55. B. Caglar, B. Afsin, A. Tabak, E. Eren, Characterization of the cation-exchanged bentonites by XRPD, ATR, DTA/TG analyses and BET measurement, Chemical Engineering Journal 200914 149242 .

56. Krasnobaeva O.N, Belomestnykh I.P, Isagulyants G.V, Nosova T.A, Elizarova T.A, Teplyakova T.D, Kondakov D.F, Danilov V.P. Synthesis of Complex Hydroxo Salts of Magnesium, Nickel, Cobalt, Aluminum, and Bismuth and Oxide Catalysts on Their Base. Russian Journal of Inorganic Chemistry 2007 52 2 141146 .

57. Krasnobaeva O.N, Belomestnykh I.P, Isagulyants G.V, Nosova T.A, Elizarova T.A, Kondakov D.F, Danilov V.P. Chromium, Vanadium, Molybdenum, Tungsten, Magnesium,and Aluminum Hydrotalcite Hydroxo Salts and Oxide Catalysts on Their Base, Russian Journal of Inorganic Chemistry 2009 54 4 495499 .

58. J. Rouquerol, F. Rouquerol, K. S. W. Sing, Adsorption by Powders and Porous Solids: Principles, Methodology and Applications, Academic Press, San Diego USA; 1999

59. Q. Liu, Y. Liu, of. Distribution, I. I. Pb, in. species, solutions. aqueous, of. Journal, Colloid, Science. Interface, 2003268 .

60. M. Kosmulski, of. P. Z. C. Compilation, I. E. P. of, soluble. sparingly, oxides. metal, from. hydroxides, Advances. literature, Colloid. in, Science. Interface, 2009152 .

61. S. V. Krishnan, I. Iwasaki, Heterocoagulation, surface precipitation in a quartz-Mg(OH)2 system, Environmental Science and Technology 19862 201224 .

62. M. Ranđelović, M. Purenović, J. Purenović, M. Momčilović, Removal, 2 Mn2+ from water by bentonite coated with immobilized thin layers of natural organic matter. Journal of Water Supply: Research and Technology-AQUA 2011; 60 8 486493 .

63. F. Ayari, E. Srasra, M. Trabelsi-Ayadi, Characterization of bentonitic clay and their use as adsorbent. Desalination 200518 185391 .

64. C. A. Kolokassidou, I. Pashalidis, C. N. Costa, A. M. Efstathiou, G. Buckau, Thermal stability of solid and aqueous solutions of humic acid. Thermochimica Acta 200745 45478 .

65. S. Ghosh, Zhen-Yu, 1 Kangl S, Bhowmik P. C, Xing B. S. Sorption and fractionation of a peat derived humic acid by kaolinite, montmorillonite, and goethite. Pedosphere 2009; 19 1 2130 .

66. Doula M.K. Removal 2 Mn2+ ions from drinking water by using clinoptilolite and a clinoptilolite Fe oxide system. Water Research 2006; 403167 .

67. J. Zhuang, G. R. Yu, Effects of surface coatings on electrochemical properties and contaminant sorption of clay minerals. Chemosphere 20024 49619 .

CHAPTER 7

Composite and Nanocomposite Metal Foams

Isabel Duarte [1,*,†] *and José M. F. Ferreira* [2,†]

[1] Department of Mechanical Engineering, TEMA, University of Aveiro, Campus Universitário de Santiago, Aveiro 3810-193, Portugal
[2] Department of Materials and Ceramics Engineering, CICECO, University of Aveiro, Campus Universitário de Santiago, Aveiro 3810-193, Portugal

ABSTRACT

Open-cell and closed-cell metal foams have been reinforced with different kinds of micro- and nano-sized reinforcements to enhance their mechanical properties of the metallic matrix. The idea behind this is that the reinforcement will strengthen the matrix of the cell edges and cell walls and provide high strength and stiffness. This manuscript provides an updated overview of the different manufacturing processes of composite and nanocomposite metal foams.

Keywords: metal foams; composite metal foams; metal matrix composites (MMC) foams; micro and nano reinforcements; ceramic particles; carbon nanotubes; hollow spheres

1. INTRODUCTION

The non-flammable, recyclable and lightweight open-cell and closed-cell metal foams have been used as functional and structural engineering applications [1,2,3,4]. The open-cell metallic foams are extensively utilized as heat exchangers, filters, electrodes, shock absorbers taking advantages of their high specific surface area and their high thermal and electrical conductivities [1,2]. The closed-cell metal foams, in particular the aluminum alloy (Al-alloy) foams, have been used in structural engineering applications (e.g., automotive, aerospace, industrial equipment and building construction) that require lightweight structures with high strength-to-weight and stiffness-to-weight ratios, high impact energy absorbing capacity and/or with a good damping of noise and vibration [3,4]. These foams, in particular the closed-cell aluminum foams, are usually applied as core and/or as filler of sandwich panels [5] and thin-walled structures [6], respectively. In these composite structures, the thin metal sheets [7] and the thin-walled structures (e.g., empty tubes) [8,9] ensure the high mechanical strength, while the foam core or filler mainly contribute to the high crashworthiness [6,10]. Although the relative low mechanical strength of these foams does not limit its range of applications, research efforts have been made towards enhancing the mechanical performance of the existing foams or/and developing high-strength Al-alloy foams or even a new class of high-strength foams in order to broaden the range of the applications of these materials. This could be achieved by strengthening the metal cell skeleton (metal-matrix) or/and by optimizing the pore structures. The most cost-effective ways currently used to enhance the mechanical properties of the existing Al-alloy foams are: (i) selecting a strong Al-alloy (in form of powder or ingot); or (ii) applying conventional heat treatment processes (e.g., age precipitation and annealing) usually used for Al-alloys [11,12,13,14]. Heat treatments promote the formation of precipitates from solid solution (precipitation hardening or age hardening) or the diffusion of alloying element into the matrix, forming a solid solution. Some authors studied the effects of these heat treatments on the mechanical properties for both open-cell [11,12] and closed-cell aluminum foams [13,14]. Zhou et $al.$ [11] investigated the effects of annealing and T6-strengthening treatments on the compressive deformation behavior of open-cell aluminum foams (10 PPI) and found that the T6-strengthneing treatment increases the compressive strength of these foams (e.g., the peak stress changed from 2.2 to 3.2 MPa). On the other hand, the thermal annealing has negative effects on the compressive strength of these foams, decreased the main mechanical parameters (the peak stress decreased from 2.2 to 1.0 MPa, after the treatment). Yamada et $al.$ [12] found a similar effect of SG91A (Al-9 wt.% Si-0.5 wt.% Mg-0.5 wt.% Fe-0.4 wt.% Mn) and AZ91 Mg (Mg9 wt.% Al1 wt.% Zn-0.2 wt.% Mn) alloys. Similar conclusions were drawn for closed-cell aluminum alloy foams. For example, the compressive strength of 6101 [13,14] and 7075 [13] Al-alloy foams could be improved by applying a suitable T6-strengthneing treatment. Nonetheless, some of these heat treatments develop cracks in the cell walls due to the thermal stresses that are responsible for the oscillation of the stress in the plateau region [14]. The mechanical performance of the existing foams could be also increased by diminishing the size of their cellular pores, demonstrated by Xia et $al.$ [15] and by Jiang et $al.$ [16]. Nonetheless, these aforementioned methodologies allow slightly increasing the mechanical strength of such foams, but not allow the fabrication of high strength foams.

Recently, great efforts have been carried out to fabricate the high strength metal foams. Most of the ideas have emerged based on the research which has been carried out to fabricate high-strength solid metals [17,18]. One strategy for strengthening of the Al-alloys is adding alloying elements (e.g., Mg, Ni) to promote the formation of intermetallics (e.g., precipitation hardening) [17]. Another strategy is incorporating micro and nano-sized reinforcement elements into the metal bulk matrix to enhance the performance of the ductile metal [18].

For example, ceramic particles, e.g., alumina (Al_2O_3), silicon carbide (SiC) [17], ceramic fibers [17], ceramic nanoparticles [19] are some of the most attractive reinforcing materials. More recently, there has been a crescent interest in exploring carbonaceous materials as reinforcing agents for metal alloys [20,21,22]. This became an attractive research field both from the scientific and industrial applications viewpoints. Considerably research activities involving the metal matrix nanocomposites have been undertaken. Till to present, research efforts have been mainly concentrated on Al-alloys reinforced with carbon nanotubes [20,22]. As it is well-known, carbon nanotubes (CNTs) have high aspect ratio (*i.e.*, length to diameter, or length to thickness ratio) and exhibit an amazing high elastic modulus and mechanical strength, as well as, excellent electrical and thermal conductivities [21]. As a result, they are considered to be the most effective reinforcing elements for fabricating composite materials for structural and functional engineering applications [23]. Although the results are promising, the incorporation of CNTs in metallic matrices is still an unsolved problem due to their high tendency to agglomerate into clusters, their poor dispersion ability in the metal-matrix, and the poor wettability of carbon by molten metal. The poor wettability derives from the large difference in surface tensions between CNTs and molten metal. The formation of interfacial reaction products leading to loss of their structural integrity is another drawback. These are the current challenges that need to be overcome in this metal matrix composite field [20]. These reinforced materials with micro- and nano-sized reinforcing elements are designated as metal matrix composites (MMC) or metal matrix nanocomposites, respectively.

This paper presents an overview of the main strategies that emerged recently aiming at fabricating composite and nanocomposite metal foams with enhanced mechanical performance, with an especial focus on Al-alloy foams that are the most industrially used ones. This review focuses processing and the properties, as well as the strengthening mechanisms associated for the reinforced foams. The reinforced foams are compared to the conventional open-cell foams and closed-cell foams.

2. COMPOSITE METAL FOAMS

2.1. Metal Foams Reinforced with Ceramic Particles

The micro-sized ceramic particles were firstly used in the metal foams field to promote the liquid foam stability and avoid the formation of non-uniform pore sizes. The ceramic particles migrate preferentially to the liquid/gas interface and stabilize the bubbles, while increase the viscosity of the melt [24]. Ceramic particles prevent drainage of metallic melt and the coalescence of bubbles, which are common causes of non-uniform structures in foams. The preferential migration of the particles to the liquid/gas interfaces is mostly related to the differences in surface tensions between the solid particles and the molten metal and the poor wetting ability. In this regard, ceramic particles in metal foams play a stabilizing role similar to that of surfactants in many other foam systems (emulsions, aqueous liquid foams, *etc.*). The presence of ceramic particles at the liquid/gas interfaces contribute to the formation of bridges between opposite liquid/gas interface, preventing them to get closer, burst the bubbles and promote their coalescence. Accordingly, they tend to retard the flow of liquid out of the foam and to hinder the growth of the bubbles, contributing to a more uniform foam structure.

Ceramic stabilizing particles (e.g., SiC and Al_2O_3) are required to fabricate closed-cell Al-alloy foams by direct foaming technique, which is one of the most common and economical methods [24,25]. The ceramic particles migrate preferentially to the liquid/gas interfaces and stabilize the bubbles, while increase the viscosity of the melt [25]. Ceramic particles prevent drainage of metallic melt and the coalescence of bubbles, which are common causes of non-uniform structures in foams. Direct foaming methods start from a molten metal containing dispersed ceramic particles, into which gas bubbles are injected directly [26], or generated chemically by the decomposition of a blowing agent (e.g., titanium hydride, calcium carbonate), or by precipitation of gas dissolved in the melt by controlling temperature and pressure [27]. When dispersed into the molten metal, ceramic particles adhere to the gas/metal interfaces of rising bubbles, avoiding their burst [25]. However, high volume fractions of ceramic particles (above 10 vol.%) are usually required to effectively the foams. For example, AlSi7Mg (A356) foams are prepared by dispersing 20 vol.% SiC particles (SiCp) in the AlSi7Mg melt containing titanium hydride as blowing agent [28]. Accordingly, the resulting reinforced foams exhibit a brittle mechanical behavior conferred by such high particle contents, preferentially located in the cell-wall where they induce localized deformations. This is the main reason why such lightweight reinforced foams are not recommended for structural applications that require a ductile behavior. This type of foam is commonly used in building construction fulfilling other roles, such as acoustic and thermal insulation [2,3]. Furthermore, the high volume fraction of ceramic particles in these foams make cutting and machining difficult (increasing of the machining time) contributing to an increase of the production costs. Reducing the size of reinforcing particles leads to a higher number of particles for a given content, and might allow decreasing the volume fraction required for an effective foam stabilization, with concomitant reductions of costs and brittleness of the foams.

Ceramic particles have been explored also to control the cellular structure (e.g., size of the cellular pores) minimizing the structural defects and imperfections to improve the quality of foams as detailed elsewhere [29]. Exploring the role of ceramic particles as reinforcing materials was also undertaken as a route to improve the mechanical properties of the metal matrix of the existing foams. In the reality, all these features are closely interrelated to each other since the cellular structure (e.g., pore size) and density gradients strongly influence the mechanical properties of the resulting metal matrix composite foams. Various micro-sized ceramic particles have been explored as both stabilizer and reinforcing elements, including Al_2O_3 [30,31,32], SiC [33,34,35,36,37,38,39,40,41], titanium diboride (TiB_2) [42], yttrium oxide (Y_2O_3) [43] and AlN [44]. Ceramic particles can be directly dispersed in the molten Al-alloy or be *in-situ* formed through chemical (e.g., oxidation, formation of oxide bi-films) and metallurgical reactions (e.g., intermetallic compounds). The extent of the *in-situ* formed particles can be tailored by controlling the atmosphere and the agitation promoted by the rise of injected gas bubbles, or by applying further mechanical stirring. Adding oxygen seeking elements of the Group 2 (beryllium, magnesium, calcium, strontium and barium) is a common way to foster internal oxidation. For example, stable Alporas foams are prepared by adding calcium to an aluminum melt under mechanical stirring and air injection enhance the viscosity of the melt (thickening) due to the *in-situ* formation of oxide particles. A blowing agent is then added into the melt. Another approach for the *in-situ* formation of stabilizing sub-micrometer carbide and boride particles in metal foams consisted in adapting the flux assisted melting method [45] already known in master alloys containing grain refiners. For example, composites consist of Al-alloys containing ceramic particles of controlled size (e.g., TiB_2, TiC) were prepared by this method using fluoride salts [45].

The findings reported in a number of published work involving different manufacturing methods are summarized in Table 1 and Table 2. Direct and indirect foaming methods have been used, in which the micro-sized particles are dispersed directly into the molten metal at high temperatures under mechanical stirring [24]; or previously mixed with metal powders based on powder metallurgy (PM) method [46], respectively. PM method [47,48], one of the most commercially exploited to fabricate the closed-cell metal foams, consists on heating of a precursor material obtained by hot compaction of a metal alloy (e.g., Al-alloy) with blowing agent powders (e.g., titanium hydride, TiH_2). Under the internal gas pressure derived from the decomposition of the blowing agent uniformly dispersed in the precursor, the metal expands and acquires a porous structure of closed-cells. A good coincidence should exist between the thermal decomposition of the blowing agent with the release of a gas (e.g., hydrogen, H_2) and the melting of the metal [48]. The liquid foam is then solidified by cooling in air to obtain solid foams consisting of closed cells, and covered by a thin external dense metal skin conferring them a good surface finish [49]. This enables producing foams with porosities between 75% and 90% [48]. The ability of PM method to produce metal foam components with different architectures (e.g., sandwich systems, filled profiles and 3D complex shaped structures) is its main advantage [49]. Furthermore, different materials or structures can be joined during the foaming step, without using chemical adhesives. For example, fasteners or standard parts used in vehicles (*i.e.*, nuts, bolts, screws, pin rivets) could be incorporated into the metallic foam during its formation [50]. PM method also allows fabricating *in-situ* foam-filled tubes in which Al foam fills empty Al-alloy tubes while expanding during the foaming process [6]. High production costs derived mostly from the precursor preparation, and the difficulty to fabricate large volume foam parts are the principal disadvantages of the PM process. Herein, the micro-sized ceramic particles (inferior to 10 vol.%) are directly dispersed in the powder mixture of metal and blowing agent using mechanical mixers (e.g., turbula). Dense reinforced precursors are obtained by hot compacting the powder mixture. PM method requires lower volume fractions of ceramic particles in comparison to direct melt foaming.

An effective dispersion of particle reinforcements in the matrix it is essential for taking full advantages of their incorporation. However, micro-sized particles are often found not well-dispersed into the metallic matrix. Their tend to aggregate and agglomerate in clusters that appear in the cell-walls has been attributed to surface tension effects, but less attention has been given to the pre-existing agglomerates in the starting ceramic powders. The easiness how the micro-sized particles react with the metal-matrix during melting and foaming to form intermetallic products that increase the viscosity of the melt is another drawback. Thickening and foam stabilization mechanisms can be assessed by quantifying the phases formed during the entire melt foaming process [51,52,53]. Phase analysis is also expected to shed light on foam fracture behavior [54,55]. However, there the impact of phase changes on the foaming process is not consensual. Some authors claiming that intermetallic phases are important, while others holding that oxides are the main responsible for foam stability [56]. Moreover, the gas released from blowing agents (e.g., titanium hydride and calcium carbonate) can accumulate at the interface between these reinforcement particles and metal matrix during the foaming process, leading to a weak interfacial bonding. This inevitably limits their reinforcing potential.

The findings summarized Table 1 and Table 2 reveal that besides the stabilizing role of ceramic particles present in metal matrix composite (MMC) foams, they also increase the stiffness and the fragile behavior of the foams. Therefore, for a given system (Al-alloy and stabilizing/reinforcing ceramic particles) the experimental variables (type of particles, mean size, content of particles, pore volume fraction, *etc.*) have to be compromised taking into account the overall changes induced in the manufacturing process of MMC foams.

However, detailed evolution trends of foams' properties as a function of a given experimental variable can hardly be drawn because of the diverse manufacturing and testing conditions used in different studies. More straightforward conclusions would require a systematic experimental approach under strictly controlled conditions, namely concerning the influence of particle size on foaming behavior and mechanical properties. In an attempt to illustrate the findings on this topic reported by Esmaeelzadeh *et al.* [34], Table 3 summarizes the data estimated from the plots published in this single work on the subject. SiC particles of different size were used to reinforce AlSi7 foams prepared by PM method, using TiH_2 as blowing agent. It can be seen that all the mechanical parameters are favored when smaller particles are used.

Table 1. Literature survey on ceramic particles reinforced foams prepared by powder metallurgy.

Reference	Metal	Ceramic Particle	Manufacturing	Test Conditions	Conclusions
Elbir et al. [33]	Al	SiC 8.6–20 vol.% Size: 22 μm	Powder Metallurgy Al-powder <74 μm TiH$_2$-powder <37 μm	Φ 20 mm × 20 mm Compression Quasi-static 0.1 mm s^{-1}	In comparison to non-reinforced Al foams, SiCp particles reduce the drainage and cell coarsening phenomena, increase linear expansion and compressive strength of Al foams, but induce fluctuations in the plateau region of stress-strain curves and accentuate the brittle behavior of composite foams.
Esmaeelzadeh et al. [34]	AlSi7	SiC up to 10 vol.% Size: 3–16 μm	Powder metallurgy Al powder: <160 μm Si powder: <150 μm TiH$_2$-powder <63 μm	Φ 30 mm × 40 mm Compression Quasi-static 1.1 × 10^{-3} s^{-1}	Increasing the added amounts of SiCp or decreasing their size reduce the drainage but lead to less homogeneous foam structures. The compressive properties and energy absorption efficiency are degraded due to an accentuation of brittleness in comparison to non-reinforced AlSi7 foams.
Kennedy and Asavavisitchai [42]	Al	TiB$_2$ Size: 10 μm 6 vol.%	Powder metallurgy Al powder: 48 μm TiH$_2$ powder: 33 μm	Φ 22 mm × 24 mm Compression Quasi-static 0.5 mm min^{-1}	TiB$_2$ particles significantly enhance the maximum foam expansion but did improve the long term stability of the foams due to their poor wetting by the molten Al, as evidenced by particles protruding the cell-walls into the gas phase. The stress-strain curves in plateau region are smooth and characterized by a slightly increasing slope, irrespective of the presence or the absence of reinforcement. The maximum yield stress is achieved for TiB$_2$-Al composite foams.
Guden and Yuksel [35]	Al	SiC 0–20 vol.% Size: 22 μm	Powder metallurgy 34.64 μm Al powder TiH$_2$ powder: <37 μm	Φ 13 mm × 13 mm Compression Quasi-static 3 × 10^{-3} s^{-1}	SiCp increase the linear foam expansion by increasing the bulk viscosities. The composite SiCp-Al foams are more brittle in comparison to with Al-foams.
Alizadeh and Mirzaei-Aliabadi [30]	Al	Al$_2$O$_3$ Size: 10 μm 0–10 vol.%	Space-holder Al-powder Carbamide: 1.2 mm Ethanol: 1–3 wt.%	Φ 25 mm × 30 mm Compression Quasi-static 0.1 mm s^{-1}	Increasing volume fractions of Al$_2$O$_3$p enhance the Young's modulus and the compressive strength of the composite foams in extends that depend on the porosity fraction. For a given porosity fraction, the plateau region of composite foams is less smooth and shorter than for the Al-foam. The plateau stress and energy absorption capacity increase with Al$_2$O$_3$p content increasing up to 2 vol.%, but this trend is reversed for higher volume fractions. However, contrarily to other literature reports [34,41] the energy absorption efficiency of the composite foams is always higher than that of non-reinforced Al-foams.
Luo et al. [36]	AlSi9Mg	SiC 4 vol.% Size: 28 μm	Infiltration process AlSi9 alloy NaCl (0.9–4 mm in size)	15 mm × 15 mm × 35 mm Compression Quasi-static 10^{-3} s^{-1}	SiCp increase yield stress and energy absorption capacity of composite foams increase. Stress-strain curves of composite foams are less smooth than as than those of non-reinforced Al-foams.
Zhao et al. [43]	Al	Y$_2$O$_3$ 0.3–1.2 wt.%, Size: 50 μm	Space holder Al powder NaCl particles: 0.66–0.90 mm	12.8 mm × 6.5 mm × 35 mm Compression Quasi-static 3 mm min^{-1}	Volume fractions of Y$_2$O$_3$p up to 0.8 wt.% enhance bending strength up to a maximum of 20.4 MPa, a trend that is reversed for further added amounts, while the maximum micro hardness is achieved within the range of 0.5–0.8 wt.%.

Table 2. Literature survey on ceramic particles reinforced foams prepared by direct foaming methods.

Reference	Metal	Ceramic Particle	Manufacturing	Test Conditions	Conclusions
Liu et al. [41]	Zn-22Al	SiC Size: 28 μm 7 vol.%	Direct melt foaming ZA22 alloy ingot CaCO$_3$; 44 μm	15 mm × 15 mm × 30 mm Compression Quasi-static (2.2 × 10^{-3} s^{-1})	SiCp accentuate the brittleness and enhanced the stress fluctuations within the plateau region of composite foams. The energy absorption capacity is slightly improved but the energy absorption efficiency is degraded in comparison to non-reinforced foams.
Luo et al. [37]	AlSi9Mg	SiC Size: 28 μm 0–20 vol.%	Direct melt foaming AlSi9Mg alloy CaCO$_3$; 44 μm	15 mm × 15 mm × 35 mm Compression Quasi-static (10^{-3} s^{-1})	The same conclusions as above [41]. At a given relative density, yield and collapsing stresses of composite foams increase with increasing SiCp volume fraction.
Yu et al. [36]	Zn-22Al	SiC Size: 28 μm 10 vol.%	Direct melt foaming ZA22-powder: 40 μm CaCO$_3$; 44 μm	15 mm × 15 mm × 30 mm Compression Quasi-static (2.2 × 10^{-3} s^{-1}) Φ 70 mm × 10 mm Damping (400 Hz)	The same conclusions as above [37/41]. The damping capacity of composite foams is slightly higher than those of ZA22 alloy and ZA22 foams.
Yu et al [39]	AlSi9Mg	SiC Size: 28 μm 10 vol.%	Direct melt foaming AlSi9Mg alloy CaCO$_3$; 44 μm	15 mm × 15 mm × 35 mm Compression Quasi-static 5 × 10^{-4}–1 × 10^{-2} s^{-1} Φ 30 mm × 10 mm High strain rate (600–1600 s^{-1}),	The same conclusions as above concerning the effects of SiCp on the mechanical properties of composite foams [37,38,41]. The yield stress depends on both relative density and strain rate, being 10 MPa and 40 MPa for quasi-static (<10^{-2} s^{-1}) and dynamic (1600 s^{-1}) loading conditions.
Daoud [32]	A359	Al$_2$O$_3$ 0–15 vol.% Size: 50–140 μm	Direct foaming CaCO$_3$	Compression Quasi-static 3 × 10^{-3} s^{-1}	Al$_2$O$_3$p enhance the uniformity of foam microstructure and the resulting compressive stress-strain curves of composite foams are smooth. The mechanical parameters increase almost linearly with increasing the volume fraction of Al$_2$O$_3$p. The energy absorbing capacity is not much sensitive to the volume fraction of Al$_2$O$_3$p up to 10 vol.% increasing for higher contents.
Song et al. [44]	Al-3.7 Pct Si-0.18 Pct Mg	AlN	Solid/liquid reaction Master ingot	10 mm × 10 mm × 10 mm Compression Quasi-static (1 mm min^{-1})	AlNp reveal an effective reinforcing role increasing the mechanical properties of Al-alloy foams. Absence of stress oscillations in the plateau region of strain-stress curves of composite foams, similarly as observed for Al$_2$O$_3$p [32] and TiB$_2$p [42] Al composite foams.

Table 3. Effect of the particle size and volume fraction of SiC particles on the compressive behavior of AlSi7 foams [34].

SiC Size	SiC (vol.%)	Yield Stress (MPa)	$\sigma_{0.1}$ (MPa)	$\sigma_{0.2}$ (MPa)	$\sigma_{0.3}$ (MPa)	$\sigma_{0.4}$ (MPa)
	0	1.13	1.13	1.38	1.41	1.50
3 μm	3	1.58	1.33	1.72	1.75	1.88
	6	1.25	1.13	1.33	1.41	1.88
16 μm	3	1.25	1.0	1.25	1.38	1.58

In an attempt to sort out some general trends from previous studies, Figure 1 and Figure 2 condense some published data about on the effects of micro-sized ceramic particles (SiCp, TiB2$_p$, and Al$_2$O$_{3p}$) on the compression behavior of different metal (ZA22, Al, AlSi9Mg and AlSi7) foams produced through different manufacturing processes. The plotted data reveal that, irrespective of the foaming method and metal matrix, the yield stress of composite foams is always higher in comparison to non-reinforced ones (Figure 1 and Figure 2). The yield stress also shows a general increasing trend with increasing contents of ceramic particles as shown in Figure 1a and Figure 2a,b and Table 1 and Table 2. However, the plateau stress strongly depends on the porosity fractions, as expected.

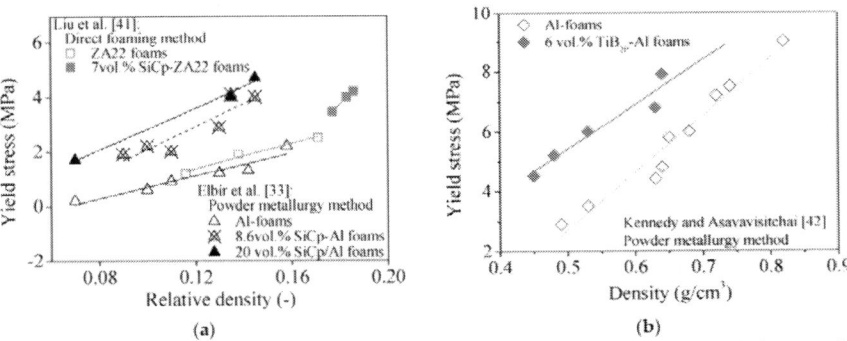

(a)   (b)

Figure 1. Effect of the volume fraction of SiC (**a**) and TiB2 (**b**) particles on the yield stress of Al based foams.

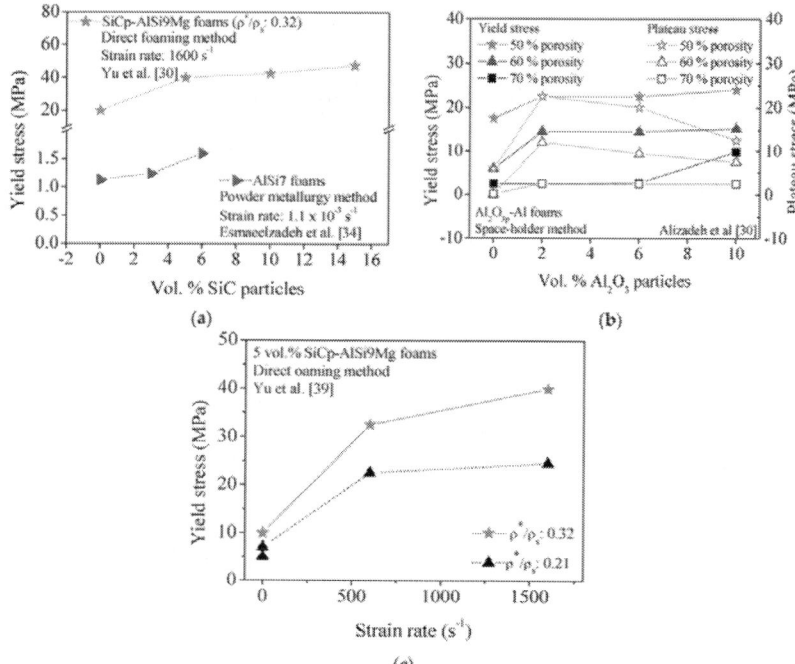

(a)   (b)

(c)

Figure 2. Effect of the volume fraction of particles on the yield stress (**a,b**) and plateau stress (**b**) of foams containing SiC (SiCp) (**a**), and Al₂O₃ particles (Al₂O₃p) (**b**). Effect of strain rate on compressive behavior of reinforced foams containing 5 vol.% SiC particles is displayed in (**c**).

Using stronger and stiffer reinforcing ceramic particles also enhance the cell walls strength. Figure 3 shows a schematic representation of typical compressive stress-strain curves often measured for metal foams under uniaxial compression. These curves usually comprise three regions (quasi-elastic, plateau and densification). Region I extends from the beginning up to the yield stress; the plateau region (II) characterized by a flat and smooth stress plateau (Curve 1, Figure 3a), sometimes with slight increasing slope (Curve 2, Figure 3a), or presenting some fluctuations (Figure 3b); and finally the densification (III) region. Furthermore, it is often difficult to precisely define the yield point (Point A in Figure 3b,c) as the curves shown in Figure 3a.

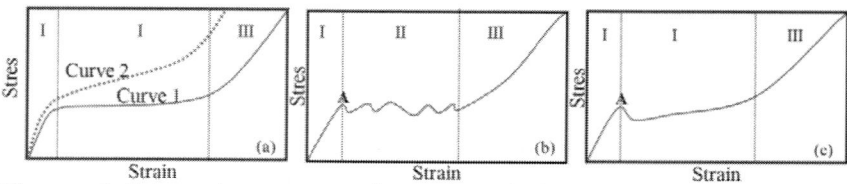

Figure 3. Schematic representation of typical stress-strain curves (**a–c**) measured for metal foam under uniaxial compression characterized by quasi-static (I); plateau (II); and densification (III) regions.

From the data published up to the present, the stress-strain curves of composite foams usually exhibit stress fluctuations within the plateau region (Figure 3b), irrespective of the manufacturing process. However, in the particular cases of foams reinforced with TiB_2p [42] produced by the PM method (Table 1) and composite foams reinforced with Al_2O_3p [32] and AlNp [44] prepared based on direct foaming method (Table 2), the absence of stress oscillations in the plateau region of strain-stress curves were found (Figure 3a,b).

2.2. Metal Foams Reinforced with Intermetallics

Stronger high-strength Al-alloy powders or ingots can be directly used for preparing high-strength Al-alloy foams. In comparison to pure Al-foams, AlZn5Mg1, AlSc0.24 [57,58], AlZn7Mg0.5 [59], AlZn10Mg0.3 [59,60] enabled stronger foams to be formed without compromising the foaming process. This makes alloying methodologies promising for the fabrication of reinforced Al-alloy foams with high strength. Alloying elements are directly added to the melt or dispersed in the powder mixture through direct and indirect melt foaming methods, respectively. Magnesium (Mg), manganese (Mn), scandium (Sc), nickel (Ni), and cobalt (Co) are some of the most studied alloying elements aiming at enhancing the mechanical properties of the existing foams [61,62,63,64,65,66,67,68,69,70,71]. The role of alloying elements on the foaming behavior of the Al-alloys, namely through the thickening and foam stabilization mechanism has also been investigated. The phases formed during the foaming melt process and the phases remaining in the solidified foams have been extensively studied due to their impact on fracture of the resulting foams [54,55]. Several strengthening mechanisms have been postulated. Some authors claim that the intermetallic phases formed are essential for foam stabilization, but others defend that formed oxides play here a more important role.

The replacement of calcium in the direct foaming method [69,70] by magnesium (Mg) as foam stabilizer was found to promote a non-uniform pore-size distribution with increasing added amounts of Mg with the resulting cell size gradient foams. This negative effect on the efficacy of metallic foam strengthening [59] might be attributed to the higher affinity of Mg towards oxygen in comparison to that of calcium. The higher affinity of Mg towards oxygen favors the formation of MgO or MgO-containing phases such as spinel [72], which, in turn might cause foaming problems.

Huang et al. [57] studied the effect of scandium on the quasi-static compressive properties of Al-alloy foams. They found that the addition of small volume fraction of scandium coupled with an appropriate heat treatment of the resulting composite solid foam, improved their compressive yield strength (Figure 4), which was attributed to the formation of precipitates (e.g., Al_3Sc). The effect tended to increase with increasing volume fractions of Sc, and with post heat treatment, as shown in Figure 4.

The stress-strain curves are smooth and, as expected, lower stresses are required for deformation with increasing porosity, and the plateau are shorter and more inclined as density increases. Moreover, the benefits of the post heat treatment that promotes a more extensive precipitation of intermetallic phase, are obvious when comparing with the heat treated (b,d) with non-treated (a,c) foams. The formation of Al3Sc precipitates at the grain boundaries of Al-Sc alloy foams was well evidenced by TEM images, which also showed atomic dislocations in <001> Al projection.

The authors also compared the Sc-containing foams with other intermetallic-reinforced foams, as well as with pure Al-foams. Except for the conventional Al-Si foams, the results (Figure 4) confirmed the superior mechanical properties of the other intermetallic-reinforced foams. Interestingly, the AlCu5Mn revealed to be generally stronger than the Sc-containing ones even after post heat treatment. Considering that Sc-containing foams are very expensive, the AlCu5Mn including cheaper metals appear as very competitive.

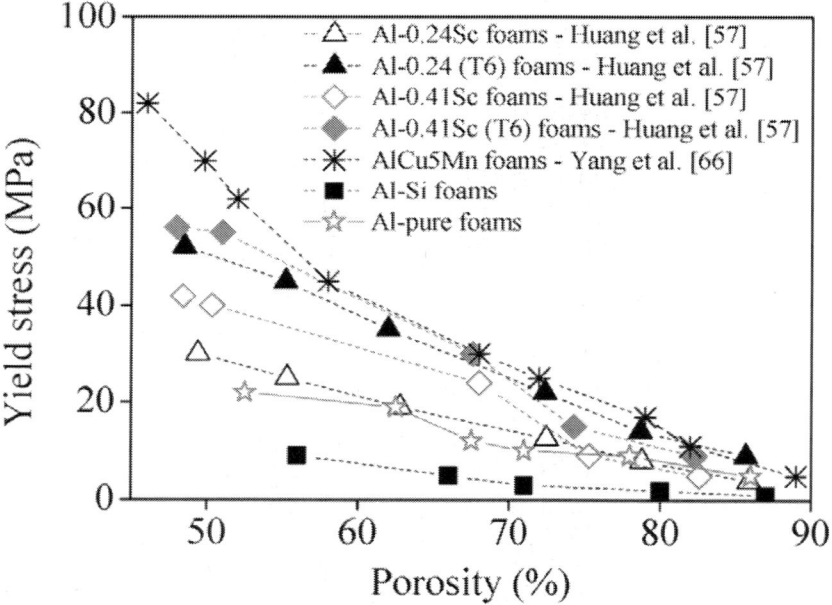

Figure 4. Comparison of different Al-based foams [57,66].

Manganese (Mn) was also shown to be an ideal element to promote reinforcement through the formation of intermetallics, enhancing micro hardness, yield stress and the plateau region of the stress-strain compressive curves of the Al-alloy foams [66]. The Mn volume fraction should be adjusted in order stabilize a good compromise between the yield strength and ductile deformation behavior [66]. The benefits of adding Mn contents ≥4 wt.% are not clear. Intermetallics-reinforced close foams were prepared by the melt foaming method using Ca as stabilizer and TiH_2 as blowing agent, varying the volume fraction of Mn powder (100 mesh). Mn was found uniformly distributed in the matrix as oxides (e.g., MnO_2), intermetallics (e.g., Al6Mn), Al-Mn solid solution (solid solution strengthening), as well as Mn particles non-dissolved in the cell wall matrix. The resulting reinforced foams possessed a much higher energy absorption capability compared to the Al-foams, which increased with increasing added amounts of Mn. It can be seen that the plateau appears gradually more inclined with increasing Mn contents.

The effects of Mn contents on mechanical properties of Mn-containing Al-foams suggest that compressive yield strength is more sensitive to the presence of Mn than micro hardness. This is likely to be related to the amount of oxides formed in the Al-Mn matrix.

Aguirre-Perales *et al.* [67] performed a systematic study on the effect of the different alloying elements (Co, Mg, Mn, Ni and Ti) on the compression behavior and the energy absorption capability of Al-3 wt.% Sn alloy. Their observations confirmed the general findings of the previous literature reports in the field that energy absorption of Al-foams and their contents in intermetallic phases increase with increasing contents of these alloying elements up to a certain limit.

In summary, the alloying elements usually enhance the mechanical properties of the metallic foams by favoring the formation of intermetallic compounds by reacting with oxygen from the atmosphere.

2.3. Composite Foams Reinforced with Hollow Spheres

Composite metallic foams incorporating lightweight filler particles such as hollow spheres are commonly designated as metallic syntactic foams (MSFs) and possess some particular features [73,74]. The term "Composite metal foam" was firstly introduced by Rabiei *et al.* [75] to designate a specific type of syntactic foams developed by PM and gravity casting techniques. The foams comprise steel hollow spheres packed into a random dense arrangement with interstitial spaces between spheres occupied with a solid metal matrix. These foams are fabricated, for example, by sintering a mixture of steel spheres and iron powder, or by infiltrating an aluminum alloy through the interstitial spaces between steel spheres. Figure 5 shows a schematic representation of an individual hollow sphere and of their arrangement in the composite syntactic foams.

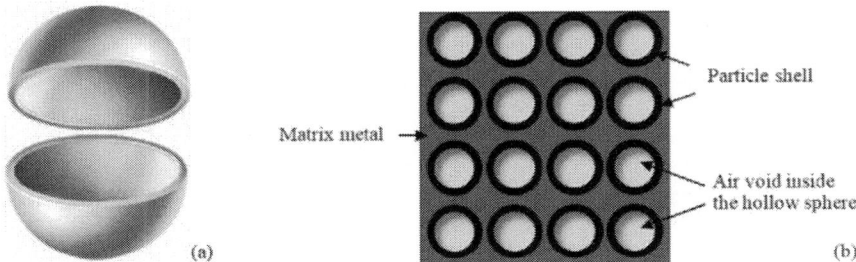

Matrix metal

Particle shell

Air void inside the hollow sphere

(a)　　　(b)

Figure 5. Scheme of a hollow particle (**a**); and its distribution on metal matrix of a syntactic foam (**b**).

The closed-cell structure is conferred by the hollow spaces instead of being derived from foaming. This allows easily controlling the cellular structure of composite foams. Furthermore, density and mechanical properties of MSFs can be easily adjusted by selecting the characteristics of both filler (e.g., volume fraction, size and wall thickness of the spheres) and the metal matrix (e.g., chemical composition). The aspect ratio, sphericity, density and chemical composition of the hollow spheres influence the mechanical behavior of the MSFs. This approach is also compatible with all the above referred strategies usually used to reinforce the alloy matrix, including the addition of micro- and nano-sized reinforcements, formation of intermetallics, oxides, thermal heat treatments, *etc*. MSFs are nowadays being considered as promising materials for commercial and military applications due to their very high level of energy absorption compared to the conventional materials.

Several alloys (e.g., magnesium [76,77,78], iron [77,78,79,80,81], steel [82,83,84,85,86,87,88,89,90], invar (FeNi$_3$6) [89,90] and titanium [91,92] alloys have been tested as matrices of MSFS, although aluminum and its alloys [72,93,94,95,96,97,98,99,100,101,102,103,104,105,106,107,108,109,110,111] were the most studied. Hollow spheres of ceramics (e.g., Al$_2$O$_3$ [93] and SiC [94,95]), perlite particles [99,100,101], metals [82,88,111,112,113,114,115,116], glasses microspheres [79,81], pumice (low-cost natural volcanic glass) [117] and even of by-products such as fly ash cenospheres [76,78], are widely used as reinforcements in different sizes (from a few microns to several mm). The selection of the filler materials is based on criteria related to their mechanical properties, the compatibility with the metal matrix and the price. Table 4 andTable 5 summarize the results of a literature survey on MSFs based on Al-alloy, and iron or steel matrices, respectively.

Some ceramic hollow particles (e.g., SiC) can be easily wetted by molten alloys, while other need to be coated with different materials (e.g., nickel and copper) to enhance their compatibility with the metal matrix [73]. Optical micrographs of an Al-alloy (A356) reinforced with SiC hollow spheres show that the particles are wetted well with the alloy, even the closely spaced particles have a layer of matrix between them, wetting of particle with the matrix alloy and mechanical interlocking at the particle-matrix interface [94].

MSFs can be fabricated by the same conventional solidification processes (e.g., casting) and powder metallurgy methods [73]. In the solidification processes, the solid reinforcements are combined with the metal matrix in liquid state, casting and then solidified in a given desired shape. One common method is the pressure infiltration in which the liquid metal is forced to flow through the voids between the hollow spheres (HS). The pressure can be applied by an inert gas, vacuum infiltration, mechanical pressure. Another common method is the stir casting in which the hollow spheres are added in the molten metal under constant agitation and stirring. Herein, this method favors the segregation and the agglomeration of the hollow spheres. In the powder metallurgy processing, the solid reinforcements are previously mixed with metal powders, pressed into a shape, outgassed, and sintered by hot compaction for obtaining near full dense parts. The dense parts can undergo post heat treatments (forging, rolling and extrusion). A dispersing agent can be used to facilitate obtaining a homogeneous distribution of the hollow spheres in the metal powder.

Table 4. Literature survey on aluminum alloy syntactic foams.

Reference	Syntactic Foam Type	Testing Conditions	Results
Licitra et al. [93]	Matrix: A356 alloy Particles: Al$_2$O$_3$, 3 mm diameter and 105 μm wall thickness	Compression Quasi-static (10^{-3} s^{-1}) High (445–910 s^{-1}) Dynamic Mechanical properties	• Young modulus, compressive strength and plateau stress of MSFs are directly proportional to density. • Particle failure initiates the specimen failure, followed by shear failure of matrix and remaining particles. • Storage modulus of A356 matrix and MSFs decreases but the loss modulus and damping parameter increase as temperature increases.
Cox et al. [95]	Matrix: A356 alloy Particles: SiC, 1 mm diameter and 70 μm wall thickness	Compression Quasi-static (10^{-3} s^{-1}) High (up to1820 s^{-1})	• Evidences of hollow spheres crushing at the end of the elastic region. • No strain rate sensitivity detected within the investigated range. • Failure at high strain rate is initiated by particle cracking and shear band formation.
Balch et al. [96]	Matrix: cp-Al, 7075alloy Particles: crystalline mullite and amorphous silica hollow microspheres	Compression Quasi-static (10^{-3} s^{-1}) High (up to2300 s^{-1})	• Pure Al MSFs show compressive strength >100 MPa with a uniform densification plateau of 60% under quasi-static conditions. • 7075-Aluminum alloy exhibit significantly higher peak strength of up to 230 MPa under quasi-static conditions. • HSR testing showed a 10%–30% increase in peak strength as compared to quasi-static testing and displayed energy absorbing capacity.
Reference	Syntactic Foam Type	Testing Conditions	Results
Orbulov et al. [97]	Matrix: Al99.5, AlSi12, AlMgSi1 and AlCu5 alloys Particles: ceramic hollow spheres with Al$_2$O$_3$, SiO$_2$ and Mullite	Compression Quasi-static (free, 10^{-2} s^{-1}) Quasi-static (Constrained, 10^{-2} s^{-1})	• Densification limit was primarily influenced by the hollow spheres' size in constrained compression. • Recoverable energy in constrained compression case is influenced by the applied heat treatment. • Overall absorbed mechanical energy is largely influenced by the compression mode (free or constrained).
Goel et al. [98]	Matrix: Al-2014 Particles: Aluminum cenospheres, 90 μm and 200 μm diameter	Compression Quasi-static (10^{-3} s^{-1}) High strain (up to1400 s^{-1})	• Syntactic foam shows about 10%–30% higher compressive strength under high strain rate conditions as compared to the quasi-static conditions. • Energy absorption capacity increases by up to 55% in the high strain rate region.
Taherishargh et al. [117]	Matrix: A356 alloy Particles: Pumice, size range: 2.8–4 mm	Compression Quasi-static (3 mm min^{-1})	• Compressive strength of pumice particles is anisotropic, showing a maximum in the direction parallel to its tubular pores. • Pumice-A356 syntactic foam is an efficient energy absorber with an average density of 1.49 g cm^{-3}, a plateau stress of 68.25 MPa, and specific energy absorption of 24.8 MJ m^{-3}.
Szlancsik et al. [133]	Matrix: Al99.5, AlSi12, AlMgSi1 and AlCu5 alloys	Compression Quasi-static (0.01 s^{-1})	• Compressive test results show plastic yielding, long and slowly ascending plateau region that ensures large EA capability. • Matrix material and the heat treatment exert strong influences on mechanical properties of MSFs.

The properties of MSFs produced by different methods and from different materials have been compared in several works and confronted with those of the conventional Al-foams [74]. From the published works, it was demonstrated that the stress-strain compressive curves of MSFs strongly depend on the material constituting the hollow spheres. The compressive strength of these MSFs is controlled by the strengths of metal matrix and of the hollow spheres. The volume fraction, structure and distribution of the hollow spheres exert considerable effects on the properties of MSFs.

Table 5. Literature survey on iron or steel matrix syntactic foams.

Reference	Syntactic Foam Type	Testing Conditions	Results
Neville and Rabiei [82]	Matrix: low carbon steel or stainless steel Particles: HS-low carbon steel (3.7–1.4 mm) or HS-stainless steel (2 mm)	Quasi-static	• EA at densification was higher for stainless steel compared to carbon steel syntactic foam.
			• Maximum energy absorption at densification was 68 MJ·m^{-3} for stainless steel syntactic foam.
Castro and Nutt [83]	Matrix: steel Particles: steel or alumina	Compression at 8×10^{-4} s^{-1}	• Low carbon and medium carbon syntactic steel foams have EA capacities of 66.45 and 122.68 MJ·m^{-3}, respectively.
			• Increasing carbon contents enhances yield strength of steel foams.
Castro and Nutt [84]	Matrix: steel Particles: steel or alumina	Compression at 8×10^{-4} s^{-1}	• Maximum EA at densification is 104.78 MJ·m^{-3}.
			• Relative density of steel foam increasing enhances compressive strength and decreases plateau stress.
			• EA capacity increased by six times per unit mass and 70 times per unit volume when compared to Al-foams.
Peroni et al. [79,80]	Matrix: 99.7% pure iron Particles: S60HS (d 30 µm) or iM30 k (d 18 µm) glass hollow particles in 5, 10 and 13 wt.%	Quasi-static (10^{-2} s^{-1}) Low (10–20 s^{-1}) High (1000–2800 s^{-1})	• Yield strength increases with strain rate, being 47% higher in comparison to that measured under quasi-static conditions.
			• Increasing glass microspheres contents reduce the strength of MSFs.
			• Strength and fracture behavior of MSFs depend on the intrinsic properties of glass microspheres used.
Weise et al. [89]	Matrix: FeNi36 Particles: S60HS (d 30 µm) glass powders	Tension	• Using fine powders is beneficial for mechanical properties.
			• Lowering density by 30% implies a 60% reduction in ultimate tensile strength.
			• Limited ductility retained under tensile load even for small additions of glass S60HS.

Reference	Syntactic Foam Type	Testing Conditions	Results
Weise et al. [55]	Matrix: AlSi6L Particles: S60HS (d 30 µm) glass hollow particles at 5.3 and 10 vol.%	Compression, tension	• High sintering temperatures lead to disintegration of glass microsphere, porosity retained, but glass phase embedded within the metal matrix rather than supporting pores as in a true syntactic foam.
			• Property scaling of QS compressive strength according to a power law with exponent 1.13, in between typical values for syntactic and non-syntactic closed-cell foams.
Peroni et al. [86]	Matrix: AISI 316L Particles: glass microspheres S60HS (d 30 µm) glass hollow particles at 40 and 60 vol.% Fillite 10% cenospheres at 40 vol.%	Compression Quasi-static (10^{-2} s^{-1}) Low (10–20 s^{-1}) High (1000–2000 s^{-1})	• Cenospheres remain intact and yield high quality syntactic foam.
			• Strength loss with decreasing density less significant for cenosphere compared to glass microsphere-based materials.
			• Compressive strength increases by 25% for glass and cenosphere-based variants with strain rate, a dependence that is attributed to lattice structure.
Brown et al. [87]	Matrix: low carbon steel or stainless steel Particles: low carbon steel or stainless steel	Three-point bending	• Flexural yield strength of 40 MPa, which is close to the compressive yield strength (42 MPa).
			• Plateau strength under compression is 80% higher than ultimate bending strength.
			• Ductile failure due to propagation of pre-existing microporosity in the matrix.
Vendra et al. [88]	Matrix: low carbon steel or stainless Steel Particles: low carbon steel or stainless steel	Compression–compression fatigue	• After 1 million cycles at fatigue load of 50% of the maximum plateau strength, stainless steel MSFs show a total strain of 8%.
			• Superior fatigue properties due to strong bonding between the hollow spheres and matrix.
Luong et al. [90]	Matrix: iron or FeNi36 Invar Particles: hollow glass microballoons (GMB)—5 and 10 wt.%	Compression Quasi-static 10^{-3} s^{-1} High (strain rate up to 2500 s^{-1})	• Yield strength decreases with increasing GMB content.
			• Quasi-static yield strengths of iron MSFs containing 5 and 10 wt.% GMB are 14% and 17% lower than that of the matrix alone.
			• Yield strength of Invar MSFs containing 5 and 10 wt.% GMB are 35% and 51% lower than that of the matrix alone. However, specific strength increases with GMB content and exceeds the respective data of other iron and steel foams.

There are two types of the stress-strain curves (Figure 6). In general, MSFs containing ceramic hollow spheres, such as alumina and silicon carbide show three main regions (Figure 6a): (I) linear elastic deformation where the stress increases linearly with strain until reach a peak stress; (II) stress drops and is followed by a plateau; (III) densification. The strong and brittle ceramic hollow spheres fail before metallic matrix being responsible by the first stress drop. Other MSFs, in particular the MSFs containing metal hollow spheres the stress-strain curves are characterized by the same three regions (Figure 6b), but the yield stress point do not have a well-defined yield point.

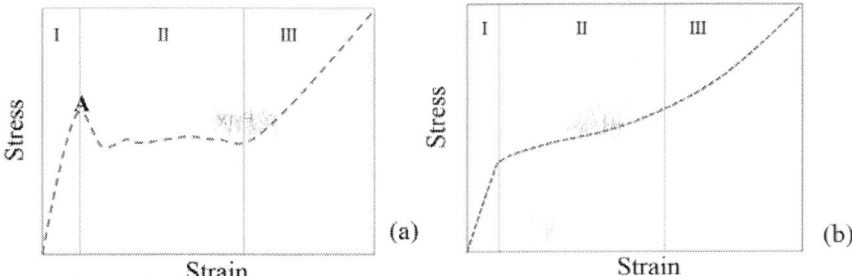

Figure 6. Representative quasi-static compressive stress-strain graphs (**a,b**) of metal matrix syntactic foams showing the linear elastic region (I), plateau region (II) and the densification strain (III).

For example, the mechanical properties (axial compressive and plateau strengths) of MSFs based on Al-alloy A356 were higher (160 MPa, [94]) when reinforced with hollow SiC spheres than when using fly ash cenospheres (75 MPa) [110]. It was also shown that the plateau stress and the energy absorption capability of MSFs were strongly dependent of the volume fraction of reinforcing material, with the mechanical properties of ceramic particles having a minor impact on the properties of bulk material.

Recently, MSFs reinforced with several kinds of ceramic microspheres and by-products have drawn a lot of attention due to their low density, heat insulating, saving energy and high mechanical features [98]. Millions tons of ceramic microsphere powders are generated in coal based thermal power plants every year and only a small portion is being utilized. Metal matrix MSFs reinforced with ceramic microspheres proved to possess better compressive properties than conventional Al-foams. The uniform distribution of ceramic microspheres in the metallic matrix MSFs prepared by melt-foaming method, the thicker pore walls and their generally higher density in comparison to conventional Al-foams are factors that contribute to the enhanced mechanical performance of MSFs.

Expanded perlite was successfully attempted as alternative reinforcement to prepare Al-alloy MSFs. Such achievement combined with the relatively low production costs of MSFs by infiltration casting stimulated the production of high amounts of expanded perlite particles that further contributed to decrease the production costs of MSFs. A few available literature reports [99,100,101] revealed that mechanical properties of perlite-MSFs increase by decreasing particle size of reinforcement and by post heat treatment. Further research is needed. The properties of MSFs have been tested for evaluating their suitability for commercial and military applications [118,119,120]. Rabiei and Vendra compared different MSFs based on Al-alloys and steel alloys with conventional Al-alloy foams (Table 6).

Table 6. Properties of composite metal foams with conventional metal foams [114].

Sample Property	MSFs				Hollow Sphere	Conventional Foams	
	PM Carbon Steel	PM SS Foam	Al-LC Cast Foam	Al-SS Cast Foam	SS HSF	Al-Foam	Al-Foam
Sphere OD (mm)	3.7	2.0	3.7	3.7	2–3	3	3
Wall thickness (mm)	0.2	0.1	0.2	0.2	0.25	–	–
Density (g·cm^{-3})	3.06	2.95	2.43	2.45	1.4	0.4	0.24
Relative density (%)	38.9	37.5	42.5	42.5	17.8	14.8	8.9
Plateau Stress (MPa)	36.2	127	60	105	23	5	2.5
Densification strain (%)	54	54	57	57	60	68	50
Strength/density	12	43.7	24.4	43	16	12.5	10
Energy absorption at densification (MJ·m^{-3})	18.9	67.8	31	51	13	2.6	1.32

In summary, compressive strength and EA efficiency of MSFs is considerably higher than the conventional closed-cell metal foams [114]. The differences relay on the higher densities and thicker cell walls of MSFs. The high weight and the high volume of MSFs, compared to the conventional foams, are the main limiting factors for their widespread structural applications.

Table 7 summarizes the data reported for the different fillers used to prepare MSFs based on Al-alloys. The following general trends are apparent: (i) stronger alloys originate stronger MSFs; (ii) Smaller size fillers favor the strength of MSFs; (iii) proper heat treatments enhance the mechanical properties of MSFs.

Table 7. Literature data on other aluminum alloy syntactic foams with different fillers.

Matrix	Filler Material	Filler Size	Filler Particle Density (g·cm^{-3})	MSFs Density (g·cm^{-3})	Plateau Stress (MPa)	References
Pure Al	Cenosphere	90–150 μm	1.00–1.74	1.52–1.43	63–42	Wu et al. [121]
A356	Cenosphere	45–125 μm	0.7	1.25–2.1	45–180	Rohatgi et al. [122]
Pure Al	Ceramic HS: 45 SiO$_2$-35 Al$_2$O$_3$-20 Mullite	100–1450 μm	0.37–0.81	1.43–1.49	77	Orbulov and Ginsztler [123]
Pure Al	Ceramic HS: 60 SiO$_2$-40 Al$_2$O$_3$-20 Mullite	250–500 μm	0.75	1.38	62	Zhang and Zhao [124]
Pure Al	Ceramic HS: 60 SiO$_2$-40 Al$_2$O$_3$-20 Mullite	75–125 μm	0.6	1.45	92	Tao and Zhao [109]
A356	Ceramic HS: SiC	1 mm	1.160	1.819	110	Luong et. al [94]
Pure Al	Ceramic HS: Alumina	3 mm		1.6–2.11	62.8	Licitra et al. [93]
Pure Al	Glass HS: 60 SiO$_2$-40 Al$_2$O$_3$-15 CaO-Na$_2$O	0.5–4 mm	0.95–0.65	1.58–1.88	42	Zhang and Zhao [124]
A35	Expanded Perlite	3–4 mm	0.18	1.05	45	Fiedler et al. [109]
A356	Pumice	2.8–4 mm	0.76–0.89	1.48–1.50	64–76	Taherishargh et al. [117]
Al 99.5	Iron (Fe pure) HS	1.92 ± 0.07 mm	0.093	1.41	35–39	Szlancski et al. [111]
AlSi12	Iron (Fe pure) HS	1.92 ± 0.07 mm	0.093	1.42	55–61	Szlancski et al. [111]
AlMgSi1	Iron (Fe pure) HS	1.92 ± 0.07 mm	0.093	1.60	54–70	Szlancski et al. [111]
AlMgSi1-T6	Iron (Fe pure) HS	1.92 ± 0.07 mm	0.093	1.60	75–91	Szlancski et al. [111]
AlCu5	Iron (Fe pure) HS	1.92 ± 0.07 mm	0.093	1.72	47–101	Szlancski et al. [111]
AlCu5-T6	Iron (Fe pure) HS	1.92 ± 0.07 mm	0.093	1.72	120–162	Szlancski et al. [111]

3. NANOCOMPOSITE METAL FOAMS

One strategy to improve the mechanical properties of conventional foams is using nano-sized reinforcements (e.g., particles, fibers, nanotubes) instead the micro-sized reinforcements. Foams reinforced with nano-sized reinforcements are called nanocomposite metal foams. The nanoscale reinforcements are much more effective in improving the desired properties in comparison to the microscale reinforcements due to their high interface-to-volume ratio. Some advantages of using nanoscale reinforcements are the much smaller volume fractions required and the negligible weight contributions for the resulting nanocomposite metal foams.

3.1. Metal Foams Reinforced with Ceramic Nanoparticles

The addition of nano-sized ceramic (e.g., alumina [125] and SiC [126]) particles was shown to enhance foam stability and homogeneity of the cellular structure without causing structural defects. Different *ex-situ* and *in-situ* strategies have been developed to uniformly disperse the nanoparticles in the powders mixture using ultrasonic methods, or incorporate them in molten metal through an *in situ* reaction. After the first screening results, several works have been conducted to study the effects of nanoparticles on the mechanical performance of the resulting nanocomposite foams prepared by using both direct and indirect foaming methods. For example, Al-foams reinforced with SiC nano-particles (nano-SiCP/Al composite foams) were prepared by mixing nanoparticles within aluminum powders using high-energy ball milling. After that, calcium carbonate (as blowing agent) was added to the initial mixture using a mechanical mixer. A dense foamable precursor material was prepared by hot pressing the mixture containing all the solids and heated at temperatures close to melting temperature to obtain the reinforced foams. The results have shown that nano-sized ceramic particles reduce the brittleness of the foams in comparison to micro-sized ones. Moreover, small additions of nanoparticles improved the foam structure, refined pore size and homogeneity of the pore distribution leading to significant enhancements of yield stress (194.5%), plateau stress, an increase of energy absorption (69.4%) in comparison to the conventional aluminum foams [126]. The pores of nanocomposites foams were much finer than those of pure Al-foams, changing from millimeter to micrometer range. In the reality, nanoparticles act as stabilizers and reinforcement agents.

Nanoscale reinforcements have been incorporated in several types of matrices to obtain nanocomposite foams, but the number of available literature reports is scarce [126,127,128,129]. A small addition (usually less than 2 wt.%) of nanoparticles can significantly strengthen the metal matrix, solving the problems of current Al-foams incorporating high loadings (generally above 10 vol.%) of microscale reinforcements required to achieve the desired levels for the mechanical properties (e.g., elastic modulus) and dimensional stability, thus compromising the weight and the toughness of the final composites.

The main difficulties to overcome in producing nanocomposites are the high surface to volume ratio and the generally low wettability of ceramic particles by aluminum. Smaller particles have a stronger tendency to agglomerate and to form micrometric clusters, losing their effectiveness in obstructing the movement of dislocations. For this reason, they cannot be prepared by simple conventional manufacturing processes. It is crucial to modify these methods.

3.2. Metal Foams Reinforced by Metal Deposition

Casting, powder metallurgy (slurry foaming and loose powder) and metallic deposition are the main manufacturing processes to fabricate these open-cell foams. An approach to improve the conventional open-cell Al-foams involves the electrodeposition into the cellular structure of metals having superior mechanical properties (e.g., nickel and copper). The aims are increasing the plateau stress, delaying the initial densification strain, and increase the energy absorption capacity. Few papers have been published in this field [130,131,132,133,134,135,136,137]. The first paper on this topic published in 2008 [130] reported on the effects of Ni-W coatings on the mechanical properties of Al-foams. The mechanical properties of the original open-cell foams were enhanced by increasing the coating thickness (density). The structure could be tailored and the maximum specific stress was substantially improved by coating with nickel. Jung *et al.* [137] reported considerable increases in plateau stress of an Al-alloy foam reinforced with nano-crystalline nickel deposition (Ni coating thickness ~250 nm). They also found that thick coatings considerably reduced the densification strain, thus compromising the energy absorption capacity of foams. Manipulating the deposition parameters (currents and times) allowed adjusting the extent of deposition. The energy absorption capacity was directly proportional to density or deposition time. However, uniform coatings could hardly be achieved for shortest deposition time while the longest one did not brought further benefits when compared to the intermediate deposition times. Such coatings also improved the resistance to corrosion which then allows using these foams even in aggressive environments. This strategy can broaden the functional applications of open-cell metallic foams, for example, as heat exchangers due to their very high thermal conductivity. The improved mechanical properties would allow producing more compact heat exchangers by partially merging the structural and heat exchanging functions.

Copper electro-deposition has been also considered to reinforce the open-cell foams (copper electrodeposited Al-foam, 10–40 PPI) [132]. Although stiffness of copper (123 GPa) is less than that of nickel (206 GPa), the first is less expensive and has a Young's modulus almost twice that of aluminum (69 GPa) [1]. The results of quasi-static systematic compression tests of nano-Cu coated open-cell Al-foams revealed enhanced energy absorption capability and plateau stress without compromising the densification strain. The overall mechanical performance was shown to be strongly affected by the foam relative density, cell topology (pore size, strut thickness, *etc.*) and the electro-deposition conditions. For example, the energy absorption capacity of Al-foams coated with a 60 nm Cu coating was 3 times greater than that of plain foams. The compressive stress–strain response of the composite samples showed no significant reduction of the densification strain compared to the uncoated foams due to the small change in the foam strut thickness and pore size. A comparison between thin Cu-coated and uncoated Al-foam samples with the same overall strut thickness (*i.e.*, same effective volume density and porosity) showed that the nano-reinforced foams had superior energy absorption capacity compared to the non-coated foams.

3.3. Metal Foams Reinforced with Carbon Nanotubes

Carbon nanotubes (CNTs) have emerged as potentially ideal nano-sized reinforcements to fabricate light weight and high-strength metal-matrix composites due to their low density and high values of aspect ratio, mechanical strength, electrical and thermal conductivities. However, the incorporation of

CNTs into the metal-matrices is not trivial because of their high tendency to form clusters, poor dispersion ability, and poor wettability of carbon by molten metal (due to a large difference of surface tensions). The formation of interfacial reaction products in molten metals is another limitation to their widespread use as reinforcements. Various processing strategies (e.g., powder metallurgy, molecular-level mixing, plasma spraying and casting) have been employed to overcome these problems, but with limited success [20]. The achievement of uniform dispersion of CNTs in the metal-matrix, the formation of strong interfacial bonding, and the retention of structural integrity of CNTs are the main challenges to overcome for successfully developing metal-matrix nanocomposites for industrial applications. These are key requirements to potentiate the homogeneous 3D reinforcing role of CNTs and to provide an efficient load transfer. As a matter of fact, CNTs in the generality of the literature reports on metallic matrix tend to appear in clusters, therefore, annulling their reinforcing potential. The research in this field is still in its infancy with only three articles published in 2015 [138,139,140,141]. The first articles in the field were published by Duarte et al. [138,139] and disclose a novel approach combining colloidal-processing (including freeze-granulation-lyophilisation) and powder PM method as schematized in Figure 7.

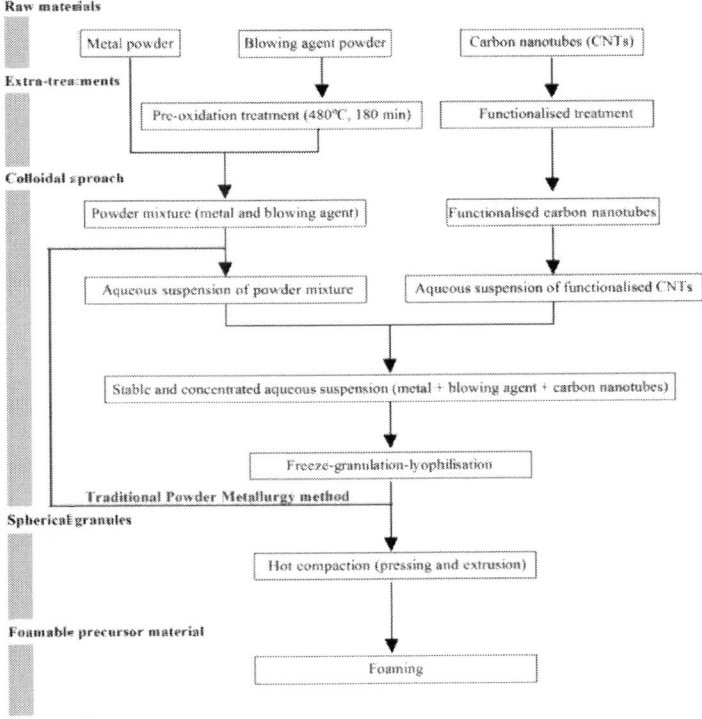

Figure 7. Schematics of the novel approach for preparing metal-foams reinforced with carbon nanotubes (CNTs) by combining the powder-metallurgy with colloidal-processing [138,139].

Figure 8 shows Al-alloy powder Al-12Si), titanium hydride powder (TiH2, as blowing agent) and –COOH-functionalized multiwall carbon nanotubes (MWCNTs-COOH, as reinforcement elements) which were the main starting raw materials used in this work.

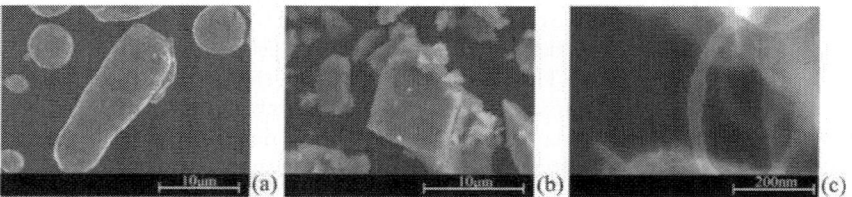

Figure 8. SEM micrographs of the starting raw materials: (**a**) Al-12 alloy; (**b**) TiH_2powder; (**c**) multiwall carbon nanotubes (MWCNTS)-COOH.

The first paper [138] reports preliminary results on the preparation of spherical granules by free granulation and preparation of precursor materials, showing evidences of the good dispersion of the MWCNTs inside the precursor materials. The second one [139] discloses the effects of the most relevant processing steps in a more systematic way, including a detailed characterization of the starting raw materials and the pre-oxidation heat treatment on the foaming agent. A detailed study of the dispersion behavior of the MWCNTs using different dispersing agents and their synergetic dispersing actions was preformed through complementary techniques (sedimentation and zeta potential measurements using diluted suspensions and rheological measurements using concentrated suspensions).

The SEM images in Figure 9 show that highly-spherical granules were obtained irrespective of the absence (Figure 9a) or the presence (Figure 9b) of MWCNTs-COOH. These SEM images also provided clear evidences about the uniform dispersion of MWCNTs-COOH and TiH_2 in the lyophilized granules (Figure 9a,b) and in the precursor materials (Figure 9c,d).

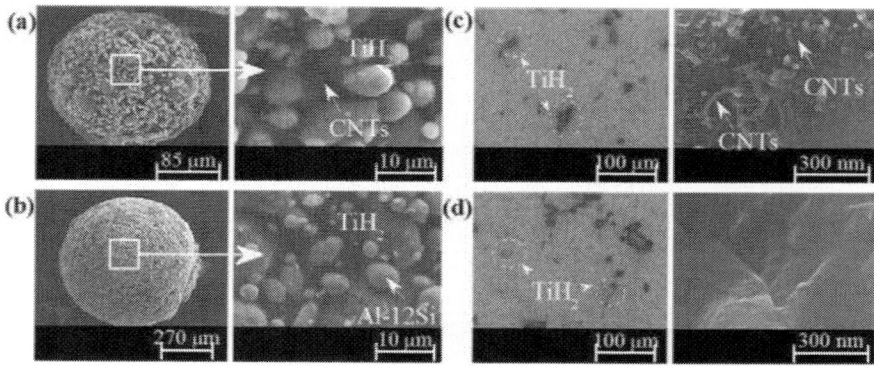

Figure 9. Highly-spherical granules prepared by freeze-granulation-lyophilisation and its surface with (**a**), and without (**b**), MWCNTs-COOH. Microstructural features of precursor materials (and of their fracture surfaces) prepared from granules with (**c**), and without (**d**), MWCNTs-COOH.

Figure 9c reveals a random distribution of individually dispersed and stretched MWCNTs-COOH aligned perpendicular direction of the hot compaction, remaining its structural integrity. This contrasts with the complete absence of MWCNTs-COOH in the non-reinforced sample (Figure 9d), as expected. As shown in Figure 9a, MWCNTs-COOH appear individually dispersed and stretched avoiding their natural tendency to form clusters that annul their reinforcing potential. Furthermore, the TiH_2 are also uniformly dispersed (dashed-circled, Figure 9a,b) ensuring the foam quality. The key for the successful preparation of Al-alloy foams reinforced with MWCNTs is ensuring a good dispersion of the MWCNTs. These highly-spherical granules were easily hot compacted to form dense precursor materials showing typical microstructures comparable to those of precursors prepared by the PM method as seen in Figure 9c and Figure 9d for precursors with and without MWCNTs-COOH, respectively.

The novel approach, a modification of the traditional powder metallurgy (PM) method by adding an advanced colloidal processing step allows: (i) achieving an effective dispersion of the functionalized MWCNTs-COOH in aqueous media and the homogeneous mixing with the other components of the system; (ii) preserving the homogeneity and structural integrity (tubular structure) of the carbon nanotubes through the process; (iii) establishing strong bonds between MWCNTs and metal matrix, which provide an efficient load transfer. Accordingly, in comparison to the non-reinforced Al-foams, the mean values of Vickers micro-hardness of reinforced ones increased within the range from 55% to 125%, depending on the neighborhood between of the indentation point and the reinforcing MWCNTs. However, further research efforts will be required to systematically investigate the effects of other relevant experimental variables not yet covered such as: (i) the reproducibility and overall quality control of the foaming process; (ii) determining the maximum allowable content MWCTs that can be incorporated without degrading the foaming process or the quality of the foams; (iii) applying the method to other Al-alloys and metallic systems; (iv) evaluating the properties and performance of the foams in high-tech applications and demonstrate their unique and competitive advantages over their existing counterparts. Individualized and stretched MWCNTs can be seen randomly distributed in the aluminum-matrix (Figure 10). This non-agglomerated condition confirms the effectiveness of the dispersion achieved in the colloidal processing step. The stretched condition is likely to be boosted upon foaming, being favorable to reinforcement.

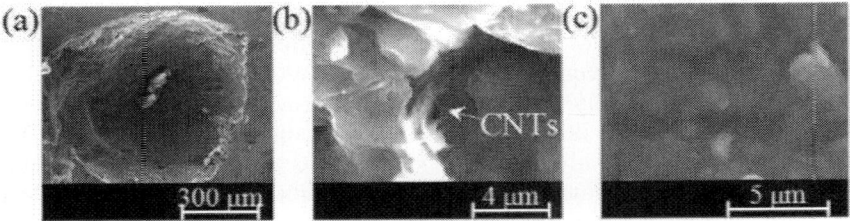

Figure 10. Cellular-pore **(a)** and magnified images of its pore-wall microstructure with **(b)** and without MWCNTs-COOH dispersed in its aluminum-matrix **(c)**.

The Vickers microhardness measured on the struts (cell-walls) changed from (60 HV ± 5.18 HV) for non-reinforced Al-foam to 93.43 ± 19.30 HV for the sample containing 0.5 wt.% MWCNTs, with some values exceeding more than double of the non-reinforced ones. The high standard deviation of reinforced foams could be attributed to: (i) the small added amount of MWCNTs; (ii) the neighborhood between of the indentation point and the reinforcing MWCNTs. This means that hardness can randomly change from one point to another depending on the closeness vicinity of a reinforcing MWCNT.

Zhang *et al.* [140] developed closed-cell Al-foams reinforced with different contents of MWCNTs by using a modified melt foaming method (Figure 11). High-energy ball-milling with adjusted the milling parameters was used to disperse the MWCNTs into the powders. According to the authors, MWCNTs existed mainly in three forms: totally embedded in cell wall; partially embedded in cell wall; and totally exposed on cell wall surface. The formation of interfacial intermetallic product (e.g., Al_4C_3) due to the reaction between the MWCNTs and the Al-alloy was reported to occur, which allegedly would lead to strong interfacial bonding between MWCNTs and aluminum matrix. Nevertheless, high-energy mixing process (e.g., ball-milling or mechanical-alloying) are likely to cause structural damage and structural-integrity loss to CNTs. Dispersion improvements of CNTs in the Al-matrix using extended ball milling times have been reported together with some structural damages. This is particularly serious for the single-walled CNTs since their tubular integrity is lost. MWCNTs may still provide the desired structural-integrity even suffering some damage in the outer-walls. The eventual formation interfacial-products (intermetallics) due to interfacial reactions between CNTs and molten metal under harsh conditions can also lead to loss of structural-integrity.

These authors studied the effect of volume fraction of MWCNTs (0.0–1.0 wt.%) on the compressive behavior and energy absorption of these nanocomposite foams. They found that the compressive yield strength increased up to 0.5 wt.% MWCNTs, while the opposite trend was noticed with further added amounts of MWCNTs. For example, the compressive yield strength of nanocomposite foams containing 1 wt.% MWCNTs was inferior to that of non-reinforced foam (without MWCNTs). Similar conclusions about the influence of the MWCNTs were drawn for the energy absorption curves that are obtained by integrating the stress-strain curves, with the highest and the lowest energy absorption capacity values being measured for the foams with 0.5 wt.% and 1.0 wt.% of MWCNTs, respectively. Although the influence of different MWCNTs contents on the mechanical properties of composites is likely to depend on the specific materials matrix, and especially on the interfacial bonding strength between the reinforcing MWCNTs and the embedding matrix, it is important underlying that the required amounts of MWCNTs for an effective reinforcement are usually small (~0.5 wt.%), provided that they are well dispersed. The probable non-fulfilment of this condition in work reported by Zhang *et al.* [140], especially for MWCNTs contents > 0.5 wt.% would explain why energy absorption capability the resulting Al-foams was even worse than for non-reinforced ones.

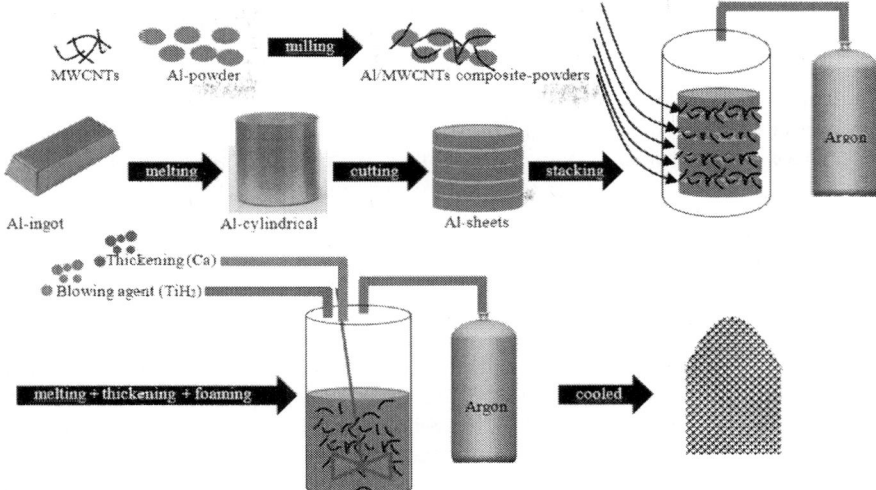

Figure 11. Schematics of the fabrication process of closed-cell Al-foams with and without MWCNTs [140].

A method for *in-situ* growing carbon nanotubes (CNTs) to reinforce Al-foams derived thereof was proposed by Wang *et al.* [141]. The method, combining chemical vapor deposition (CVD), ball milling and space-holder technique, is schematized in Figure 12. The CNTs were grown onto the surface of Al-alloy particles by *in-situ* chemical vapor deposition and the Al/CNTs powders were then used to fabricate the nanocomposites. The authors claim that the proposed method allows CNTs to be well dispersed and integrated in Al powders and fosters good interfacial bonding between CNTs and Al matrix, while morphology and size of pores could be well controlled by the space-holder technique.

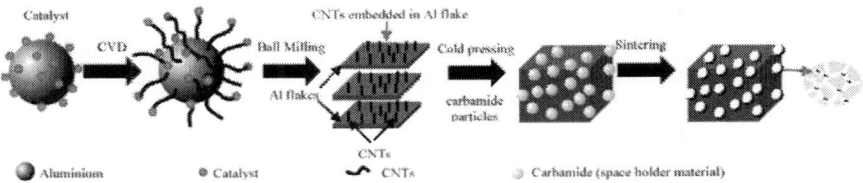

Figure 12. Schematic diagram of the procedures used to fabricate CNT/Al [141].

The SEM images of the resulting nanocomposite foams show some intercommunicating pores derived from the mutual contacts among the carbamide spheres, but also some closed pores. The type of foams obtained (closed cell *vs.* open cell) strongly depends on the added volume fraction of the pore generating agent. Below the percolation threshold, the pores tend to appear separated from each other and closed cell foams are formed. From this perspective, the intercommunicating voids could even be considered as structural imperfections. The authors [141] also investigated the effects of ball milling on the compressive behavior and energy absorption capacity of non-ball-milled and ball-milled composite foams prepared by compacting the Al/CNTs powders without or after being ball milled, respectively. The properties of the resulting nanocomposite foams were compared to the non-reinforced Al-foams (without CNTs) obtained using the conventional space-holder technique. The stress-strain and energy absorption capacity curves reveal that the nanocomposites exhibited typical stress behavior of Al-foams characterized by three regions (linear elastic, plateau, densification). It is clear that ball-milling the Al/CNTs powder is essential for profiting from this *in-situ* chemical vapor deposition approach of growing CNTs onto the surface of Al-particles, as the overall properties of Al-foams made from non-ball-milled nanocomposite powders were worse than those of non-reinforced ones. Besides the superior mechanical properties of ball-milled composite foams, their stress-strain and energy absorption capacity curves are smoother. The compressive yield strength of the ball-milled 2 wt.%-CNT/Al composite foams was reported to be 25% and 67% superior in comparison to pure Al-foams and non-ball-milled nanocomposite foams (with 2 wt.%-CNTs), respectively. The authors postulated that for CNTs to effectively act as reinforcing agents they have to be uniformly distributed and without structural damage for improving the interfacial bonding between the CNTs and Al matrix, and acting as bridges to restrict the cell wall deformations.

For taking full advantages of CNTs, it is vital to understand how they act to reinforce a composite. Fortunately, this issue has been tackled recently by [142]. The researchers used a powder metallurgy route to fabricate an Al-matrix composites reinforced with 0.6 wt.% MWCNTs produced by chemical vapor deposition and performed advanced *in-situ* tensile tests by operating the tensile stage with a CNTs/Al sample inside a FE-SEM chamber. This *in-situ* SEM approach provides a direct and easy method to investigate the mechanical behavior of MWCNTs in composites, which is essentially regulated through a load transfer strengthening mechanism. When a force is applied to the composite, the MWCNTs initially act like a bridge to suppress crack growth. As further force is applied, the outer walls of the nanotubes in contact with the Al matrix start to break. The inner walls then fracture, either breaking vertically or unpeeling to expose the next inner walls, and so on. The SEM images of the completely fractured composite surface showed clear evidences of ruptured MWCNTs.

Several strengthening mechanisms have been already proposed for MWCNTs in metal-matrix composites (MMCs) [142,143,144,145,146,147] including load transfer from matrix to the MWCNTs [146]; grain refining [144] and texture strengthening [145] by pinning effect of MWCNTs; dispersion strengthening of MWCNTs [146]; solution strengthening of carbon atoms [147]; strengthening of in-situ formed or participant carbide particles [147]; and thermal mismatch between MWCNTs and matrix [6]. However, the composite strength might be a synergetic result of several strengthening mechanisms although the specific contribute of each one is not easy to discriminate from these previous reports [142,143,144,145,146,147]. In the parallel work, Chen *et al.* [147] also examined the failure behaviors of MWCNTs (produced by chemical vapor deposition) in an Al metal matrix composite reinforced with 0.6 wt.% MWCNTs in an attempt to shed further light on this issue using the same *in-situ* tensile tests reported elsewhere [147]. The tensile sample was prepared by extrusion and machining. This *in-situ* advanced tensile testing technique enabled them concluding that the mechanical behavior of MWCNTs in composites is essentially controlled by a load transfer strengthening mechanism. There was an effective load transfer between the MWCNTs and the matrix and between the cell walls in the composites during the tensile failure.

Zhendong *et al.* [148] and Wang et.al. [149] patented an innovative method to prepare a foam metal-CNTs and metal-graphene composite materials comprising a metal foam substrate and a graphene film layer positioned on the substrate. The metal-graphene composite foam was prepared by means of electrophoresis. Specifically, the preparation method included the following steps: removing greasy dirt and oxides from the surface of the metal foam substrate, preparing graphene by the oxidation-reduction method, modifying graphene, and performing electrophoretic deposition of graphene onto the surface of the metal foam substrate. Within certain of electromagnetic waveband, the metal-graphene foam composite material has the structural advantages of light weight and porosity, large specific surface area, and good conductivity. On the other hand, the composite material integrates excellent electrical conductivity and high dielectric constant, a capacity of being more conductive to absorbing electromagnetic waves due to a large amount of defects and functional group residues and other properties of the self-made graphene. The composite material was claimed to have excellent electromagnetic shielding performance.

3.4. Metal Foams Reinforced with Short Fibres

Short ceramic (Al_2O_3) fibers are more effective than ceramic particles in enhancing the viscosity of metallic melts due to their high aspect ratio. Therefore, their use as stabilizing agents for the fabrication of metal foams has been attempted. However, only a few reports on this topic are available in open literature.

Liu *et al.* [150,151] fabricated closed-cell Zn-22Al composite foams reinforced with 3 vol.% of short Al_2O_3 fibers by the conventional direct foaming method melt using $CaCO_3$ as blowing agent. Zn-22Al matrix was melted at 590 °C in a graphite crucible in an electric furnace, and then $CaCO_3$ powder as blowing agent was added into the melt under mechanical stirring (900 rpm) for 2 min. The temperature was increased to 700 °C–720 °C and held for several minutes to allow the release gas bubbles from the decomposition of the blowing agent. Finally, the composite foams were cooled down.

The distributions of the short fibers in the composite foams were observed by SEM and the compressive properties of the composite foams were investigated in quasi-static condition. The short Al_2O_3 fibers in the composite foams were mostly distributed in two locations: some uniformly dispersed in cell edges/walls; and others are penetrating through the cells. The Zn-22Al foams reinforced with short Al_2O_3 fibers exhibit higher compressive yield stress and energy absorption capacity than the non-reinforced Zn-22Al alloy foams. Moreover, the compressive curves of composite foams are smoother without any dentate collapse plateau region, and increase more rapidly than those of Zn-22Al alloy foams.

3.5. Metal Foams Reinforced with Spinels

Synthesizing or growing reinforcing nano- or sub-micron particles inside the matrix during manufacturing process through an *in situ* reaction is an alternative approach. Such *in situ* reactions can also be used to make grain refiners in melts. For example, SiO_2 dispersed into the AlMg5 alloy melt form 3.4 vol.% of spinel particles that were sufficient for an efficient foam stabilization [152]. *In-situ* fabrication enables accomplishing homogeneous distribution of the reinforcements and their good wetting by the matrix at lower costs in comparison to the *ex-situ* method.

Guo *et al.* [153] applied to the first time a method for *in-situ* generating the reinforcing elements. $MgAl_2O_3$ nano-wiskers reinforced metallic foams were fabricated via sintering and dissolution processes using sodium chloride particles as a space holder material (Figure 13). Such spinel nano-wiskers (50–300 nm and the aspect ratio of 10–50) are ideal reinforcement for Al composite foams. The $MgAl_2O_3$ spinel whiskers are generated in the cell wall and might exists in three forms: (i) entirely embedding in the cell walls; (ii) partially protruding through the cell walls; (iii) penetrating through the micropores. The improvements brought by the *in-situ* $MgAl_2O_3$ spinel whiskers in compressive properties and energy absorption suggest the usefulness of the method to prepare Al composite foam with excellent properties.

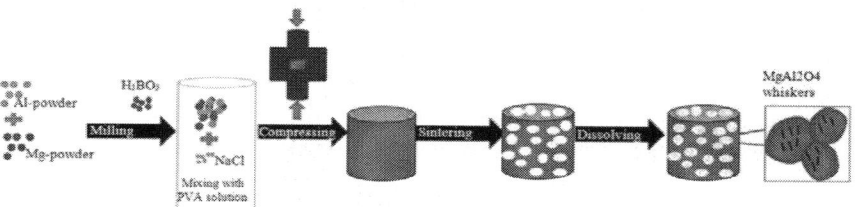

Figure 13. Schematic illustration of procedures to fabricate Al composite foams reinforced by$MgAl_2O_4$ spinel whiskers [153].

4. FUTURE DIRECTIONS

The paper summarizes all important manufacturing routes of MMCs and describes the role of particles in regard of foam stability and mechanical strengthening comprising. As this topic is of great interest for industry and academia—as the number of publications verifies—this review article is indeed a needed task. Besides the summarizing efforts of the already published perceptions of other works, attempts were also made to draw some general conclusions.

The most important functional properties of foams include low density, essential for for light weight applications; specific energy absorption capacity required for crash safety applications; mechanical strength for the most demanding structures under load conditions; acoustic and heat insulation properties capacity for noise and heat management; and refractoriness for fireproof applications. A number of traditional processing approaches of composites and nanocomposite foams are based on both direct and indirect foaming methods. Each method offers several advantages and drawbacks, which need to be balanced according to the intended properties for any specific application. As in any other technological field, the final properties of foams derived from each manufacturing method are strongly dependent of the quality of raw materials, their right combination and the processing details that might require specific skills and knowhow.

There are at least two main methodological approaches for further enhancing the properties of metallic foams, depending on the intended set of final properties: The first one involves exerting a closer control over all the relevant processing steps and experimental parameters of the relatively low density foams manufactured by the traditional methods, including the search for more effective reinforcing agents (particles, carbon nanotubes, graphene, short fibers, etc.) and their uniform distribution in the matrix to maximize the mechanical benefits. Developing experimental methodologies for achieving uniform dispersions of the nano- and micro-sized reinforcing elements and avoiding their strong agglomeration into cluster are some of the main challenges. Achieving such targets and promoting a proper (strong) interfacial bonding to delay premature failure while avoiding interfacial reaction between the reinforcements and the metal melt and the consequent formation of undesirable interfacial products will be the future power research lines in the field. Obtaining uniform dispersion of the nano-sized reinforcing elements is perhaps the biggest challenge. Efforts should be made towards developing reinforced foams exhibiting long flat plateau, having higher yield stresses (close to the stress plateau), and possessing higher specific energy absorbing capacities. The second approach is to further explore the new concept of syntactic foams, especially when the light weight aspect can be sacrificed to favor high demanded mechanical properties. Syntactic foams are less porous but enable a close control of porosity fraction and its spatial distribution, and consequently obtain superior mechanical properties when compared to conventional metallic foams. In this last case, mass production still requires the search for suitable fillers (hollow spheres or porous particles) that should be available at an economically acceptable cost. However, the minimum achievable density and cost of MSFs is limited by the filler material. Reinforcing the metallic matrix of MSFs with nano- and micro-sized reinforcing elements would certainly be a very interesting direction to follow taking advantages of the specificities of the two main approaches described above. In all kinds of foams, a better level of understanding of the strengthening mechanisms is needed to make supported progresses in the field.

REFERENCES

1. Ashby, M.F.; Evans, A.G.; Fleck, N.A.; Gibson, L.J.; Hutchinson, J.W.; Wadley, H.N.G. *Metal Foams. A Design Guide*, 1st ed.; Butterworth Heinemann: Woburn, MA, USA, 2000.

2. Degischer, H.-P.; Kriszt, B. *Handbook of Cellular Metals: Production, Processing, Applications*; Wiley-VCH Verlag GmbH & Co. KgaA: Weinheim, Germany, 2002.

3. Banhart, J. Manufacture, Characterization and Application of Cellular Metals and Metal Foams. *Prog. Mater. Sci.* **2001,***46*, 559–632.

4. Gibson, L.J.; Asbhy, M.F. *Cellular Solids: Structure and Properties*, 2nd ed.; Cambridge University Press: Cambridge, UK, 1997.

5. Banhart, J.; Seeliger, H.-W. Aluminium Foam Sandwich Panels: Manufacture, Metallurgy and Applications. *Adv. Eng. Mater.* **2008**, *10*, 793–802.

6. Lehmhus, D.; Busse, M.; Herrmann, A.S.; Kayvantash, K. *Structural Materials and Processes in Transportation*, 1st ed.; Wiley-VCH Verlag GmbH & Co. KGaA: Weinheim, Germany, 2013.

7. Duarte, I.; Teixeira-Dias, F.; Graça, A.; Ferreira, A.J.M. Failure Modes and Influence of the Quasi-static Deformation Rate on the Mechanical Behavior of Sandwich Panels with Aluminum Foam Cores. *Mech. Adv. Mater. Struct.* **2010**, *17*, 335–342.

8. Duarte, I.; Vesenjak, M.; Krstulović-Opara, L.; Anžel, I.; Ferreira, J.M.F. Manufacturing and bending behaviour of *in situ* foam-filled aluminium alloy tubes. *Mater. Des.* **2015**, *66*, 532–544.

9. Nia, A.A.; Hamedani, J.H. Comparative analysis of energy absorption and deformations of thin walled tubes with various section geometries. *Thin Walled Struct.* **2010**, *48*, 946–954.

10. Duarte, I.; Vesenjak, M.; Krstulović-Opara, L. Static and dynamic axial crush performance of *in-situ* foam-filled tubes.*Compos. Struct.* **2015**, *124*, 128–139.

11. Zhou, J.; Gao, Z.; Cuitino, A.M.; Soboyejo, W.O. Effects of heat treatment on the compressive deformation behavior of open cell aluminum foams. *Mater. Sci. Eng. A* **2004**, *386*, 118–128.

12. Yamada, Y.; Shimojima, K.; Sakaguchi, Y.; Mabuchi, M.; Nakamura, M.; Asahina, T. Effects of heat treatment on compressive properties of AZ91 Mg and SG91A Al foams with open-cell structure. *Mater. Sci. Eng. A* **2000**, *280*, 225–228.

13. Campana, F.; Pilone, D. Effect of heat treatments on the mechanical behavior of aluminium alloy foams. *Scr. Mater.***2009**, *60*, 679–82.

14. Lehmhus, D.; Banhart, J. Properties of heat-treated aluminium foams. *Mater. Sci. Eng. A* **2003**, *349*, 98–110.

15. Xia, X.C.; Chen, X.W.; Zhang, Z.; Chen, X.; Zhao, W.M.; Liao, B.; Hur, B. Effects of porosity and pore size on the compressive properties of closed-cell Mg alloy foam. *J. Magnes. Alloy.* **2013**, *1*, 330–335.

16. Jiang, B.; Wang, Z.; Zhao, N. Effect of pore size and relative density on the mechanical properties of open-cell aluminum foams. *Scr. Mater.* **2007**, *56*, 169–172.

17. Everett, R. *Metal Matrix Composites: Processing and Interfaces*; Elsevier: San Diego, CA, USA, 1991.

18. Clyne, T.W.; Whiters, P.J. *Metal Matrix Composites*; Cambridge University Press: Cambridge, UK, 1993.

19. Casati, R.; Vedani, M. Metal Matrix Composites Reinforced by Nano-Particles—A Review. *Metals* **2014**, *4*, 65–83.

20. Tjong, S. Recent progress in the development and properties of novel metal matrix nanocomposites reinforced with carbon nanotubes and graphene nanosheets. *Mater. Sci. Eng. Rep.* **2013**, *74*, 281–350.

21. Harris, P. Carbon nanotube composites. *Int. Mater. Rev.* **2004**, *49*, 31–43.

22. Agarwal, A.; Bakshi, S.; Lahiri, D. *Carbon Nanotubes-Reinforced Metal Matrix Composites*; Taylor and Francis Group: Boca Raton, FL, USA, 2011.

23. Thostenson, E.T.; Ren, Z.; Chou, T.-W. Advances in the science and technology of carbon nanotubes and their composites: A review. *Compos. Sci. Technol.* **2001**, *61*, 1899–1912.

24. Jin, I.; Kenny, L.D.; Sang, H. Lightweight Foamed Metal and Its Production. International Patent Application WO91/03578, 21 March 1991.

25. Kaptay, G. Interfacial Criteria for ceramic stabilised metallic foams. In Proceedings of the 1st International Conference on Metal Foams and Porous Metal Structures (MetFoam'99), Bremen, Germany, 14–16 June 1999; Banhart, J., Ashby, M.F., Fleck, N.A., Eds.; MIT Verlag: Bremen, Germany, 1999; pp. 141–146.

26. Thomas, M.; Kenny, L.D. Production of Particle-Stabilized Metal Foams. PCT Patent WO 94/017218, 21 January 1994.

27. Zeppelin, F.; Hirscher, M.; Stanzick, H.; Banhart, J. Desorption of hydrogen from blowing agents used for foaming metals. *Compos. Sci. Technol.* **2003**, *63*, 2293–2300.

28. Gui, M.C.; Wang, D.B.; Wu, J.J.; Yuan, G.J.; Li, C.G. Deformation and damping behaviors of foamed Al-Si-SiC$_p$composite. *Mater. Sci. Eng. A* **2000**, *286*, 282–288.

29. Banhart, J. Light-metal foams-history of innovation and technological challenges. *Adv. Eng. Mater.* **2013**, *15*, 82–111.

30. Alizadeh, M.; Mirzaei-Aliabadi, M. Compressive properties and energy absorption behavior of Al-Al$_2$O$_3$ composite foam synthesized by space-holder technique. *Mater. Des.* **2012**, *35*, 419–424.

31. Mahmutyazicioglu, N.; Albayrak, O.; Ipekoglu, M. Effects of alumina (Al₂O₃) addition on the cell structure and mechanical properties of 6061 foams. *J. Mater. Res.* **2013**, *28*, 2509–2519.

32. Daoud, A. Compressive response and energy absorption of foamed A359-Al₂O₃ particle composites. *J. Alloy. Compd.***2009**, *486*, 597–605.

33. Elbir, S.; Yilmaz, S.; Toksoy, A.K.; Guden, M.; Hall, I.W. SiC-particulate aluminum composite foams Produced by powder compacts: Foaming and compression behavior. *J. Mater. Sci.* **2003**, *38*, 4745–4755.

34. Esmaeelzadeh, S.; Simchi, A.; Lehmhus, D. Effect of ceramic particle addition on the foaming behavior, cell structure and mechanical properties of P/M AlSi7 foam. *Mater. Sci. Eng. A* **2006**, *424*, 290–299.

35. Guden, M.; Yuksel, S. SiC-particulate aluminum composite foams produced from powder compacts: Foaming and compression behavior. *J. Mater. Sci.* **2006**, *41*, 4075–4084.

36. Luo, Y.; Yu, S.; Liu, J.; Zhu, X.; Luo, Y. Compressive property and energy absorption characteristic of open-cell. SiCp/AlSi9Mg composite foams. *J. Alloy. Compd.* **2010**, *499*, 227–230.

37. Luo, Y.; Yu, S.; Li, W.; Liu, J.; Wei, M. Compressive behavior of SiCp/AlSi9Mg composite foams. *J. Alloy. Compd.* **2008**,*460*, 294–298.

38. Yu, S.; Liu, J.; Luo, Y. Compressive behavior and damping property of ZA22/ SiCp composite foams. *Mater. Sci. Eng. A* **2007**, *457*, 325–328.

39. Yu, S.; Luo, Y.; Liu, J. Effects of strain rate and SiC particle on the compressive property of SiCp/AlSi9Mg composite foams. *Mater. Sci. Eng. A* **2008**, *487*, 394–399.

40. Prakash, O.; Sang, H.; Embury, J.D. Structure and properties of Al-SiC foam. *Mater. Sci. Eng. A* **1995**, *199*, 195–203.

41. Liu, J.; Yu, S.; Zhu, X.; Wei, M.; Li, S.; Luo, Y.; Liu, Y. Correlation between ceramic additions and compressive properties of Zn-22 Al matrix composite foams. *J. Alloy. Compd.* **2009**, *476*, 220–225.

42. Kennedy, A.R.; Asavavisitchai, S. Effects of TiB₂ particle addition on the expansion, structure and mechanical properties of PM Al foams. *Scr. Mater.* **2004**, *50*, 115–119.

43. Zhao, N.Q.; Jiang, B.; Du, X.W.; Li, J.J.; Shi, C.S.; Zhao, W.X. Effect of Y₂O₃ on the mechanical properties of open-cell aluminium foams. *Mater. Lett.* **2006**, *60*, 1665–1668.

44. Song, Y.H.; Tane, M.; Ide, T.; Seimiya, Y.; Hur, B.Y.; Nakajima, H. Fabrication of Al-3.7 Pct Si-0.18 Pct Mg Foam Strengthened by AlN Particle Dispersion and its Compressive Properties. *Metall. Mater. Trans. A* **2010**, *41*, 2104–2111.

45. Heim, K.; Vinod-Kumar, G.S.; García-Moreno, F.; Rack, A.; Banhart, J. Stabilisation of aluminium foams and films by the joint action of dispersed particles and oxide film. *Acta Mater.* **2015**, *99*, 313–324.

46. Banhart, J. Metal foams: Production and stability. *Adv. Eng. Mater.* **2006**, *8*, 781–794.
47. Baumgärtner, F.; Duarte, I.; Banhart, J. Industrialization of powder compact foaming process. *Adv. Eng. Mater.* **2000**, *2*, 168–174.
48. Duarte, I.; Banhart, J. A study of aluminium foam formation—kinetics and microstructure. *Acta Mater.* **2000**, *48*, 2349–2362.
49. Duarte, I.; Oliveira, M. Chapter 3. Aluminium Alloy Foams: Production and Properties. In *Powder Metallurgy*; Katsuyoshi, K., Ed.; InTech: Rijeka, Croatia, 2012; pp. 47–72.
50. Duarte, I.M.A.; Banhart, J.; Ferreira, A.J.M.; Santos, M.J.G. Foaming around fastening elements. *Mater. Sci. Forum* **2006**, *514–516*, 712–717.
51. Kadoi, K.; Nakae, H. Relationship between Foam Stabilization and Physical Properties of Particles on Aluminum Foam Production. *Mater. Trans.* **2011**, *52*, 1912–1919.
52. Simone, A.E.; Gibson, L.J. Aluminum foams produced by liquid-state processes. *Acta Mater.* **1998**, *46*, 3109–3123.
53. Babcsán, N.; Leitlmeier, D.; Banhart, J. Metal foams—High temperature colloids Part I. Ex situ analysis of metal foams. *Colloid. Surf. A Physicochem. Eng. Asp.* **2005**, *261*, 123–130.
54. Markaki, A.E.; Clyne, T.W. The effect of cell wall microstructure on the deformation and fracture of aluminium-based foams. *Acta Mater.* **2001**, *49*, 1677–1686.
55. Amsterdam, E.; de Hosson, J.T.M.; Onck, P.R. Failure mechanisms of closed-cell aluminum foam under monotonic and cyclic loading. *Acta Mater.* **2006**, *54*, 4465–447.
56. Gergely, V.; Clyne, B. The FORMGRIP Process: Foaming of Reinforced Metals by Gas Release in Precursors. *Adv. Eng. Mater.* **2000**, *2*, 175–178.
57. Huang, L.; Wang, H.; Yang, D.H.; Ye, F.; Lu, Z.P. Effects of scandium additions on mechanical properties of cellular Al-based foams. *Intermetallics* **2012**, *28*, 71–76.
58. Huang, L.; Wang, H.; Ye, F.; Lu, Z.P.; Yang, D.H. Mechanical properties of cellular Al-0.2 wt.% Sc foams. In Proceedings of the 7th International Conference on Porous Metals and Metallic Foams (MetFoam2011), Busan, Korea, 18–21 September 2011; Hur, B.-Y., Kim, B.-K., Kim, S.-E., Hyun, S.-K., Eds.; GS Intervision: Seoul, Korea, 2012; pp. 301–306.
59. Miyoshi, T.; Hara, S.; Mukai, T.; Higashi, K. Development of a Closed-cell Aluminum Alloy Foam with Enhancement of the Compressive Strength. *Mater. Trans.* **2001**, *42*, 2118–2123.
60. Miyoshi, T.; Nishi, S.; Furuta, S.; Hamada, T.; Yoshikawa, K. Current activities and new technologies of aluminium foam by melt route. In Proceedings of the 4th International Conference on Porous Metals and Metal Foaming Technology (MetFoam2005), Kyoto, Japan, 21–23

September 2005; Nakajima, H., Kanetake, N., Eds.; Japan Institute of Metals (JIMIC-4): Sendai, Japan, 2006; pp. 255–260.

61. Shang, X.L.; Zhang, B.; Han, E.H.; Ke, W. The effect of 0.4 wt.% Mn addition on the localized corrosion behaviour of zinc in a long-term experiment. *Electrochimica Acta* **2012**, *65*, 294–304.

62. Hwang, J.Y.; Doty, H.W.; Kaufman, M.J. The effects of Mn additions on the microstructure and mechanical properties of Al–Si–Cu casting alloys. *Mater. Sci. Eng. A* **2008**, *488*, 496–504.

63. Xia, X.; Feng, H.; Zhang, X.; Zhao, W. The compressive properties of closed-cell aluminum foams with different Mn additions. *Mater. Des.* **2013**, *51*, 797–802.

64. Davydov, V.G.; Rostova, T.D.; Zakharov, V.V.; Filatov, Y.A.; Yelagin, V.I. Scientific principles of making an alloying addition of scandium to aluminum alloys. *Mater. Sci. Eng. A* **2000**, *280*, 30–36.

65. Aguirre-Perales, L.Y.; Robin, A.L.; Jung, I-H. The Effect of In Situ Intermetallic Formation on Al-Sn Foaming Behavior. *Metall. Mater. Trans. A* **2014**, *45*, 3714–3727.

66. Yang, D.-H.; Yang, S.-R.; Ma, A.-B.; Jiang, J.-H. Compression properties of cellular AlCu5Mn alloy foams with wide range of porosity. *J. Mater. Sci.* **2009**, *44*, 5552–5556.

67. Aguirre-Perales, L.Y.; Jung, I.-H.; Drew, R.A.L. Foaming Behaviour of Metallurgical Al-Sn Foams. *Acta Mater.* **2012**, *60*, 759–769.

68. Park, S.-H.; Hur, B.-Y.; Kim, S.-Y.; Ahn, D.-K; Ha, D.-I. A study on the viscosity and surface tension for Al foaming and the effects of addition elements. In Proceedings of the 65th World Foundry Congress, Gyeongju, Korea, 20–24 October 2002; Hong, C.-P., Kim, D.-H., Kim, K.-Y., Eds.; pp. 515–523.

69. Gokhale, A.A.; Sahu, S.N.; Kulkarni, V.K.; Sudhakar, B.; Rao, N.R. Materials issues in foaming of liquid aluminium. In Proceedings of the 4th International Conference on Porous Metals and Metal Foaming Technology (MetFoam2005), Kyoto, Japan, 21–23 September 2005; Nakajima, H., Kanetake, N., Eds.; Japan Institute of Metals (JIMIC-4): Sendai, Japan, 2006; pp. 95–100.

70. Suzuki, S.; Murakami, H.; Kadoi, K.; Saiwai, T.; Nakae, H.; Babcsán, N. Aluminum foam fabrication through the melt route by adding Mg and Bi. In Proceedings of the 7th International Conference on Porous Metals and Metallic Foams (MetFoam2011), Busan, Korea, 18–21 September 2011; Hur, B.-Y., Kim, B.-K., Kim, S.-E., Hyun, S.-K., Eds.; GS Intervision: Seoul, Korea, 2012; pp. 3–6.

71. Kim, S.Y.; Kang, K.H.; Um, Y.S.; Hur, B.Y. Foaming characteristics of Al-Mg alloy foam. In Proceedings of the 4th International Conference on Porous Metals and Metal Foaming Technology (MetFoam2005), Kyoto,

Japan, 21–23 September 2005; Nakajima, H., Kanetake, N., Eds.; Japan Institute of Metals (JIMIC-4): Sendai, Japan, 2006; pp. 115–118.

72. Helwig, H.-M.; Garcia-Moreno, F.; Banhart, J. A study of Mg and Cu additions on the foaming behaviour of Al-Si alloys. *J. Mater. Sci.* **2011**, *46*, 5227–5236.

73. Gupta, N.; Rohatgi, P.K. *Metal Matrix Syntactic Foams: Processing, Microstructure, Properties and Applications*; DEStech Publications, Inc.: Lancaster, PA, USA, 2015.

74. Ochsner, A.; Augustin, A. Manufacturing, Properties and Application. In *Multifunctional Metallic Hollow Sphere Structures*; Springer-Verlag: Berlin, Germany, 2009.

75. Rabiei, A.; O'Neill, A.T.; Neville, B.P. Processing and development of a new high strength metal foam. In Prpceedings of the Materials Research Society Symposium-(2004 MRS Fall Meeting & *Exhibit*), Boston, MA, USA, 28 November–2 December 2004; Chipara, M., Edwards, D.L., Benson, R.S., Phillips, S., Eds.; Cambridge University Press: Cambridge, UK, 2005; Volume 851, pp. 517–526.

76. Rohatgi, P.K.; Daoud, A.; Schultz, B.F.; Puri, T. Microstructure and mechanical behavior of die casting AZ91D-fly ash cenosphere composites. *Compos. Part A Appl. Sci. Manuf.* **2009**, *40*, 883–896.

77. Gupta, N.; Luong, D.D.; Cho, K. Magnesium matrix composite foams-density, mechanical properties, and applications. *Metals* **2012**, *2*, 238–252.

78. Luong, D.; Gupta, N.; Rohatgi, P. The high strain rate compressive response of Mg-Al alloy/fly Ash cenosphere composites. *JOM J. Miner. Met. Mater. Soc.* **2011**, *63*, 48–52.

79. Peroni, L.; Scapin, M.; Avalle, M.; Weise, J.; Lehmhus, D. Dynamic mechanical behavior of syntactic iron foams with glass microspheres. *Mater. Sci. Eng. A* **2012**, *552*, 364–375.

80. Peroni, L.; Scapin, M.; Avalle, M.; Weise, J.; Lehmhus, D.; Baumeister, J.; Busse, M. Syntactic iron foams—on deformation mechanisms and strain-rate dependence of compressive properties. *Adv. Eng. Mater.* **2012**, *14*, 909–918.

81. Weise, J.; Baumeister, J.; Yezerska, O.; Salk, N.; Silva, G.B.D. Syntactic iron foams with integrated microglass bubbles produced by means of metal powder injection moulding. *Adv. Eng. Mater.* **2010**, *12*, 604–608.

82. Neville, B.P.; Rabiei, A. Composite metal foams processed through powder metallurgy. *Mater. Des.* **2008**, *29*, 388–396.

83. Castro, G.; Nutt, S.R. Synthesis of syntactic steel foam using mechanical pressure infiltration. *Mater. Sci. Eng. A* **2012**, *535*, 274–280.

84. Castro, G.; Nutt, S.R. Synthesis of syntactic steel foam using gravity-fed infiltration. *Mater. Sci. Eng. A* **2012**, *553*, 89–95.

85. Weise, J.; Lehmhus, D.; Baumeister, J.; Kun, R.; Bayoumi, M.; Busse, M. Production and properties of 316L stainless steel cellular materials and syntactic foams. *Steel Res. Int.* **2014**, *85*, 486–497.

86. Peroni, L.; Scapin, M.; Fichera, C.; Lehmhus, D.; Weise, J.; Baumeister, J.; Avalle, M. Investigation of the mechanical behavior of AISI 316L stainless steel syntactic foams at different strain-rates. *Compos. Part B Eng.* **2014**, *66*, 430–442.

87. Brown, J.; Vendra, L.; Rabiei, A. Bending properties of Al-steel and steel-steel composite metal foams. *Metall. Mater. Trans. A* **2010**, *41*, 2784–2793.

88. Vendra, L.; Neville, B.; Rabiei, A. Fatigue in aluminum-steel and steel-steel composite foams. *Mater. Sci. Eng. A* **2009**, *517*, 146–153.

89. Weise, J.; Salk, N.; Jehring, U.; Baumeister, J.; Lehmhus, D.; Bayoumi, M.A. Influence of powder size on production parameters and properties of syntactic invar foams produced by means of metal powder injection moulding. *Adv. Eng. Mater.* **2013**, *15*, 118–122.

90. Luong, D.D.; Shunmugasamy, V.C.; Gupta, N.; Lehmhus, D.; Weise, J.; Baumeister, J. Quasi-static and high strain rates compressive response of iron and Invar matrix syntactic foams. *Mater. Des.* **2015**, *66*, 516–531.

91. Xue, X.; Zhao, Y. Ti matrix syntactic foam fabricated by powder metallurgy: Particle breakage and elastic modulus. *JOM J. Miner. Met. Mater. Soc.* **2011**, *63*, 43–47.

92. Mondal, D.P.; Datta-Majumder, J.; Jha, N.; Badkul, A.; Patel, A.; Gupta, G. Titanium cenosphere syntactic foam made through powder metallurgy route. *Mater. Des.* **2012**, *34*, 82–89.

93. Licitra, L.; Luong, D.D.; Strbik, O.M., III; Gupta, N. Dynamic properties of alumina hollow particle filled aluminum alloy A356 matrix syntactic foams. *Mater. Des.* **2015**, *66*, 504–515.

94. Luong, D.D.; Strbik, O.M., III; Hammond, V.H.; Gupta, N.; Cho, K. Development of high performance lightweight aluminum alloy/SiC hollow sphere syntactic foams and compressive characterization at quasi-static and high strain rates. *J. Alloy. Compd.* **2013**, *550*, 412–422.

95. Cox, J.; Luong, D.D.; Shunmugasamy, V.C.; Gupta, N.; Strbik, O.M.; Cho, K. Dynamic and thermal properties of aluminumalloyA356/silicon carbide hollow particle syntactic foams. *Metals* **2014**, *4*, 530–548.

96. Balch, D.K.; O'Dwyer, J.G.; Davis, G.R.; Cady, C.M.; GrayIII, G.T.; Dunand, D.C. Plasticity and damage in aluminum syntactic foams deformed under dynamic and quasi-static conditions. *Mater. Sci. Eng. A* **2005**, *391*, 408–417.

97. Orbulov, I.N.; Májlinger, K. Compressive properties of metal matrix syntactic foams in free and constrained compression. *J. Miner. Met. Mater. Soc.* **2014**, *66*, 882–891.

98. Goel, M.D.; Peroni, M.; Solomos, G.; Mondal, D.P.; Matsagar, V.A.; Gupta, A.K.; Larcher, M.; Marburg, S. Dynamic compression behavior of cenosphere aluminum alloy syntactic foam. *Mater. Des.* **2012**, *42*, 418–423.

99. Sulong, M.A.; Taherisharsgh, M.; Belova, I.V.; Murch, G.E.; Fiedler, T On the mechanical anisotropy of the compressive properties of aluminium perlite syntactic foam. *Comput. Mater. Sci.* **2015**, *109*, 258–265.

100. Fiedler, T.; Taherishargh, M.; Krstulović-Opara, L.; Vesenjak, M. Dynamic compressive loading of expanded perlite/aluminum syntactic foam. *Mater. Sci. Eng. A* **2015**, *626*, 296–304.

101. Taherishargh, M.; Belova, I.V.; Murch, G.E.; Fiedler, T. On the mechanical properties of heat-treated expanded perlite-aluminium syntactic foam. *Mater. Des.* **2014**, *63*, 375–383.

102. Mondal, D.P.; Jha, N.; Gull, B.; Das, S.; Badkul, A. Microarchitecture and compressive deformation behavior of Al-alloy (LM13)—Cenosphere hybrid Al foam prepared using $CaCO_3$ as foaming agent. *Mater. Sci. Eng. A* **2013**, *560*, 601–610.

103. Guo, R.Q.; Rohatgi, P.K.; Nath, D. Preparation of aluminium-fly ash particulate composite by powder metallurgy technique. *J. Mater. Sci.* **1997**, *32*, 3971–3974.

104. Ramachandra, M.; Radhakrishna, K. Synthesis-microstructure-mechanical properties-wear and corrosion behavior of an Al-Si (12%)-fly ash metal matrix composite. *J. Mater. Sci.* **2005**, *40*, 5989–5997.

105. Tao, X.F.; Zhang, L.P.; Zhao, Y.Y. Al matrix syntactic foam fabricated with bimodal ceramic microspheres. *Mater. Des.* **2009**, *30*, 2732–2736.

106. Kiser, M.; He, M.Y.; Zek, F.W. The mechanical response of ceramic microballoon reinforced aluminum matrix composites under compressive loading. *Acta Mater.* **1999**, *47*, 2685–2694.

107. Xia, X.; Chen, X.; Zhang, Z.; Chen, X.; Zhao, W.; Liao, B.; Hur, B. Compressive properties of closed-cell aluminum foams with different contents of ceramic microspheres. *Mater. Des.* **2014**, *56*, 353–358.

108. Orbulov, I.N. Compressive properties of aluminium matrix syntactic foams. *Mater. Sci. Eng. A* **2012**, *555*, 52–56.

109. Tao, X.F.; Zhao, Y.Y. Compressive behavior of Al matrix syntactic foams toughened with Al particles. *Scr. Mater.* **2009**, *61*, 461–464.

110. Dou, Z.Y.; Jiang, L.T.; Wu, G.H.; Zhang, Q.; Xiu, Z.Y.; Chen, G.Q. High strain rate compression of cenosphere-pure aluminium syntactic foams. *Scr. Mater.* **2007**, *57*, 945–948.

111. Szlancsik, A.; Katona, B.; Bobor, K.; Májlinger, K.; Orbulov, I.N. Compressive behavior of aluminum matrix syntatic foams reinforced by iron hollow spheres. *Mater. Des.* **2015**, *83*, 230–237.

112. Vendra, L.J.; Rabiei, A. A study on aluminum-steel composite metal foam processed by casting. *Mater. Sci. Eng. A* **2007**, *465*, 59–67.

113. Rabiei, A.; O'Neill, A. A study on processing of a composite metal foam via casting. *Mater. Sci. Eng. A* **2005**, *404*, 159–164.

114. Rabiei, A.; Vendra, L.J. A comparison of composite metal foam's properties and other comparable metal foams. *Mater. Lett.* **2009**, *63*, 533–536.

115. Vendra, L; Rabiei, A. Evaluation of modulus of elasticity of composite metal foams by experimentaland numerical techniques. *Mater. Sci. Eng. A* **2010**, *527*, 1784–1790.

116. Rabiei, A.; Garcia-Avila, M. Effect of various parameters on properties of composite steel foams under variety of loading rates. *Mater. Sci. Eng. A* **2013**, *564*, 539–547.

117. Taherishargh, M.; Belova, I.V.; Murch, G.E.; Fiedler, T. Pumice/aluminum syntactic. *Mater. Sci. Eng. A* **2015**, *635*, 102–108.

118. Chen, S.; Bourham, M.; Rabiei, A. Neutrons attenuation on composite metal foams and hybrid open-cell Al foam.*Radiat. Phys. Chem.* **2015**, *109*, 27–39.

119. Garcia-Avila, M.; Portanova, M.; Rabiei, A. Ballistic performance of composite metal foams. *Compos. Struct.* **2015**, *125*, 202–211.

120. Chen, S.; Bourham, M.; Rabiei, A. Attenuation efficiency of X-ray and comparison to gamma ray and neutrons in composite metal foams. *Radiat. Phys. Chem.* **2015**, *117*, 12–22.

121. Wu, G.H.; Dou, Z.Y.; Sun, D.L.; Jiang, L.T.; Ding, B.S.; He, B.F. Compression behaviors of cenosphere-pure aluminum syntactic foams. *Scr. Mater.* **2007**, *56*, 221–224.

122. Rohatgi, P.K.; Kim, J.K.; Gupta, N.; Alaraj, S.; Daoud, A. Compressive characteristics of A356/fly ash cenosphere composites synthesized by pressure infiltration technique. *Compos. Part A Appl. Sci. Manuf.* **2006**, *37*, 430–437.

123. Orbulov, I.N.; Ginsztler, J. Compressive behaviour of metal matrix syntactic foams. *Acta Polytech. Hung.* **2012**, *9*, 43–56.

124. Zhang, L.P.; Zhao, Y.Y. Mechanical Response of Al Matrix Syntactic Foams Produced by Pressure Infiltration Casting.*J. Compos. Mater.* **2007**, *41*, 2105–2117.

125. Prabhu, B.; Suryanarayana, C.; An, L.; Vaidyanathan, R. Synthesis and characterization of high volume fraction Al-Al_2O_3 nanocomposite powders by high-energy milling. *Mater. Sci. Eng. A* **2006**, *425*, 192–200.

126. Du, Y.; Li, A.B.; Zhang, X.X.; Tan, Z.B.; Su, R.Z.; Pu, F.; Geng, L. Enhancement of the mechanical strength of aluminum foams by SiC nanoparticles. *Mater. Lett.* **2015**, *148*, 79–81.

127. Casati, R.; Vedani, M. Metal matrix composites reinforced by nanoparticles—A review. *Metals* **2014**, *4*, 65–83.

128. Casati, R.; Fabrizi, A.; Timelli, G.; Tuissi, A.; Vedani, M. Microstructural and Mechanical Properties of Al-Based Composites Reinforced with In-Situ and Ex-Situ Al_2O_3 Nanoparticles. *Adv. Eng. Mater.* **2015**.

129. Casati, R.; Fabrizi, A.; Tuissi, A.; Xia, K.; Vedani, M. ECAP consolidation of Al matrix composites reinforced with in-situ γ-Al_2O_3 nanoparticles. *Mater. Sci. Eng. A* **2015**, *648*, 113–122.

130. Boonyongmaneerat, Y.; Schuh, C.A.; Dunand, D.C. Mechanical properties of reticulated aluminum foams with electrodeposited Ni-W coatings. *Scr. Mater.* **2008**, *59*, 336–339.

131. Bouwhuis, B.A.; McCrea, J.L.; Palumbo, G.; Hibbard, G.D. Mechanical properties of hybrid nanocrystalline metal foams. *Acta Mater.* **2009**, *57*, 4046–4053.

132. Wang, W.; Burgueño, R.; Hong, J.-W.; Lee, I. Nano-deposition on 3-d open-cell aluminium foam materials for improved energy absorption capacity. *Mater. Sci. Eng. A* **2013**, *572*, 75–82.

133. Sun, Y.; Burgueño, R.; Wang, W.; Lee, I. Effect of annealing on the mechanical properties of nano-copper reinforced open-cell aluminum foams. *Mater. Sci. Eng. A* **2014**, *613*, 340–351.

134. Sun, Y.; Burgueño, R.; Vanderklok, A.J.; Tekalur, S.A.; Wang, W.; Lee, I. Compressive behaviour of aluminum/copper hybrid foams under high strain rate loading. *Mater. Sci. Eng. A* **2014**, *592*, 111–120.

135. Antenucci, A.; Guarino, S.; Tagliaferri, V.; Ucciardello, N. Improvement of the mechanical and thermal characteristics of open-cell aluminum foams by the electrodeposition of Copper. *Mater. Des.* **2014**, *59*, 124–129.

136. Devivier, C.; Tagliaferri, V.; Trovalusci, F.; Ucciardello, N. Mechanical characterization of open-cell aluminium foams reinforced by nickel electro-deposition. *Mater. Des.* **2015**, *86*, 272–278.

137. Jung, A.; Chen, Z.; Schmauch, J.; Motz, C.; Diebels, S. Micromechanical characterisation of Ni/Al hybrid foams by nano- and microindentation coupled with EBSD. *Acta Mater.* **2016**, *102*, 38–48.

138. Duarte, I; Ventura, E.; Olhero, S.; Ferreira, J.M.F. A novel approach to prepare aluminium-alloy foams reinforced by carbon-nanotubes. *Mater. Lett.* **2015**, *160*, 162–166.

139. Duarte, I.; Ventura, E.; Olhero, S.; Ferreira, J.M.F. An effective approach to reinforced closed-cell Al-alloy foams with multiwalled carbon nanotubes. *Carbon* **2015**, *95*, 589–600.

140. Zhang, Z; Ding, J.; Xia, X.; Sun, X.; Song, K.; Zhao, W.; Liao, B. Fabrication and characterisation of closed-cell aluminium foams with differents contents of multi-walled carbon nanotubes. *Mater. Des.* **2015**, *88*, 359–365.

141. Wang, J.; Yang, X.; Zhang, M.; Li, J.; Shi, C.; Zhao, N.; Zou, T. A novel approach to obtain in-situ growth carbon nanotube reinforced aluminum foams with enhanced properties. *Mater. Lett.* **2015**, *161*, 763–766.

142. George, R.; Kashyap, K.T.; Rahul, R.; Yamdagni, S. Strengthening in carbon nanotube/aluminium (CNT/Al) composites. *Scr. Mater.* **2005**, *53*, 1159–1163.

143. Nam, D.H.; Cha, S.I.; Lim, B.K.; Park, H.M.; Han, D.S.; Hong, S.H. Synergistic strengthening by load transfer mechanism and grain refinement of CNT/Al-Cu composites. *Carbon* **2012**, *50*, 2417–2423.

144. Wei, H.; Li, Z.; Xiong, D.-B.; Tan, Z.; Fan, G.; Qin, Z.; Zhang, D. Towards strong and stiff carbon nanotube-reinforced high-strength aluminum alloy composites through a microlaminated architecture design. *Scr. Mater.* **2014**, *75*, 30–33.

145. Yoo, S.J.; Han, S.H.; Kim, W.J. Strength and strain hardening of aluminum matrix composites with randomly dispersed nanometer-length fragmented carbon nanotubes. *Scr. Mater.* **2013**, *68*, 711–714.

146. Li, S.; Sun, B.; Imai, H.; Kondoh, K. Powder metallurgy Ti-TiC metal matrix composites prepared by *in situ* reactive processing of Ti-VGCFs system. *Carbon* **2013**, *61*, 216–228.

147. Chen, B.; Li, S.; Imai, H.; Jia, L.; Umeda, J.; Takahashi, M.; Kondoh, K. Load transfer strengthening in carbon nanotubes reinforced metal matrix composites via *in-situ* tensile tests. *Compos. Sci. Technol.* **2015**, *113*, 1–8.

148. Dong, Z.; Keju, J.; Huihui, Z. Foam Metal-Carbon Nanotube Composite Material and Preparation Method Thereof. Patent CN 103434207 A, 11 December 2013.

149. Wang, Y.; Chin, Y.-H.; Gao, Y.; Aardahl, C.L.; Stewart, T.L. Carbon Nanotube-Containing Structures, Methods of Making, and Processes Using Same. Patent US 7011760 B2, 14 March 2006.

150. Liu, J.; Yu, S.; Zhu, X.; Wei, M.; Li, S.; Luo, Y.; Liu, Y. Effect of Al_2O_3 short fiber on the compressive properties of Zn-22Al foams. *Mater. Lett.* **2008**, *62*, 3636–3638.

151. Liu, J.; Yu, S.; Zhu, X.; Wei, M.; Li, S.; Luo, Y.; Liu, Y. Deformation and energy absorption characteristic of Al_2O_{3f}/Zn-Al composite foams during compression. *J. Alloy. Compd.* **2010**, *506*, 620–625.

152. Kumar, G.S.; Chakraborty, M.; Garcia-Moreno, F.; Banhart, J. Foamability of $MgAl_2O_4$ (Spinel)-reinforced aluminum alloy composites. *Metall. Mater. Trans. A* **2011**, *42*, 2898–2908.

153. Guo, C.; Zou, T.; Shi, C.; Yang, X.; Zhao, N.; Liu, E.; He, C. Compressive properties and energy absorption of aluminum composite foams reinforced by in-situ generated $MgAl_2O_4$ whiskers. *Mater. Sci. Eng. A* **2015**, *645*, 1–7.

CHAPTER 8

Bond Strength of Composite CFRP Reinforcing Bars in Timber

Marco Corradi [1,3,], Luca Righetti [2] and Antonio Borri [3]*

[1] Department of Mechanical and Construction Engineering, Northumbria University, 212 Wynne-Jones Building, Newcastle upon Tyne NE1 8ST, UK
[2] Department of Mechanical and Construction Engineering, Northumbria University, 209 Wynne-Jones Building, Newcastle upon Tyne NE1 8ST, UK
[3] Department of Engineering, University of Perugia, 92 Via Duranti, Perugia 06125, Italy

ABSTRACT

The use of near-surface mounted (NSM) fibre-reinforced polymer (FRP) bars is an interesting method for increasing the shear and flexural strength of existing timber members. This article examines the behaviour of carbon FRP (CFRP) bars in timber under direct pull-out conditions. The objective of this experimental program is to investigate the bond strength between composite bars and timber: bars were epoxied into small notches made into chestnut and fir wood members using a commercially-available epoxy system. Bonded lengths varied from 150 to 300 mm. Failure modes, stress and strain distributions and the bond strength of CFRP bars have been evaluated and discussed. The pull-out capacity in NSM CFRP bars at the onset of debonding increased with bonded length up to a length of 250 mm. While CFRP bar's pull-out was achieved only for specimens with bonded lengths of 150 and 200 mm, bar tensile failure was mainly recorded for bonded lengths of 250 and 300 mm.

Keywords: timber; composite materials; carbon fibre; epoxy resin; bonding; testing

1. INTRODUCTION

With increasing focus on the development of sustainable construction systems, the reinforcement of existing wood members and the use of timber in new constructions is at present receiving much attention.

In order to upsurge the useful life of wood structural elements, it is necessary to afford suitable retrofitting techniques. Timber, when used as a structural material, is constantly exposed to several agents of deterioration (insects assault, moisture variation, aging, biological attack, *etc.*), which reduce the strength and stiffness, and reinforcement interventions are often necessary to increase the capacity or to reduce flexural deflections. Many innovative techniques are available in the literature, which consider the use of traditional materials, such aluminium and steel rods, composite materials, such as carbon and glass fibre-reinforced polymers (FRPs), and, more recently, natural-based composite materials, such basalt FRP. In many cases these reinforcements are in the form of rods or bars.

In the first studies, the use of steel rods [1,2] glued on glulam beams produced interesting increases in capacity and stiffness. Recent research programs involving the reinforcement of wood beams have also examined the use of FRP bars applied on the tension side. FRP materials have excellent mechanical properties and exhibit very good characteristics in relation to long-term behaviour [3,4]. FRPs have been employed either to improve flexural and shear characteristics of existing structures or to reduce the dimension of new timber structures. Currently, the restricted data are accessible on the bond behaviour of FRP rods in timber, and design guidelines provided in Eurocodes and national standards for steel bars cannot be properly used for this purpose due to essential differences in surfaces deformations and mechanical properties.

Bars are usually glued into grooves realized along the direction parallel to the beams' fibres using epoxy resins. The use of near-surface mounted (NSM) FRP bars as a replacement for steel has been encouraged, because it implicates higher mechanical properties, ease of application, a high stiffness-to-weight ratio (10- to 15-times higher than the steel) and better long-term behaviour. Several studies [3,4] have been carried out in order to analyse the mechanical characteristics of FRP bars, in particular their tensile stress, Young's modulus, ultimate strain and creep behaviour, and results confirmed the good properties of the material that could constitute a suitable solution instead of the steel bar to strengthen concrete [5,6,7], masonry [8,9] and timber members [10,11,12,13].

Thus, current research on wood reinforcement has focused on the use of FRP strips or bars epoxy bonded to wood solid or glulam beams. However the response of the interface bond timber-epoxy-bar under loading is not yet fully defined, and additional information is needed to develop specifications and design values for reinforcement of timber with FRP. Gentile*et al.* [14] carried out several bending tests on a large number of half-scale timber beams reinforced with different diameter glass fibre-reinforced polymer (GFRP) bars, bonded with epoxy resin inside grooves realized in different position on the tensile surface. The results showed an increase of ultimate strength between 18% and 46% in the reinforced beams. Borri *et al.* [15] used carbon FRP (CFRP) bars, applied on the tensile side using epoxy resins, to reinforce solid timber beams, which produced an increase of the capacity up to 52%. GFRP and CFRP rods have been also used, with encouraging results, also for the reinforcement of glulam beams [16,17,18,19,20].

However, the increasing production costs of FRPs are significantly widening the field of research, especially toward natural materials, which readily available and considerably more economical. Recently, Raftery *et al.* [21] tested bending timber beams strengthened with basalt fibre-reinforced polymer (BFRP) bars bonded in notches realized on the tensile zone of Irish spruce beams exhibiting an ultimate capacity over 23% of the unreinforced beams. Several pull-out tests were carried out to investigate the bond capacity of FRP bars glued in timber specimens [22,23,24,25], in particular used for the connection of timber element. The results showed that the main failure mode was longitudinal splitting and pull-out of the rods along with a timber volume surrounding the bonded length; however, capacity increases with the growth of bonded length and notched size.

The above cited experimental results show that a debonding failure of the FRP may occur in some cases because of the push-off of the split timber near the beam midspan. However, it is unclear if this depends on the grade of the timber material, on the type of resin used to apply the FRP bars or on the position of the bar reinforcement. From recent studies on the behaviour of bonded FRP bars to solid and glulam wood, it is apparent that the problem is quite complicated, both experimentally and analytically, and more experimental data are necessary to address the problem. The bond behaviour of different wood species is another aspect to consider: in Northern Europe, the common use of faster-grown wood species, mainly softwood (fir, larch, *etc.*), which produce lower grade timber, can particularly benefit from the reinforcement with FRP bars, but it may determine problems at the timber-epoxy interface, compared to the use of hardwood (oak, chestnut, *etc.*).

This article examines the behaviour of carbon FRP bars in soft (fir) and hardwood (chestnut) under direct pull-out conditions. Results were previously partially presented in [26]. The objective of this experimental program is to investigate the bond strength between composite bars and timber: bars were epoxied into small notches made into chestnut and fir wood members using a commercially-available epoxy system with bonded lengths varying from 150 to 300 mm.

There are many different experimental setups for determining the bond behaviour of the FRP substrate, amongst which single shear tests, double shear pull and push tests and shear bending tests are the most common (Figure 1). Since FRP bars are usually applied for flexural strengthening of beams, bending creates a tension zone; the stresses in the bond line between FRP and timber are much more complex than when pure tension tests are used. Thus, bending bond tests are more likely to represent the actual conditions than the direct pull-out tests, but significant limitations are also present for this setup [27].

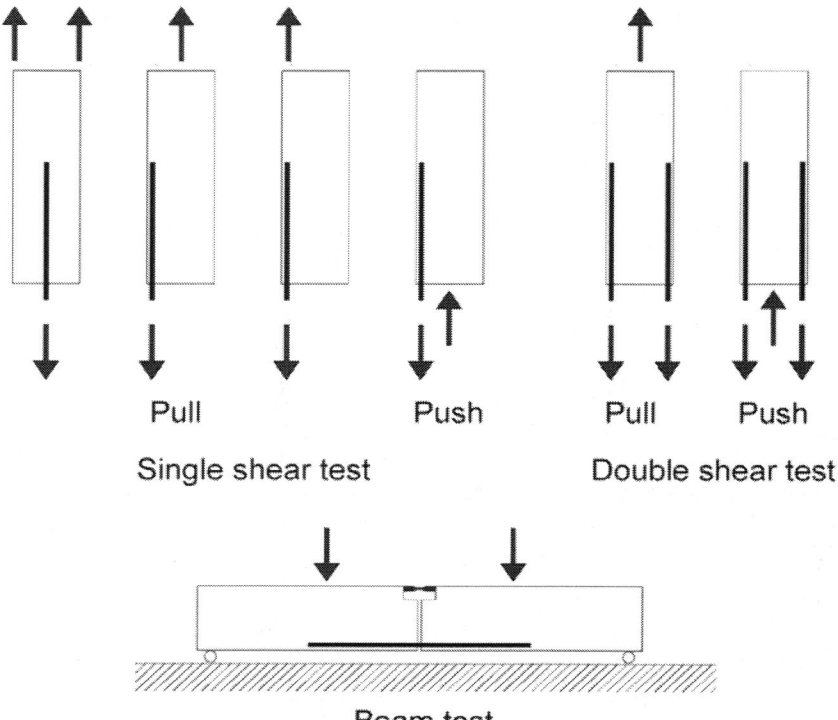

Figure 1. Different test setups.

In this experimental work, the double shear push test has been used as a result of its simplicity. However, it must be pointed out that numerical and experimental investigations have demonstrated that different test setups can produce different results, and small variations in setup may have significant effects [28].

2. MATERIALS

2.1. Timber

Tests were carried out on prism specimens in fir (*Abies alba*) (Figure 2a) and chestnut wood (*Castanea sativa*) (Figure 2b). Specimen dimensions were 200 mm × 200 mm × 500 mm and 220 mm × 220 mm × 500 mm with a density of 453.6 and 448 kg/m³ for fir and chestnut wood, respectively. Moisture content was evaluated according to the EN 13183-1 standard [29]; the average value was 10.9% for fir and 17.12% for chestnut wood. Two notches, with cross-section of 14 mm × 15 mm, were realized into the side surfaces of the timber specimens, parallel to the grain, with a circular saw with different lengths (150, 200, 250 and 300 mm).

The timber material used in this experimental campaign was classified according to the EN 338 standard [30] in C24 and D24, according to [31], for fir and chestnut wood, respectively. In order to verify the quality of the timber, four-point-bending tests were carried out on four timber beams (two for each type of wood). The fir and chestnut wood bending strengths were 32.55 and 34 N/mm², respectively. The limited number of characterization tests performed must be considered given the high natural variability of timber and the local character of the bonding tests.

(a)   **(b)**

Figure 2. (a) Softwood prism (fir); (b) hardwood prism (chestnut).

2.2. CFRP Bars

Tests were performed to characterize the mechanical properties of the CFRP materials used in this investigation. CFRP unidirectional pultruded bars (Figure 3) were produced by MAC SpA (a product commercially known as "Leonardo"). NSM rods were 7.5 mm-diameter CFRP bars having a sandblasted surface to improve the bond characteristics and a deformed, helically-wound surface produced by fibre wraps. These rods are comparable to steel bars used in reinforced concrete in nominal dimensions; nevertheless, the main difference is the pultrusion process by which the bars are manufactured and the stress-strain behaviour. The pultrusion process produces bars with a reasonably constant cross-section. In the experimental campaign, bars with indentations were used; those were realized by wrapping a carbon fibre string around the bar before the resin dried. The stress-strain behaviour of CFRP bars is linear at all stress levels up to collapse, without showing any yielding behaviour.

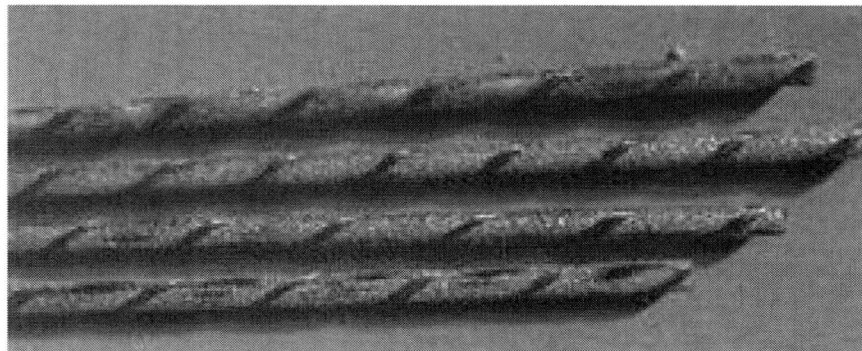

Figure 3. Carbon fibre-reinforced polymer (CFRP) bars: bars are sandblasted and superficially deformed.

The mechanical characterization of the CFRP bars was conducted on nine coupons tested in tension according to the ASTM D3039 standards [32], with a crosshead speed of 1 mm/min (displacement control mode). To avoid damaging the CFRP bars by the compression stresses introduced by the loading shoes of the test machine, the end of each specimen was inserted into a steel cylindrical pipe and fixed with epoxy resin. Test results are shown in Table 1.

Table 1. Mechanical properties of CFRP bars.

Nominal Diameter (mm)	Number of Specimens	Failure Load (kN)	Tensile Strength (N/mm^2)	Strain at Failure (%)	Modulus of Elasticity E_b (N/mm^2)
7.5	9	46.49 (1.82)	1053 (41.22)	0.69	151030

In parentheses: Standard deviation.

2.3. Epoxy System

The epoxy system is constituted of two epoxy components: primer and saturant. Both of them are bi-component epoxy resins with a weight ratio of epoxy-to-curing agent of 3:1. The primer was initially applied on the wood surface to facilitate bonding. Notches were then filled with the saturant resin. Both components were mixed in ratio of 3:1 by volume and cured for 10 days at room temperature. The system was manufactured by MAC SpA. Mechanical characteristics of the saturant and primer were evaluated by testing 5 specimens in compression according to the ASTM D695 standard [33] and 5 specimens in tension according to the ASTM D638 standard [34]. Test results are shown in Table 2.

Table 2. Mechanical properties of the epoxy system.

	Saturant (N/mm²)	Primer (N/mm²)
Sample size	5	5
Compression strength	56.54	26.15
Sample size	5	5
Tensile strength	23.43	12.69
Modulus of elasticity	4510	426

Microscopic analysis has been performed on chestnut wood samples. The primer and saturant were strained on the samples in order to analyse the penetration of the two resins into the wood. Microscopic analysis showed that the penetration of the primer in the chestnut wood was about 50 μm (Figure 4a), but that the penetration of the saturant was completely negligible (Figure 4b). This explains the necessity to use the epoxy primer, whose application was made by a brush, with a bond-line thickness varying between 0.02 and 0.06 mm [35].

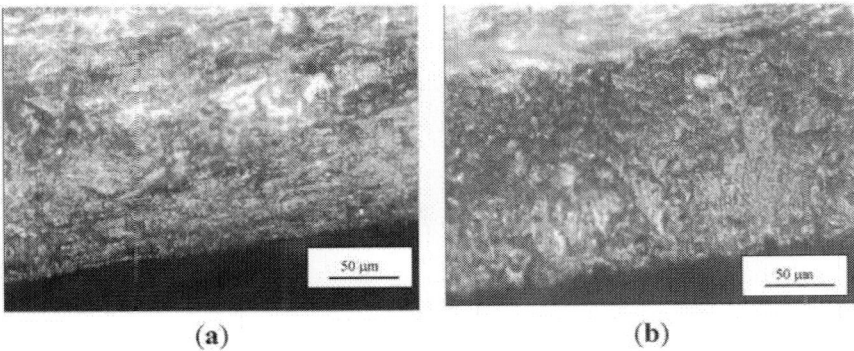

(a)   (b)

Figure 4. Microscopic analysis of the bonding: (a) primer-wood interface; (b) saturant-wood interface.

3. TEST SETUP

Twelve fir (Figure 5a) and twelve chestnut wood pull-out specimens (Figure 5b) have been obtained joining two square-section timber prisms with two unidirectional NSM CFRP bars positioned inside notches and secured in place using the epoxy system. The total number of pull-out specimens was 24. Fir and chestnut wood specimens are identified by the letters "SF" and "SC", respectively.

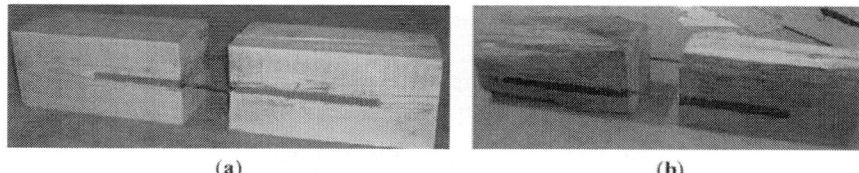

Figure 5. (a) Fir wood specimen; (b) chestnut wood specimen.

Tests were divided into two categories. The first category of pull-out tests investigated the effect of different bonded lengths. Figure 6a shows the test schematic arrangement with bonded-in CFRP bars. Reinforcement bars were tested with bonded lengths of 150, 200, 250 and 300 mm. The pull-out specimens had a uniform rectangular notch size of 14 mm (width) × 15 mm (height) (Figure 6b). Based on the dimensions of the notch and the CFRP bar diameter, the glue-line thickness is approximately 3.5 mm: this value, relatively thick, was chosen to facilitate the penetration of the epoxy resin into the notch and to improve the bonding strength [35].

The curing time before testing was 15 days at room temperature. A mutual distance of 115 mm between the two prismatic elements was used to allow the allocation of the test equipment (hydraulic jack, steel plates).

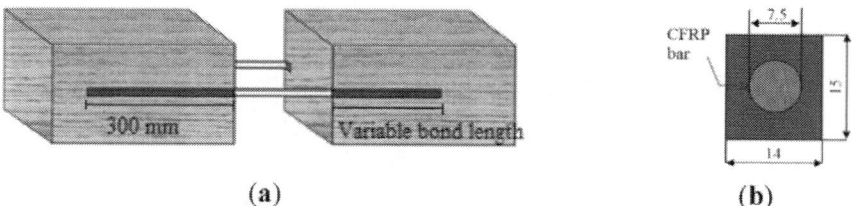

Figure 6. (a) CFRP bar schematic arrangement; (b) section of the notch (dimensions in mm).

4. EXPERIMENTAL CAMPAIGN

Twenty-four specimens were tested to study the bonding behaviour of the CFRP bars with timber. The analysed parameters were the failure mode, maximum load (pull-out capacity), average bond stress on the lateral bar's surface and on the interface between bar and epoxy resin. Furthermore, strain gauges were applied on the bar surface of four of the above specimens (SF_2, SF_4, SC_2 and SC_3), to evaluate the stress distribution along the CFRP bars.

Timber species and bonded length were the test variables considered in this work. Tests were carried out placing between the prismatic wood elements an Enerpac 20-ton hydraulic jack (Figure 7) with a stroke of 50 mm actuated by a 700-bar manual pump; the pushing cylinder of the jack was placed in contact with the timber prism surfaces. To avoid the crushing of the timber and shear along the grain, due to the application of the jack on a small surface, two square (110 mm × 110 mm) bearing steel plates were inserted between the jack and the timber specimen (Figure 7). While preventing timber local failure, the bearing plates had a clear distance from the notches to allow for timber failure near the epoxy-wood interface. The gradient of the pressure manually applied with the pump to the specimen was approximately 3 bar/s.

Figure 7. Pull-out test arrangement (tests without strain gauges).

4.1. Tests without Strain Gauges

The objective of these pull-out tests was to measure the bond strength for different bonded lengths. Twelve fir wood and twelve chestnut wood specimens were tested to investigate the behaviour of the CFRP bars epoxied into timber elements. Tensile stress σ in the bar could be computed from the externally-applied load using equilibrium. By assuming an equal distribution of the load between the CFRP bars and that shear stress was constant over the bar-epoxy interface, the bond strength τ_b of the single bar can be easily determined by dividing the poll-out load by the lateral bar surface.

Three different failure modes were recorded during the tests: (1) pull-out of the CFRP reinforcement from the epoxy substrate; (2) timber shear failure; and (3) tensile failure of the CFRP bars.

In detail, the first mode was characterized by the bar-epoxy interface failure and the subsequent pull-out of the CFRP bar from the epoxy resin (Figure 8a). This was the most frequent mode of failure for bonded lengths of 150 and 200 mm. The longitudinal micro-cracking that appeared was due to the compressive forces radiating out in an inclination that varies with rib surface. This cracking slowly propagated up with the increase of the load until a noticeable cracking of the epoxy resin and the subsequent slipping out of the bar occurred.

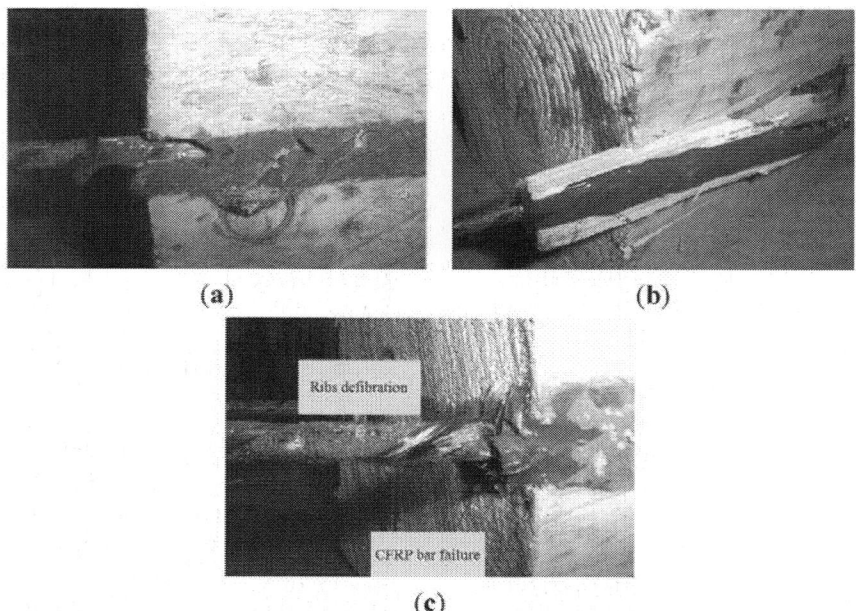

(a)   (b)

(c)

Figure 8. Failure modes: (**a**) bar pull-out; (**b**) timber shear failure; (**c**) CFRP bar tensile failure.

The second failure mode was frequent for soft wood (fir) specimens with small bonded lengths (150, 200 and 250 mm), and this mainly involved the wood material. Wood failure occurred when the failure surface was within the wood adjacent to the notch. After developing small shear cracks primarily located in the timber, specimens failed for the pull-out of the CFRP bar and a portion of the timber materials surrounding the bar (Figure 8b).

The third failure mode was characterized by the rupture of the FRP reinforcement (Figure 8c). The majority of pull-out specimens with a bonded length of 300 mm failed according to this mode. This failure was not instantaneous: the carbon fibres applied helicoidally over the CFRP bars failed first, followed by the tensile rupture of longitudinal carbon fibres.

As expected, the failure load increased with the bonded length. The average normal stress at failure σ, bond stresses at interface bar-resin τ_b and at interface resin-timber τ_{ew} were 530.3, 6.63 and 3.55 N/mm^2 respectively, for fir wood specimens with a bonded length of 150 mm; similar results were measured for chestnut wood having the same bonded length (Table 3).

Table 3. Test results. SF, fir specimen; SC, chestnut specimen.

Index	Bonded Length (mm)	Maximum Load F_{max} (kN)	Tensile Normal Stress σ (N/mm²)	Bond Stress τ_b (N/mm²)	Bond Stress τ_{ew} (N/mm²)	Failure Mode
SF_1	150	23.96	542.6	6.78	3.63	Bar pull-out
SF_2	150	20.52	464.7	5.81	3.11	Timber shear failure
SF_7	150	25.77	583.6	7.30	3.90	Bar pull-out
		23.42 (2.67)	530.3 (60.4)	6.63 (0.75)	3.55 (0.40)	
SF_3	200	31.95	723.6	6.78	3.63	Timber shear failure
SF_4	200	33.78	765.0	7.17	3.84	Bar pull-out
SF_8	200	32.45	734.9	6.89	3.69	Timber shear failure
		32.73 (0.95)	741.2 (21.4)	6.95 (0.20)	3.72 (0.11)	
SF_5	250	47.23	1070	8.02	4.29	Timber shear failure
SF_6	250	47.92	1085	8.14	4.36	Bar tensile failure
SF_9	250	44.65	1011	7.58	4.06	Timber shear failure
		46.60 (1.72)	1055 (39.0)	7.92 (0.29)	4.24 (0.16)	
SF_10	300	48.08	1089	6.81	3.64	Bar tensile failure
SF_11	300	48.77	1104	6.90	3.69	Bar tensile failure
SF_12	300	44.95	1018	6.36	3.41	Timber shear failure
		47.27 (2.04)	1070 (46.1)	6.69 (0.29)	3.58 (0.15)	
SC_1	150	26.01	589.0	7.36	3.94	Timber shear failure
SC_2	150	26.74	605.6	7.57	4.05	Bar pull-out
SC_7	150	23.45	531.1	6.64	3.55	Bar pull-out
		25.40 (1.73)	575.2 (39.1)	7.19 (0.49)	3.85 (0.26)	
SC_3	200	28.7	650.0	6.09	3.26	Bar pull-out
SC_4	200	32.37	733.1	6.87	3.68	Bar pull-out
SC_8	200	32.12	727.4	6.82	3.65	Bar pull-out
		31.06 (2.05)	703.5 (46.4)	6.60 (0.44)	3.53 (0.23)	
SC_5	250	49.31	1117	8.38	4.48	Bar tensile failure
SC_6	250	47.91	1085	8.14	4.36	Bar tensile failure
SC_9	250	46.56	1054	7.91	4.23	Bar tensile failure
		47.93 (1.38)	1085 (31.1)	8.14 (0.23)	4.36 (0.13)	
SC_10	300	48.56	1100	6.87	3.68	Bar tensile failure
SC_11	300	49.56	1122	7.01	3.75	Bar tensile failure
SC_12	300	47.9	1085	6.78	3.63	Bar tensile failure
		48.67 (0.84)	1102 (18.9)	6.89 (0.12)	3.69 (0.06)	

In parentheses: Standard deviation.

By increasing the CFRP bar bonded length of 33% (from 150 to 200 mm), an almost consistent increment of the maximum load (pull-out capacity) and stresses was measured and calculated (31.1%). The average normal stress and bond stresses increased 39.8 and 22.3%, respectively, for softwood (fir) and hardwood (chestnut) specimens. Specimens with 200-mm bonded lengths mainly exhibited a failure due to pull-out of the bar from the notches.

An interesting observation can be underlined for specimens with a longer bonded length. For both timber species, bar pull-out or timber shear failure modes were prevented when a bonded length of 250 mm was used. The bar's failure appeared in correspondence to an average load F_{max} of 47.26 kN and an average bond stress at bar-epoxy interface of 8.03 N/mm². The bonded length is the most influential parameter on the test; the differences between fir wood and chestnut wood specimens are very low for all the three different bonded length in terms of maximum load F_{max} (Figure 9), axial strength σ and bond strengths τ_b and τ_{ew}.

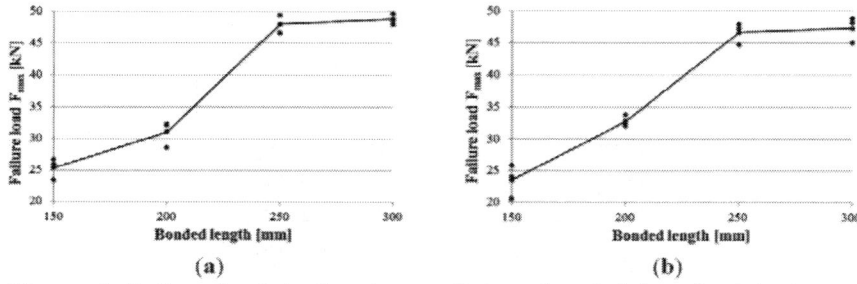

Figure 9. Failure load (pull-out capacity) *vs.* bonded length: (**a**) chestnut wood; (**b**) fir wood.

Since the number of specimens tested was limited, results should be confirmed by a larger experimental programme. However, the emerging line seems quite correct: using bonded lengths greater than 250 mm does not cause an increase in the pull-out capacity in the CFRP bars. This was evident by comparing the results of specimens with a bonded length of 250 mm with the ones with 300 mm: fir wood specimens exhibited a capacity of 46.6 and 47.27 kN for 250 mm and 300 mm bonded lengths, respectively. For chestnut specimens, pull-out capacity increased from 47.93 to 48.67 kN.

4.2. Tests with Strain Gauges

In order to evaluate the distribution of stress and strain, four pull-out tests [SF_2, SF_4, SC_2 and SC_3 (Table 3)] were carried out with the use of strain gauges. Strain gauges were produced by Micro-Measurements under the commercial name "CEA-06-125UN-350" (gage factor 2.085, resistance 350 Ohms, length 4.57 mm). Three strain gauges (Figure 10a) were fixed to the specimens with a bonded length of 150 mm and four strain gauges (Figure 10b) to the specimens with a 200-mm bonded length. Strain gauges were applied at a distance of 10 mm from the loaded end and with a mutual centre-to-centre distance of 50 mm (Figure 11).

Test results with the use of strain gauges were used to plot graphs in terms of strain *versus* location. The strain in the CFRP bar along the bonded length is plotted for different values of the load, indicated as a percentage of the failure load (10%, 20%, 50%, 75% and 100%, respectively). All points of the graphs were plotted from the strain gauge readings, with the exception of the strain at the unloaded end of the bars, which was assumed equal to zero and the loaded end (stress calculated from the axial load).

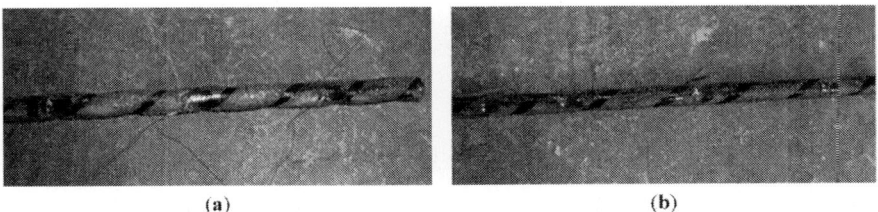

(a)   (b)

Figure 10. Strain gauge arrangement on FRP bar: (**a**) for a bonded length of 150 mm; (**b**) for a bonded length of 200 mm.

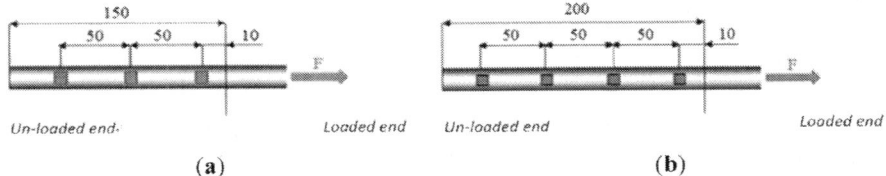

(a)   (b)

Figure 11. (**a**) Strain gauge arrangement along reinforcement (150-mm bonded length); (**b**) strain gauge arrangement along reinforcement (200-mm bonded length).

From the strain-location data, much useful information can be drawn. Strain distributions exhibit an approximately linear behaviour for low load levels. When the load increases, strain distribution along the different positions show an almost non-linear trend. This could mean that, as the axial load rises, the bond stresses become more evenly distributed along the bonded length as a consequence of variations in the characteristics of the bond. For low values of the axial load, the primary bond mechanism seems to be governed by the chemical adhesion due to the epoxy resin, but when the load increases, the primary bond mechanism changes from a chemical adhesion to mechanical friction mechanism between the bar's indentation and the epoxy resin in the interface between the materials. This could be noted in Figure 8a: the cracks are parallel to the CFRP bar and particularly large near the bar indentations.

Strain readings obtained from strain gauges were used also to evaluate the axial stress σ_b values on the lateral surface of the CFRP bars using Hooke's law:

$$\sigma_b = E_b \times \varepsilon_b \tag{1}$$

where E_b is Young's modulus of the CFRP bar and ε_b the normal strain of the bar.

Results are reported in Figure 12. Due to the effect of the application of the load using the hydraulic jack, timber material was mainly in compression, and both CFRP bars epoxied on the notches were in tension. The distribution of the tensile axial stresses on the CFRP bars is linear near the unloaded end and parabolic at the loaded end. The maximum value of the tensile stress is always located near the loaded end. The non-linear behaviour is more evident for values of axial loads of 100%, 75% and 50% of the maximum load.

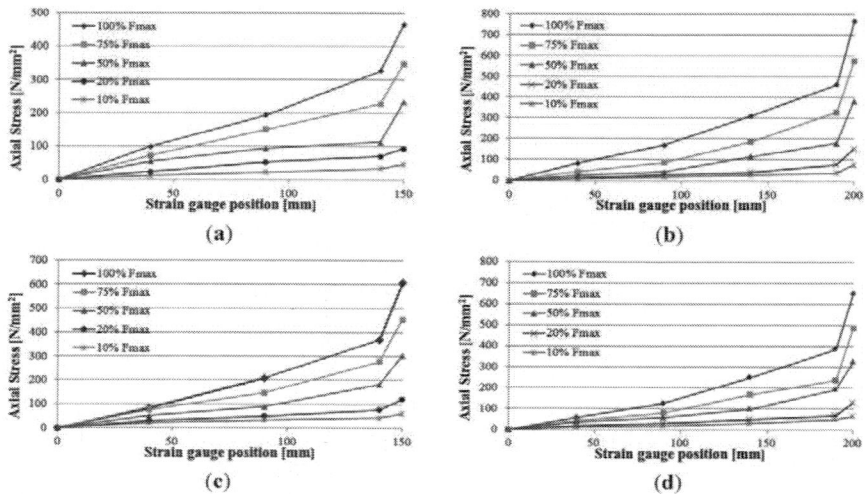

Figure 12. Normal stress *versus* position: (**a**) SF_2 test; (**b**) SF_4 test; (**c**) SC_2 test; (**d**) SC_4.

Data obtained from tests with the use of strain gauges finally were used to evaluate the bond stress τ_b between the CFRP bar and epoxy resin. The equilibrium of a CFRP bar's element of length dx, with the hypothesis of linearly-elastic behaviour, is defined by the following:

$$\tau_b(x) = \frac{d_b}{4} E_b \frac{d\varepsilon_b(x)}{dx} \tag{2}$$

where: d_b = bar's diameter; E_b = Young's modulus of the bar; $d\varepsilon_b$ = strain of the bar element length dx. Because strain measurements are available at discrete points along the bonded length and indicating with ε_{bi} the strain reading at the location expressed by the coordinate x_i and with ε_{bj} the strain reading at the coordinate x_j, Equation (2) could be approximated as:

$$\tau_b\left(\frac{x_i + x_j}{2}\right) = \frac{d_b}{4} E_b\left(\frac{\varepsilon_{bj} - \varepsilon_{bi}}{x_j - x_i}\right) \tag{3}$$

Experimental results obtained for the specimens equipped with strain gauges, using Equation (3), are shown in the following Figure 13. Due to the surface deformation of CFRP bars, the primary bond mechanism is the mechanical interlocking, while chemical adhesion of the epoxy resin is a secondary bond mechanism. During the tests, the radial components of the bond stresses generate micro-cracks in the epoxy resin and the consequent slip between bars and adhesive. This was observed for the SF_4, SC_2 and SC_3 tests. Micro-cracking and the consequent slip between the materials tend to cause the bond stress to be more evenly distributed. Figure 13 shows that, for low lcad levels, the bond stresses at the unloaded end is close to 0 N/mm². As the load increases, the peak of the bond stress gradually shifts towards the loaded end, and it mainly contributes to resisting the exterral force applied by the jack.

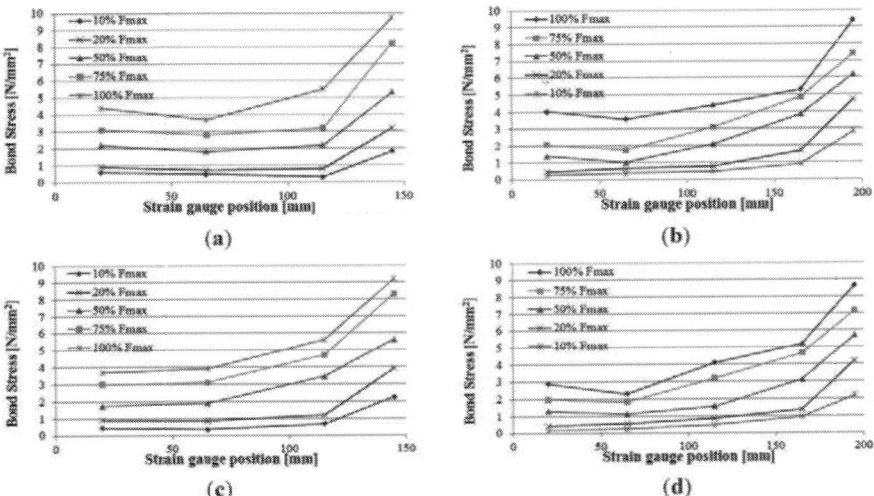

Figure 13. Bond stress *versus* position: (**a**) SF_2 test; (**b**) SF_4 test; (**c**) SC_2 test; (**d**) SC_4.

4.3. Discussion and Comparison with Previous Research

Several bending tests carried out in the past by the same authors [15] have shown that the most frequent failure mechanism of CFRP-reinforced beams is due to cracking of timber on the tension side without an important plasticization of the timber in the compression side, depending on the quality of the wood. Wood yield is interrupted from beam failure due to the appearance of cracks, particularly for softwood (fir) beams.

Beams reinforced with CFRP bars revealed less ductile behaviour compared to that of those reinforced with CFRP sheets [15]. The CFRP reinforcement caused an increase of 52% for a double CFRP bar reinforcement. Load *vs.* deflection curves show that the beams exhibited a more basically linear behaviour up to failure. The positive effect induced by the presence of the bars does not suffice to confine local ruptures and bridge local defects. Moreover, the grooves cut into the beams in order to insert the CFRP bars produce some limited damage.

Results of this research on the bond strength of CFRP bars to timber show that an adequate stress transfer is possible both for soft and hardwood using limited bonded lengths (approximately 250 mm) and standard epoxy resins. However, the local character of the bar reinforcement is not able to prevent failure of the timber in the tension un-reinforced areas, especially at the beam edges. According to this, it could be advised that a "more uniform" reinforcement is to be preferred: numerous small FRP bars inserted in small notches produce a smaller damage and contribute to reducing stress and strain in the timber material.

5. CONCLUSIONS

This paper investigates the bond behaviour of NSM CFRP bars installed in softwood (fir) and hardwood (chestnut) timber. A series of 24 specimens, 12 in fir and 12 in chestnut wood, were tested to analyse the effect of different bonded lengths and different timber species on the bonded strength of the bars. The following conclusion can be drawn:

1. The pull-out capacity in NSM CFRP bars at the onset of debonding increased with bonded length up to a length of 250 mm. The CFRP bar's pull-out was achieved only for specimens with bonded lengths of 150 and 200 mm.
2. The test results identified three basic modes of failure: one related to the parent material (*i.e.*, timber); and the other two associated with the reinforcing composite material (CFRP bar pull-out and tensile failure, *i.e.*, rupture of CFRP bar or cracking of the epoxy system). For high bonded lengths (250 and 300 mm), timber failure was observed to be the controlling mode. For small bonded lengths (150 and 200 mm), either timber cracking or CFRP pull-out was observed, the latter being the most common.
3. For 250- and 300-mm bonded lengths, rupture was always initiated by the failure of the carbon filaments, which are not parallel to the bar and produced a radial component of the bond stress. This component caused micro-cracks that slowly propagated up to determine a noticeable cracking of the epoxy resin and the subsequent slipping-out of the bar as the load increased.
4. Timber type did not affect the bond behaviour: the different test results between fir wood (softwood) and chestnut wood (hardwood) specimens were small for all four different bonded lengths in terms of pull-out capacity and bond strength.

Further experimental investigation taking into account different bonded lengths, epoxy resins and types of FRP bars will be necessary to address the problem.

ACKNOWLEDGMENTS

The authors would like to acknowledge the support of the Structural Engineering Laboratory at Perugia University for the use of test and measurement equipment critical to the collection and evaluation of the data presented. The experimental program was carried out with the help of

Gianfelice Montani and Marco Canduzzi, undergraduate students, and Alessio Molinari, Graduate Research Assistant. CFRP bars were supplied by MAC SpA, Treviso, Italy.

AUTHOR CONTRIBUTIONS

M.C. and A.B. conceived and designed the experiments; L.R. performed the experiments; M.C. and L.R. analysed the data; L.R. wrote the paper.

REFERENCES

1. Lantos, G. The flexural behavior of steel reinforced laminated timber beams. *Wood Sci.* **1970**, *2*, 136–143.
2. Bulleit, W.M.; Sandberg, L.B.; Woods, G.J. Steel-reinforced glued laminated timber. *J. Str. Eng.* **1989**, *115*, 433–444.
3. Castro, P.F.; Carino, N.J. Tensile and nondestructive testing of FRP bars. *J. Compos. Constr.* **1998**, *2*, 17–27.
4. Kocaoz, S.; Samaranayake, V.; Nanni, A. Tensile characterization of glass FRP bars. *Compos. B Eng.* **2005**, *36*, 127–134.
5. Larralde, J.; Silva-Rodriguez, R. Bond and slip of FRP rebars in concrete. *J. Mater. Civ. Eng.* **1993**, *5*, 30–40.
6. Micelli, F.; Nanni, A. Tensile characterization of FRP rods for reinforced concrete structures. *Mech. Compos. Mater.* **2003**, *39*, 293–304.
7. Ceroni, F.; Cosenza, E.; Gaetano, M.; Pecce, M. Durability issues of FRP rebars in reinforced concrete members. *Cem. Concr. Compos.* **2006**, *28*, 857–868.
8. Turco, V.; Secondin, S.; Morbin, A.; Valluzzi, M.R.; Modena, C. Flexural and shear strengthening of un-reinforced masonry with FRP bars. *Compos. Sci. Technol.* **2006**, *66*, 289–296.
9. Galati, N.; Tumialan, G.; Nanni, A. Strengthening with FRP bars of URM walls subject to out-of-plane loads. *Constr. Build. Mater.* **2006**, *20*, 101–110.
10. Pleviris, N.; Triantafillou, T.C. FRP-reinforced wood as structural material. *J. Mater. Civ. Eng.* **1992**, *4*, 330–317.
11. Plevris, N.; Triantafillou, T.C. Creep behavior of FRP-reinforced wood members. *J. Str. Eng.* **1995**, *12*, 174–186.
12. Triantafillou, T.C. Shear reinforcement of wood using FRP materials. *J. Mater. Civ. Eng.* **1997**, *9*, 65–69.
13. Borri, A ; Corradi, M.; Grazini, A. FRP reinforcement of wood elements under bending loads. In Proceedings of the 10th International Conference on Structural Faults + Repair, London, UK, 1–3 July 2003.

14. Gentile, C.; Svecova, D.; Rizkalla, S.H. Timber beams strengthened with GFRP bars: Development and applications. *J. Compos. Constr.* **2002**, *6*, 11–20.

15. Borri, A.; Corradi, M.; Grazini, A. A method for flexural reinforcement of old wood beams with CFRP materials.*Compos. B Eng.* **2005**, *39*, 143–153.

16. Miceli, F.; Scialpi, V.; la Tegola, A. Flexural reinforcement of glulam timber beams and joints with Carbon Fiber-Reinforced Polymer rods. *J. Compos. Constr.* **2005**, *9*, 337–347.

17. Gilfillan, J.; Gilbert, S.; Patrick, G. The use of FRP composites in enhancing the structural behavior of timber beams.*J. Reinf. Plast. Compos.* **2003**, *22*, 1373–1388.

18. Bainbridge, R.; Mettem, C.; Harvey, K.; Ansell, M. Bonded-in rod connections for timber structures-development of design methods and test observations. *Int. J. Adhes. Adhes.* **2002**, *22*, 47–59.

19. Raftery, G.M.; Whelan, C. Low-grade glued laminated timber beams reinforced using improved arrangements of bonded-in GFRP rods. *Constr. Build. Mater.* **2014**, *52*, 209–220.

20. Borri, A.; Corradi, M.; Speranzini, E. Reinforcement of wood with natural fibers. *J. Compos. B Eng.* **2013**, *53*, 1–8.

21. Raftery, G.M.; Kelly, F. Basalt FRP rods for reinforcement and repair of timber. *Compos. B Eng.* **2015**, *70*, 9–19.

22. Scialpi, V.; de Lorenzis, L.; la Tegola, A. Bond of CFRP rods in glulam members. In Proceedings of International Conference on Composites in Constructions, Rende, Italy, 16–19 September 2003.

23. Horeczy, G.; Svecova, D.; Rizkalla, S. The bond behavior of timber reinforced with GFRP bars. In Proceedings of 2rd Annual Conference on Advanced Engineered Wood Composites, Sunday River, Bethel, ME, USA, 14–16 August 2001.

24. Yeboah, D.; Taylor, S.; McPolin, D.; Gilfillan, R. Pull-out behavior of axially loaded Basalt Fibre Reinforced Polymer (BFRP) rods bonded perpendicular to the grain of glulam elements. *Constr. Build. Mater.* **2013**, *38*, 962–969.

25. Ling, Z.; Yang, H.; Liu, W.; Lu, W.; Zhou, D.; Wang, L. Pull-out strength and bond behavior of axially loaded rebar glued-in glulam. *Constr. Build. Mater.* **2014**, *65*, 440–449.

26. Righetti, L.; Corradi, M.; Borri, A. Analytical study on bond behavior of CFRP bars epoxied into timber elements. In Proceedings of 3rd International Conference on Structural Health Assessment of Timber Structures, Wroclaw, Poland, 9–11 September 2015.

27. Yao, J.; Teng, J.G.; Chen, J.F. Experimental study on FRP-to-concrete bonded joints. *Compos. B Eng.* **2005**, *36*, 99–113.

28. Chen, J.F.; Yang, Z.J.; Holt, G.D. FRP or steel plate-to-concrete bonded joints: Effect of test methods on experimental bond strength. *Steel Compos. Str.* **2001**, *1*, 231–244.

29. *Moisture Content of A Piece of Sawn Timber*; BS EN 13183–1:2002; British Standards Institute: London, UK, 2002.

30. *Structural Timber. Strength Classes*; BS EN 338:2009; British Standards Institute: London, UK, 2009.

31. *Eurocode 5 Design of Timber Structures Part 1-1: General-Common Rules and Rules for Buildings*; EN 1995-1-1: 2004; European Committee for Standardization: Brussels, Belgium, 2004.

32. *Standard Test Method For Tensile Properties of Polymer Matrix Composite Material*; ASTM D3039; ASTM: West Conshohocken, PA, USA, 2014.

33. *Standard Test Method For Compressive Properties of Rigid Plastics*; ASTM D695; ASTM: West Conshohocken, PA, USA, 2008.

34. *Standard Test Method for Tensile Properties of Plastics*; ASTM D638; ASTM: West Conshohocken, PA, USA, 2014.

35. Raftery, G.M.; Harte, A.M.; Rodd, P.D. Bonding of FRP materials to wood using thin epoxy gluelines. *Int. J. Adhes. Adhes.* **2009**, *29*, 580–588.

CHAPTER 9

Fiber Reinforced Polymer and Polypropylene Composite Retrofitting Technique for Masonry Structures

Saleem Muhammad Umair [1,*], *Muneyoshi Numada* [2], *Muhammad Nasir Amin* [1] *and Kimiro Meguro* [2]

[1] Department of Civil and Environmental Engineering, College of Engineering, King Faisal University, Al-Ahsa 31982, Saudi Arabia
[2] International Center for Urban Safety Engineering, Institute of Industrial Science, The University of Tokyo, 4-6-1 Komaba, Meguro-ku, Tokyo 153-8505, Japan

ABSTRACT

In the current research work, an attempt is made to increase the seismic capacity of unreinforced masonry (URM) structures by proposing a new composite material which can improve shear strength and deformation capacity of URM wall systems. Fiber Reinforced Polymer (FRP) having high tensile and shear stiffness can significantly increase in-plane and out-of-plane strength of masonry walls, but, inherently, FRP strengthened wall systems exhibit brittle failure under extreme seismic loading. Polypropylene (PP-band) is a low cost material with sufficient ductility and deformation capacity. Keeping in view the behavior of FRP and PP-band, a composite of FRP and PP-band is proposed for retrofitting of URM walls. Mechanical behavior of the proposed composite material is assessed by carrying out an in-plane diagonal compression test and an out-of-plane bending test on twenty-five 1/4-scaled masonry wall panels. Experimental plan for each panel, URM, PP-band retrofitted, FRP retrofitted and FRP + PP-band retrofitted masonry, is diagonal compression test and three-point bending test. Experimental results have determined that FRP + PP-band composite increased, not only the initial peak strength, but also the ductility, deformation capacity and residual strength of URM wall systems.

Keywords: rehabilitation; masonry; composite materials; polypropylene; fiber reinforced polymer

1. INTRODUCTION

Masonry is a historical construction material and has served as the first choice of people for many centuries. Pleasing aesthetics with good sound and heat insulation properties are some of the inherent advantages of masonry. Local availability of masonry raw materials and having lower cost of manufacturing has made it a popular construction material. Even in the recent times masonry structures are higher in number than concrete, steel and wooden structures [1]. However, historically unreinforced masonry (URM) structures have suffered extensive damage during earthquakes. Therefore, to increase the seismic capacity of low earthquake-resistant masonry structures is one of the key issues for earthquake disaster mitigation and for reduction of casualties. Seismic retrofitting reduces not only casualties and damage to buildings during earthquakes, but also the costs of first aid activities, rescue, rubble removal and permanent residential reconstruction to help and support re-establishment of daily life [2]. URM buildings can fail due to deficient strength of walls when loaded in in-plane and out-of-plane directions [3]. Seismic capacity of URM buildings depends on the ability of in-plane walls to effectively transfer lateral forces to the foundations [4,5].

Failure of in-plane walls occurs due to formation of diagonal shear cracks, by sliding of a portion of the wall generally along a bed joint (sliding shear deformation), by rocking about the wall toe or by crushing of the wall toe [6]. URM structures are much weaker in out-of-plane direction as compared to in-plane direction and final collapse of masonry buildings is due to out-of-plane failure of URM walls [7].

In order to avoid diagonal shear failure and out-of-plane failure of URM walls, different retrofitting procedures have been adopted by different researchers [8]. Some of the retrofitting methods include the seismic retrofit of URM walls using externally-bonded or near surface mounted Fiber Reinforced Polymer (FRP) laminates, bars and fabrics. Experiments on various patterns and layouts of FRP have validated that FRP can significantly increase in-plane and out-of-plane strength of URM walls [7,9,10,11,12,13,14,15,16, 17,18,19,20,21,22]. Higher strength to weight ratio, ease of application and corrosion resistance are some of well-known advantages of FRP retrofitting technique over conventional retrofitting methods. On the other side, FRP is a costly material and exhibits a brittle failure [11,14,23].

In addition to aforementioned retrofitting methods, Mayorca and Meguro [24] proposed polypropylene band (PP-band) mesh retrofitting technique. PP-band is normally used for packing and also available worldwide. PP-band is a low cost material with large deformation capacity [25]. A mesh of PP-band can be applied to URM structures to hold the masonry components into a single unit and to prevent the collapse of masonry structures. After carrying out a series of experiments ranging from small-scale model to full-scale masonry house, it was found that PP-band retrofitted walls can withstand much stronger input ground motion without collapse [2,24,26,27,28,29,30].

2. RESEARCH OBJECTIVES

In this study, a new composite retrofitting material is proposed, which consists of FRP and PP-band. This composite retrofitting technique can increase the shear strength and bending strength of URM walls and can serve satisfactorily to hold the structural system by providing sufficient deformation capacity. FRP is a brittle material having ultimate tensile strain ranging from only 2% to 4% [31]. PP-band cannot increase initial strength, but it can increase deformation capacity and energy dissipation capacity of masonry wall systems [30]. In order to get holding effect, if FRP is used for brick masonry, then it is required to apply FRP on whole wall surface, which will tremendously increase the retrofitting cost and still the ductile failure of masonry wall system is not guaranteed. On the other hand, PP-band mesh is not only a fairly ductile and deformable, but can also be wrapped around to the whole wall system, because of very low retrofitting cost.

Main objective of this study was to investigate the effect of PP-band and FRP composite on increase in strength and deformation capacity of masonry wall panels. In order to achieve aforementioned objective, diagonal compression tests and three-point bending tests using 1/4-scale retrofitted and non-retrofitted masonry panels were carried out.

3. EXPERIMENTAL RESULTS

Experiments were conducted in three main phases, material test, diagonal compression tests and out-of-plane bending tests on 1/4-scale retrofitted and non-retrofitted masonry panels.

3.1. Diagonal Compression Tests Results

Figure 1a shows the stress-strain of three non-retrofitted masonry panels, which are abbreviated as URM-1, URM-2 and URM-3. All of the URM masonry panels have shown different peak strength values. Stress-strain curve of PP-band retrofitted masonry panel along with URM masonry panels is shown in Figure 1b. PP-band retrofitted panel has shown same initial stiffness as that of URM's, with a peak stress value of 0.126 MPa. After the initial peak stress, there was sudden drop in peak value due to failure of masonry panel, but as the applied load further increases PP-band comes into action and restrains the further movement of failed masonry. This phenomenon can be seen by increase in load carrying capacity of panel after the strain value of 1.5%. There are some rises and falls in load carrying capacity of masonry panel, which is due to subsequent failure of masonry and PP-band restraining action. After the strain value of 8%, the load carrying capacity has even increased from the initial peak load and keeps on increasing up to 12.1% strain. Beyond 12.1%, bricks and mortar started crushing due to very high compressive displacements.

To understand behavior of retrofitted and non-retrofitted masonry panels under diagonal compression loading, stress-strain curves of non-retrofitted are presented along with retrofitted masonry panels in Figure 2. Figure 2a shows the stress-strain curves of Carbon Fiber Reinforced Polymer (CFRP) and CFRP + PP-band retrofitted masonry panels along with URM, and PP-band retrofitted masonry panels. CFRP retrofitted masonry panel has increased the initial peak load of URM from average value of 0.11 to 0.79 MPa and failure strain from 0.7% to 3.7%. After the initial peak, there is a sudden drop in strength of CFRP retrofitted masonry panel. Proposed CFRP + PP-band retrofitted masonry panel has increased the initial peak strength as that of CFRP retrofitted masonry panel but it has also increased the strain from 0.7% to 20%, which is closer to the failure strain of PP-band retrofitted masonry panel. In addition to this, the panel has shown a high residual strength after the initial peak load. This increase in residual strength was greater than the residual strength of only PP-band retrofitted masonry panel. Failure of CFRP + PP-band retrofitted masonry panel was also gradual with sufficient ductility.

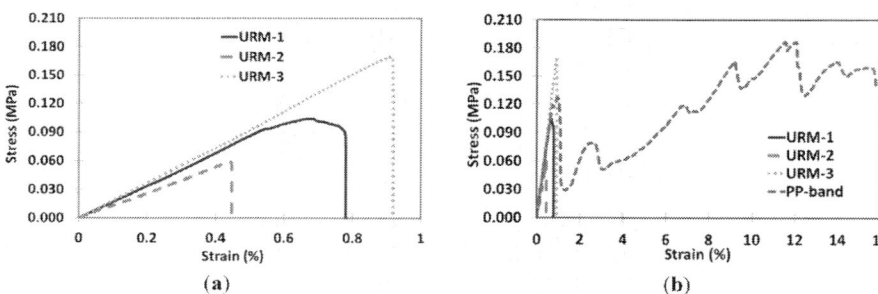

Figure 1. Stress-strain curves of masonry panels under diagonal compression test, (**a**) stress-strain curves of non-retrofitted (URM) masonry panels and (**b**) stress-strain curves of URM and PP-band retrofitted masonry panels.

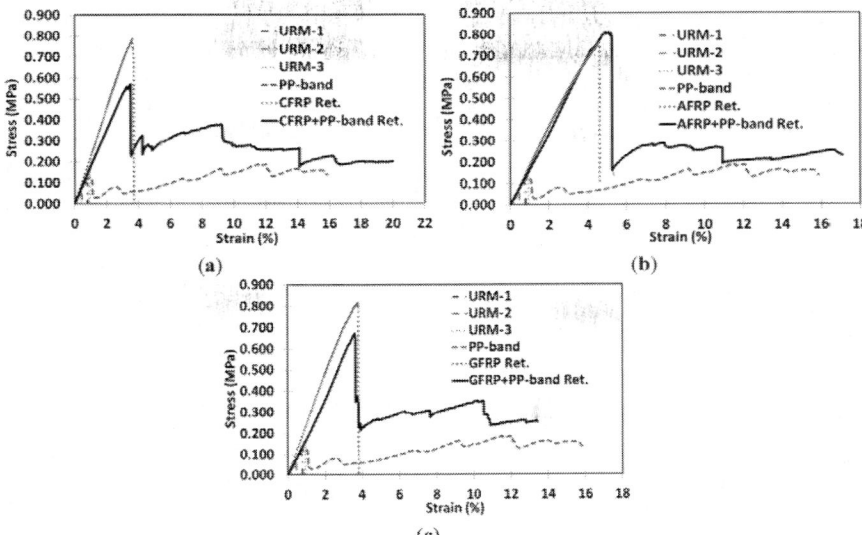

Figure 2. Stress-strain curves of URM and PP-band retrofitted masonry panels along with, (**a**) CFRP and CFRP + PP-band retrofitted masonry panels, (**b**) AFRP and AFRP + PP-band retrofitted masonry panels, and (**c**) GFRP and GFRP + PP-band retrofitted masonry panels.

Stress-strain curves of Aramid Fiber Reinforced Polymer (AFRP) and AFRP + PP-band retrofitted masonry panels are shown in Figure 2b. As that of CFRP retrofitted masonry panel, AFRP retrofitted masonry panel has increased initial shear strength but has shown a highly sudden and brittle failure as compared to PP-band retrofitted masonry panel. Increase in initial strength in the case of AFRP retrofitted masonry panels is almost the same as that of CFRP retrofitted masonry panel. Like CFRP + PP-band retrofitted masonry panel, AFRP + PP-band retrofitted masonry panel has also increased the initial shear strength, residual strength and the displacement carrying capacity of masonry panel. In AFRP + PP-band retrofitted case, residual strength was comparatively less than the residual strength of CFRP + PP-band retrofitted masonry panel. This effect could be due to loosen PP-band mesh but still the residual strength is even higher than the initial peak strength of URM panels. Similar type of stress-strain behavior was observed in the case of Glass Fiber Reinforced Polymer (GFRP) and GFRP + PP-band retrofitted masonry panels, as shown in Figure 2c.

Failure Modes

Failure modes and cracking pattern of different types of masonry panels are shown in Figure 3. Figure 3a,b show the cracking pattern of URM and PP-band retrofitted masonry panel. In both of the masonry panels, failure was initiated by a diagonal slip of masonry panel through the mortar joints. Figure 3c,e,g show the cracking pattern of CFRP, AFRP and GFRP retrofitted masonry panels, respectively. All types of FRP retrofitted panels have exhibited a brittle failure with less ductility and having no warning before failure. In none of the FRP retrofitted panels, fracture of FRP strips was witnessed. In most of the cases, failure was initiated due to detachment of FRP along with a thin layer of brick from the masonry panel surface.Figure 3c clearly shows the detachment of CFRP with thin layer of brick from masonry panel surface. Similar type of failure was witnesses by Bahman *et al.* who performed single lap shear test to evaluate the deboning behavior of FRP from brick surface by using Digital Image Correlation (DIC) technique [32,33]. DIC technique is becoming popular among researchers for full-field measurement of strains and deformation by comparing the image features at different mechanical states. Ghorbani *et al.* also used DIC technique to predict the cracking pattern and deformation behavior of confined masonry walls [34]. Figure 3d,f,h show the failure pattern of different FRP + PP-band retrofitted masonry panels.

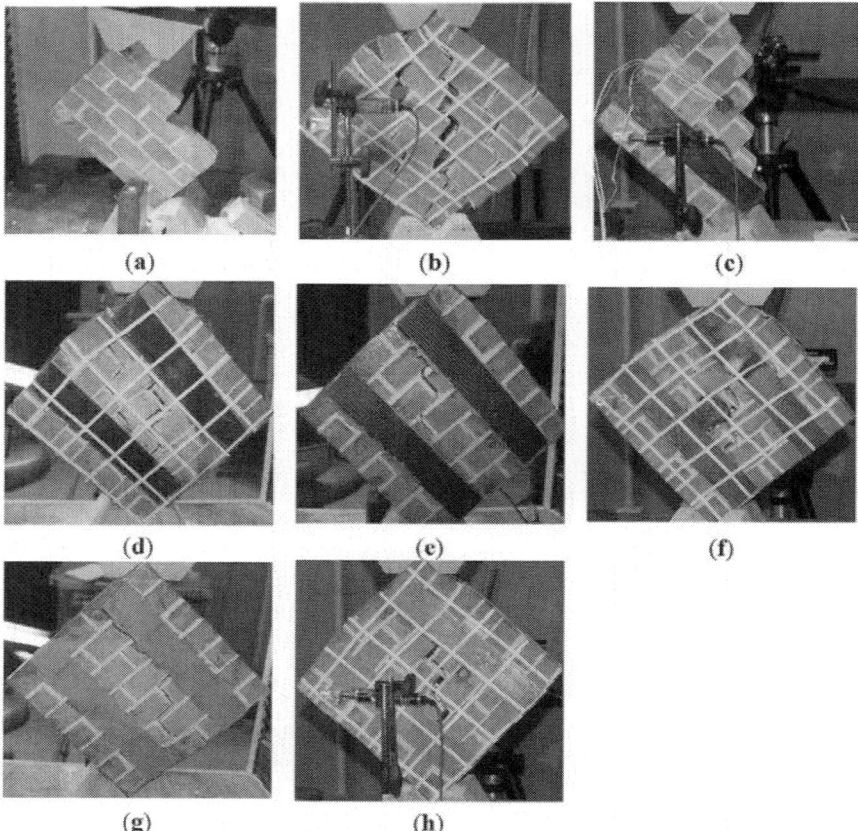

Figure 3. Failure mode and cracking pattern of masonry panels under diagonal compression test for (**a**) URM, (**b**) PP-band retrofitted, (**c**) CFRP retrofitted, (**d**) CFRP + PP-band retrofitted, (**e**) AFRP retrofitted, (**f**) AFRP + PP-band retrofitted, (**g**) GFRP retrofitted, and (**h**) GFRP + PP-band retrofitted.

3.2. Out-of-Plane Bending Test Results for Type-1 Masonry Walls

Figure 4a,b show the moment-deflection curves of three URM and PP-band retrofitted masonry panels, whereas corresponding shear stress is plotted on the secondary vertical axis. URM panels have shown a variety of peak strengths with an average value of 0.08 kN-m with almost similar initial stiffness. Behavior of panels was almost linear up to the peak load, as shown in Figure 4a. Figure 4b shows the moment-deflection curve of PP-band retrofitted masonry panel. It has shown peak strength of 0.085 kN-m, which is almost the same as that of URM masonry panels. In the case of PP-band retrofitted masonry panel after the deflection of 1.2 mm, it again started taking load and has gone up to a value of 57.7 mm with a moment value of 0.097 kN-m, as shown in Figure 4b. PP-band retrofitted masonry panel has shown a fairly long deflection and deformation as compared to URM masonry panels.

Figure 5a shows the moment-deflection curve of CFRP, AFRP and GFRP retrofitted masonry panels under out-of-plane bending test. Application of FRP has significantly increased the out-of-plane bending strength of masonry panel. CFRP has increased initial peak strength of URM masonry panel from 0.08 to 0.88 kN-m, whereas AFRP and GFRP retrofitted masonry panel has shown a peak strength of 0.59 and 0.33 kN-m, respectively. FRP has increased the failure displacement of URM masonry panels but no residual strength of masonry panels was witnessed. All FRP's retrofitted masonry panels have shown a brittle failure. Figure 5a shows an initial drop in strength of FRP retrofitted masonry panels at a displacement range of 0.5 to 0.9 mm. This sudden drop in load was due to separation and falling of a layer of bricks at the inner edge of the masonry panel. After this drop, panels, again, started taking load and reached their corresponding peak strengths.

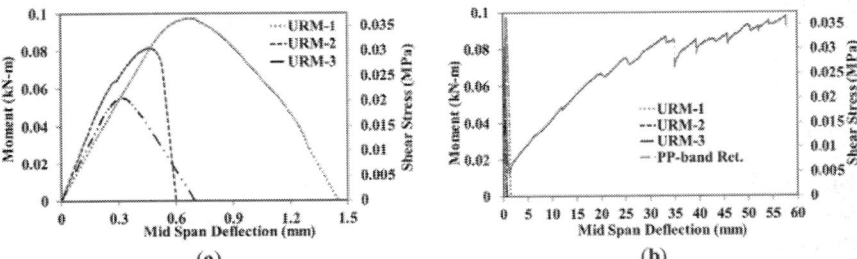

Figure 4. Moment-deflection curves of masonry panels under bending test for **(a)** URM and **(b)** PP-band retrofitted.

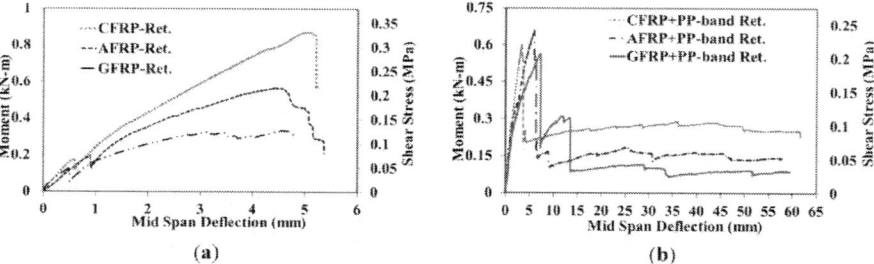

Figure 5. Moment-deflection curves under out-of-plane bending test for (**a**) CFRP, AFRP and GFRP retrofitted masonry panels and (**b**) CFRP + PP-band, AFRP + PP-band and GFRP + PP-band retrofitted masonry panels.

Figure 5b shows moment-deflection curves for CFRP + PP-band, AFRP + PP-band and GFRP + PP-band retrofitted masonry panels. Application of FRP has increased the initial strength of URM masonry panel compared to that of only FRP retrofitted masonry panels. This increase in strength was variable depending upon the type of FRP used and its application on the surface of masonry panel. Presence of PP-band has increased the displacement carrying capacity of masonry panel as that of only PP-band retrofitted masonry panel. In addition to these two predictive effects, FRP + PP-band retrofitted masonry panels have shown a higher residual strength compared to only PP-band retrofitted masonry panel. In the case of CFRP + PP-band and GFRP + PP-band retrofitted masonry panels, this residual strength was even greater than the strength of URM masonry panel. Figure 5b also shows that, the initial drop of strength due to detachment of brick layers as seen inFigure 5a was either vanished or moved to higher load levels due to the presence of PP-band on the surface of masonry panels. All FRP + PP-band retrofitted masonry panels have increased the initial strength, residual strength and deformation capacity of masonry panels.

Failure Modes for Type-1 Walls

Figure 6 shows that failure pattern of different type of masonry panels under out-of-plane bending test. Figure 6a shows a pure bending type of failure with major bending cracks and spalling of bricks from the central part of the panels. There were no signs of shear failure near supports. Figure 6b shows that PP-band retrofitted masonry panel failure. Failure of PP-band retrofitted masonry panels was the same as that of URM masonry panels but the presence of PP-band kept the brick masonry as a single unit. PP-band has not provided any increase in bending strength but after the initial drop of strength due to masonry failure, PP-band enabled the panel to take load after the initial failure. Figure 6c,e,g show the failure of CFRP, AFRP and GFRP retrofitted masonry panels. All FRP retrofitted masonry panels have shown a stepwise failure initiated with the fall of inner and outer bricklayers. Brittle failure of panels was mainly due to detachment and debonding of FRP from brick surface. No FRP rupture was witnessed in any of CFRP, AFRP and GFRP retrofitted masonry panel, failure was due to delamination of FRP from the panel surface. Failure patterns of CFRP + PP-band, AFRP + PP-band and GFRP + PP-band retrofitted masonry panels are shown in Figure 6d,f,h, respectively. Initial failure of all three FRP + PP-band retrofitted masonry panels was similar as that of only FRP retrofitted masonry panels but presence of PP-band did not allow the masonry panel to dismember and kept it as one unit which enabled it to undergo further displacements.

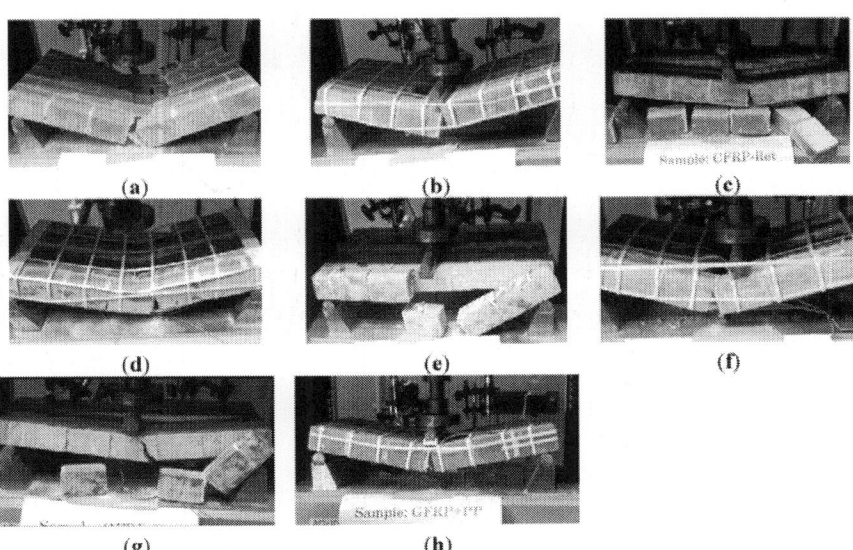

Figure 6. Failure mode and cracking pattern of Type-1 masonry panels under out-of-plane bending test for (**a**) URM, (**b**) PP-band retrofitted, (**c**) CFRP retrofitted, (**d**) CFRP + PP-band retrofitted, (**e**) AFRP retrofitted, (**f**) AFRP + PP-band retrofitted, (**g**) GFRP retrofitted, and (**h**) GFRP + PP-band retrofitted.

3.3. Out-of-Plane Bending Test Results for Type-2 Masonry Walls

Figure 7 shows the moment-deflection curves of URM, PP-band retrofitted, CFRP and CFRP + PP-band retrofitted masonry panels under out-of-plane bending test for Type-2 walls, whereas their corresponding shear stress values are plotted on secondary vertical axis. Figure 7a shows the moment-deflection curves of URM masonry panels. Both of URM masonry panels have shown a variety of peak strengths ranging from 0.0066 to 0.0154 kN-m, which is significantly less than peak strength of URM panels in Type-1 as shown in Figure 4a. Figure 7b shows the moment-deflection curve for PP-band retrofitted masonry panel along with URM masonry panels. PP-band has shown almost same peak strength as that of URM masonry panels but it has shown a long deformation capacity. Figure 7c shows that CFRP has increased the URM strength from 0.0154 to 0.77 kN-m, but failure displacement was far less than

PP-band retrofitted masonry panel. CFRP + PP-band retrofitted masonry panel has increased the peak strength from 0.0154 to 0.79 kN-m and failure displacement from 0.17 to almost 44 mm as shown in Figure 7c. Type-2 masonry panels have shown similar type of behavior as that of their corresponding Type-1 masonry panels but in Type-2 masonry panels, the residual strength has vanished by stepwise degradation, as shown in Figure 7b,c. This phenomenon could be due to gradual separation of bricks from mortar as all bed joints are parallel to the plane of loading.

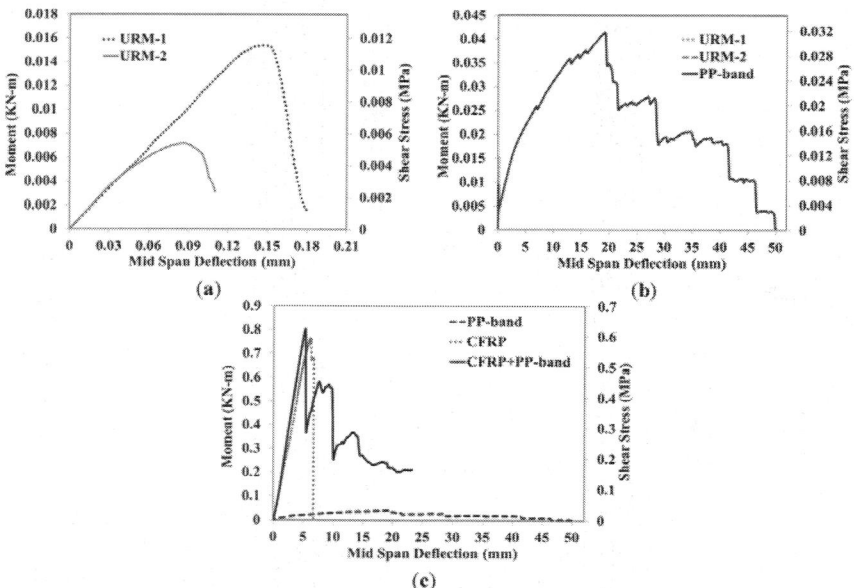

Figure 7. Moment-deflection curves of Type-2 masonry panels under out-of-plane load for (**a**) URM, (**b**) URM and PP-band retrofitted and (**c**) PP-band, CFRP and CFRP + PP-band retrofitted.

Failure Modes for Type-2 Walls

Figure 8a–d shows the failure pattern of masonry panel under out-of-plane loading for Type-2 walls. Failure patterns were also similar as that of Type-1 masonry panels. URM has shown a highly brittle failure with separation of wall segments at bed joints as shown in Figure 8a. In the case of PP-band retrofitted masonry panel, failure was gradual and panel did not break into different segments as shown in Figure 8b. Figure 8c shows failure of CFRP retrofitted masonry panel. CFRP retrofitted masonry panels failed by delamination of CFRP from the bottom face at a section located near the supports, which shows a kind of shear failure near the support. Once the delamination of CFRP started, it did not stop and, finally, the panel broke into several pieces. Figure 8d shows the failure of CFRP + PP-band retrofitted masonry panel. Failure started by the delamination of CFRP from the panel surface near the support, but PP-band hindered the complete delamination and breaking of panel into small pieces, as shown in Figure 8d.

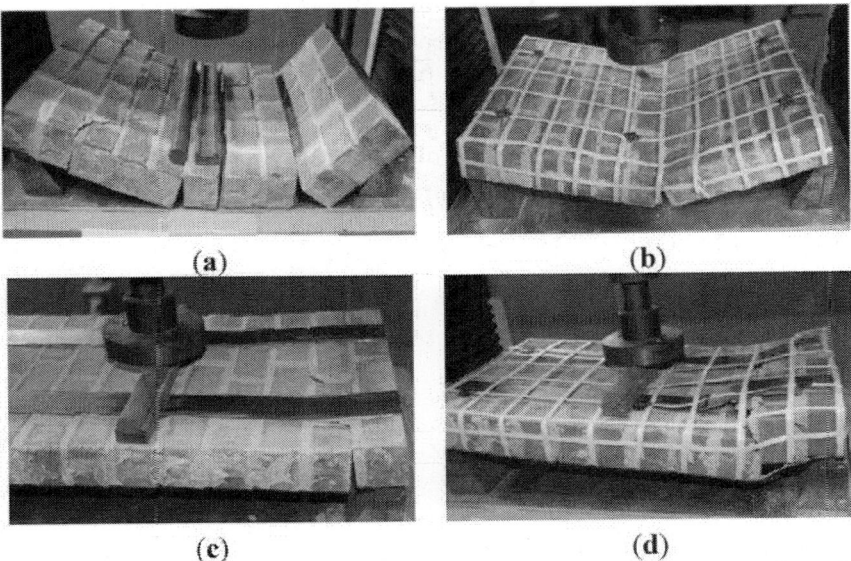

(a)   (b)

(c)   (d)

Figure 8. Failure mode and cracking pattern of Type-2 masonry panels under out-of-plane bending test for **(a)** URM, **(b)** PP-band retrofitted, **(c)** CFRP retrofitted, and **(d)** CFRP + PP-band retrofitted.

4. RETROFITTING PROCEDURE

For in-plane diagonal compression test, all FRP and FRP + PP-band retrofitted masonry panel were retrofitted by 50 mm × 0.5 mm FRP strips. Whereas for out-of-plane bending test, FRP strip width of 40 mm with fabric thickness of 0.5 mm was used. In the case of FRP and FRP + PP-band retrofitted masonry panels, quantity of FRP was decided based upon the FRP reinforcement ratios used in experiments conducted in the past [12,16,18,35,36,37,38]. The surfaces of masonry walls were cleaned with the help of a wet cloth and then FRP was applied on both faces of wall using strong epoxy glue.

In the case of PP-band and FRP + PP-band retrofitted masonry panels, PP-band mesh pitch of 50 mm was used. Mesh pitch was decided based upon the parametric study carried out by Sathiparan (2005) and Mayorca and Meguro (2008) [26,39]. PP-band mesh making process is very simple and do not require any skilled labor. Unbonded and untensioned PP-bands are arranged in a mesh fashion and connected at their intersection points using portable ultrasonic welder. Mesh of PP-bands with any desired pitch is prepared commercially by using ultrasonic welding machine in which multiple ultrasonic welders are arranged in rows. Holes are drilled through the wall at approximately four times the mesh pitch for an existing masonry wall and small straws/pipes are left embedded at the joints in the case of new construction [39]. Figure 9 shows some of the features of PP-band retrofitting method. Figure 9a,b show Polypropylene bands and mesh making process, whereas Figure 9c shows

PP-band meshes prepared by ultrasonic welding machine. Once PP-band mesh is ready, it can be applied on both faces of wall with the help of out-of-plane connectors, as shown in Figure 9d. These out-of-plane connectors could be a steel wire, PP string or PP-band itself. PP-band is attached on walls with the help of out-of-plane connectors and no epoxy is applied on wall surface. At wall edges or corners PP-bands are connected with the help of portable ultrasonic welders, which are commercially available in markets and could be easily brought to the site. After applying the PP-band mesh on masonry walls, a layer of surface finish is applied, which gives an appearance just like an ordinary masonry house as shown in Figure 10 [28].

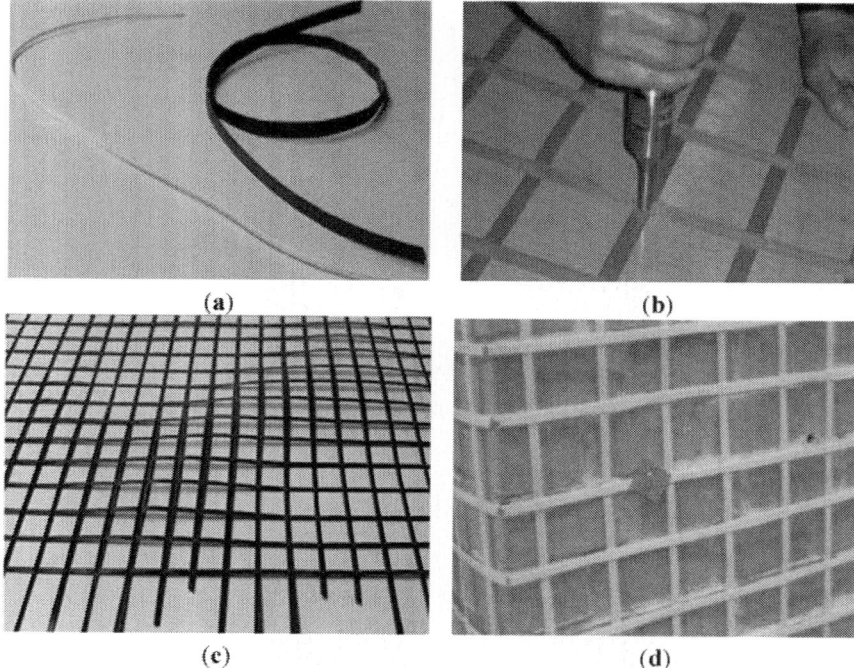

Figure 9. Retrofitting of masonry walls using PP-band mesh: (**a**) Unbonded and untensioned PP-band, (**b**) ultrasonic welding of PP-bands, (**c**) PP-band mesh prepared by ultrasonic welding machine, and (**d**) out-of-plane connection of PP-band mesh.

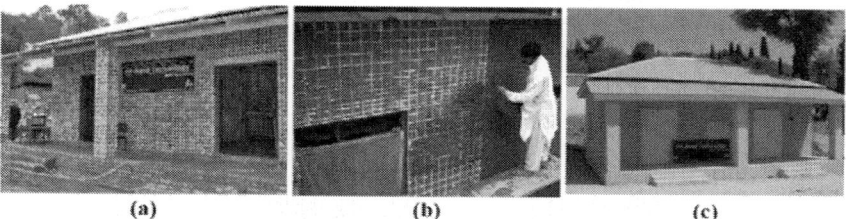

Figure 10. Application of PP-band retrofitting technology in Pakistan, (**a**) PP-band retrofitted house, (**b**) plastering of PP-band retrofitted house, and (**c**) surface Finished PP-band retrofitted house [28].

5. EXPERIMENTAL PROGRAM

Current experimental study is carried out to evaluate the alternate strengthening techniques for URM masonry structures. Many researchers have tried to evaluate the performance of retrofitted masonry using full-scale brick units with size of masonry panels ranging from 1070 to 1975 mm but it requires large size testing facilities and a large amount of research funds [10,18,19,38]. Structural tests of scaled models are also an alternative to understand the behavior of masonry wall system. Many researchers have carried out experiments using scaled bricks and scaled size of masonry panels (ranging from 300 to 600 mm) based upon available testing facilities and number of units to be tested [12,16,40,41].

In most of developing countries, length of brick ranges from 200 to 300 mm, whereas thickness of mortar ranges from 13 to 20 mm. Keeping in mind aforementioned aspects, length of the brick was scaled down to 75 mm for better handling, laying and based upon available manufacturing facilities, whereas depth of brick was kept half of its length. Thickness of brick was kept 50 mm to provide sufficient out-of-plane stability during application of in-plane loads. Thickness of mortar was kept 5 mm to control the consistency of the wall panels and to allow the failure, either inside the mortar or at brick mortar interface.

5.1. Test Specimens and Setup

5.1.1. Diagonal Compression Test

Figure 11 shows test setup of masonry panels under diagonal compression test. Same test setup was used for all masonry panels. All panels were tested with the same initial rate of loading up to the first initial peak load; after that, different loading rates were used depending upon the type of masonry panel. In the case of the URM masonry panel, and all FRP retrofitted masonry panels, a loading rate of 0.15 mm/min was used up to the failure of masonry panels. For PP-band retrofitted and FRP + PP-band retrofitted masonry panels, loading rate of 0.15 mm/min was used up to initial peak, but after the initial failure peak, loading rate was increased to 1.00 mm/min to evaluate the performance of PP-band and FRP + PP-band retrofitted panels under higher diagonal displacements. Diagonal force over the corners of masonry panels was applied with specially designed strong wooden wedges placed at top and bottom platens of 100 kN Universal Testing Machine (UTM) manufactured by Shimadzu Scientific Instruments (SSI), Kyoto, Japan, as shown in Figure 11a,b. In-plane displacements of masonry panels were measured with linear variable displacement transducers (LVDT) of 500×10^{-6} mm sensitivity.

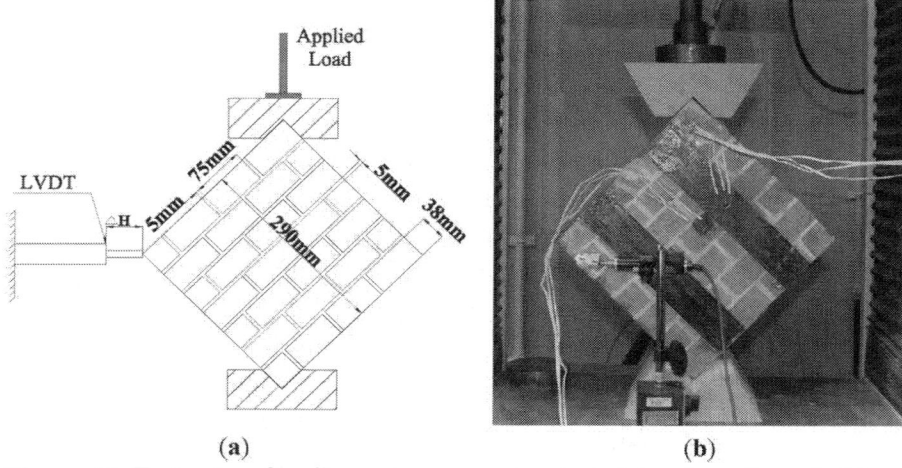

(a)   (b)

Figure 11. Test setup for diagonal compression test: (**a**) schematic and (**b**) experimental.

Recorded horizontal and vertical deformation and diagonal loads are converted to corresponding stress and strains by the following Equations (1) and (2):

$$v = \frac{P \cos \alpha}{((L+W)/2)t} \tag{1}$$

$$\gamma = \frac{(\Delta H + \Delta V)}{\sqrt{L^2 + W^2})} \tag{2}$$

where P = load applied, α = angle between the load and wallet top surface, L = length of wallet, W = height of wall, t = thickness of wall, ΔV = vertical deformation and ΔH = horizontal deformation.

Panel retrofitting scheme for in-plane diagonal compression test is shown in Figure 12. In the case of three FRP retrofitted panels, each one of the panels was retrofitted with CFRP, AFRP and GFRP. Similarly, out of three FRP + PP-band retrofitted masonry panels, the first panel was CFRP + PP-band retrofitted, the second was AFRP + PP-band retrofitted, and the third was GFRP + PP-band retrofitted masonry panel. Dimensions of each masonry panel were 280 mm × 290 mm × 50 mm, made with 75 mm × 38 mm × 50 mm brick units, as shown in Figure 12. Mortar mixed proportion of cement, lime and sand (140 g:1110 g:2800 g) with 0.14 water/cement ratio was used for in-plane diagonal compression test. Five-millimeter mortar thickness was used for both vertical and horizontal joints.

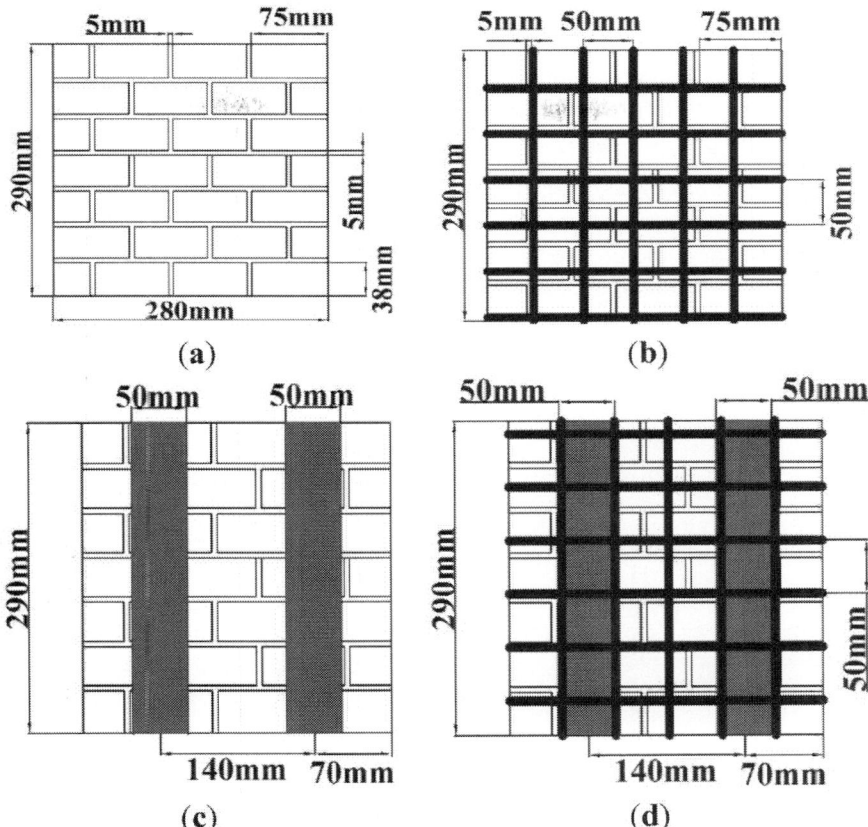

Figure 12. Panel retrofitting scheme for in-plane diagonal compression test: (a) non-retrofitted masonry panel, (b) PP-band retrofitted masonry panels, (c) FRP retrofitted masonry panel, and (d) FRP + PP-band retrofitted masonry panel.

5.1.2. Out-of-Plane Load Test

In out-of-plane bending test, masonry panels were tested like a simply support beam having center-to-center span of 440 mm with a line load applied at the center of span throughout the width of the masonry panel, as shown in Figure 13a,b. Different displacement loading rates were selected depending upon the failure displacement and duration of test. All masonry panels were tested at initial loading rate of 0.1 mm/min up to the initial peak, whereas PP-band and FRP + PP-band retrofitted panels had 2 mm/min loading rate after the initial failure. Two LDTV displacement transducers with 500 × 10^{-6}mm sensitivity were attached at quarter span to measure deflections. Wall specimens were simply supported on two specially designed steel rollers and applied displacement was transferred from machine to specimen with the help of steel roller and a cap plate as shown in Figure 13b. A small initial displacement is applied to assure the full contact of the panel and loading arrangement.

Masonry walls under lateral loads can have either only uniaxial bending or biaxial bending depending upon the aspect ratio (length/height) of wall and its boundary conditions (presence of other walls perpendicular to wall under consideration). In order to consider this effect, two types of masonry panels were constructed. Type-1 corresponds to a condition where wall is long and tall (length/height > 0.5) having bending about longitudinal and vertical axis of the wall, whereas Type-2 represents a condition when walls are tall (aspect ratio < 0.5) having bending only about the longitudinal axis of the wall. For Type-2 walls, it was very difficult to test walls under biaxial bending, so in order to simplify the testing plan, two types of walls were constructed for each case, as shown in Figure 14. In Type-1 masonry panels, bed joints of wall were kept parallel to horizontal axis, whereas in Type-2 bed joint were kept perpendicular to horizontal axis.

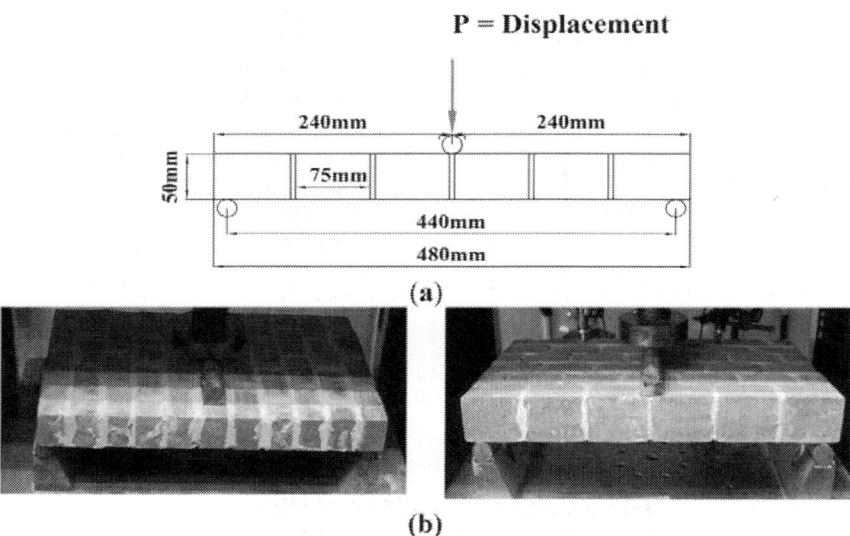

Figure 13. Test setups of masonry panels under out-of-plane bending test: (**a**) schematic and (**b**) experimental.

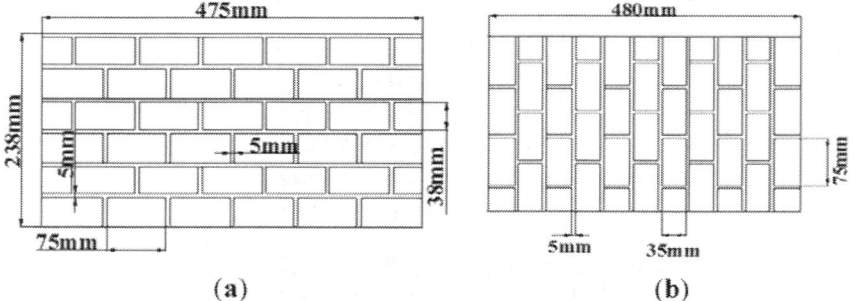

Figure 14. Layout of Type-1 and Type-2 masonry panels for out-of-plane bending test: (**a**) Type-1 and (**b**) Type-2.

Figure 15 shows the testing plan for out-of-plane load tests for Type-1 and Type-2 walls. Ten masonry panels were tested in out-of-plane direction for Type-1 walls. Out of ten masonry panels, three were non-retrofitted, named URM-1, URM-2 and URM-3; one was PP-band retrofitted; three were CFRP, AFRP and GFRP retrofitted; and three were CFRP + PP, AFRP + PP and GFRP + PP-band retrofitted masonry panels. Size of each masonry panel was 475 mm × 238 mm × 50 mm. Five wall panels were constructed for Type-2 walls, out of five masonry panels, two were non-retrofitted, one was PP-band retrofitted, one was CFRP retrofitted, and one was CFRP + PP-band retrofitted. All masonry wall panels were constructed, cured and tested under similar conditions.

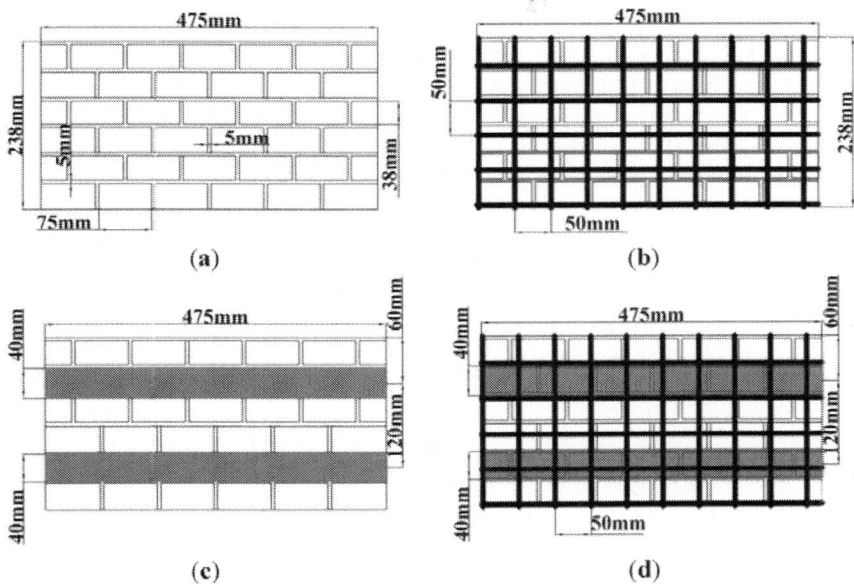

Figure 15. Panels retrofitting scheme for out-of-plane bending test: (**a**) non-retrofitted masonry panel, (**b**) PP-band retrofitted masonry panels, (**c**) FRP retrofitted masonry panel, and (**d**) FRP + PP-band retrofitted masonry panel.

5.2. Material Properties

5.2.1. FRP, Polypropylene Band (PP-band) and Epoxy
Figure 16 shows the stress-strain curve of three PP-bands samples (PP-1, PP-2 and PP-3) of same polypropylene material and three different types of FRPs (CFRP, AFRP and GFRP) supplied by Sekisui Jushi Strapping Co. Milton Keynes, UK and Sankyo Manufacturers, Yokohama, Japan, respectively. PP-bands and FRPs were tested under uniaxial tensile conditions. Figure 16 shows that all three PP-band samples have shown a bi-linear behavior having initial and residual modulus of elasticity of 6.92 and 1.98 GPa. Three different types of FRP materials were used in order to find the best economical and efficient type of FRP to be used with the PP-band. For this purpose, CFRP, AFRP and GFRP were used. Out of these three, CFRP was found to be the most expensive and

GFRP was the cheapest among all. All FRPs carry a biaxial type of fiber layout having fabric thickness of 0.5 mm. CFRP, AFRP and GFRP have shown a linear elastic behavior with tensile modulus of elasticity of 37, 24 and 18 GPa, respectively. Figure 16 shows that FRP has high stiffness and tensile strength with poor elongation capacity, whereas PP-bands have comparatively low strength, but have higher strain values compared to CFRP, AFRP and GFRP. Failure behavior and strength increment are mainly function of FRP strength, epoxy used and the surface on which it is applied. Two types of epoxies are available, the first adhesive is epoxy resin and other is a flexible polyurethane polymer. Flexible polyurethane polymer with higher ductility and lower tensile modulus can ensure lesser brittle failure of FRP systems but strength increment could not be sufficient due to movability of applied composites. According to Bhaman *et al.*, due to higher tensile modulus, epoxy resin can contribute much higher in terms of strength as compared to flexible polymer resins [32]. In this study, an epoxy resin Epoxy Bond E-250 was used as it has strong bonding properties with lower debonding strain values. Table 1 summarizes the properties of epoxy resin (Epoxy Bond E-250) used to apply FRP over the masonry wall surface. All epoxy strength parameters were provided by the epoxy supplier Konishi Chemical Ind., Wakayama, Japan and examined at a temperature of 20 ± 1 °C after curing time of seven days.

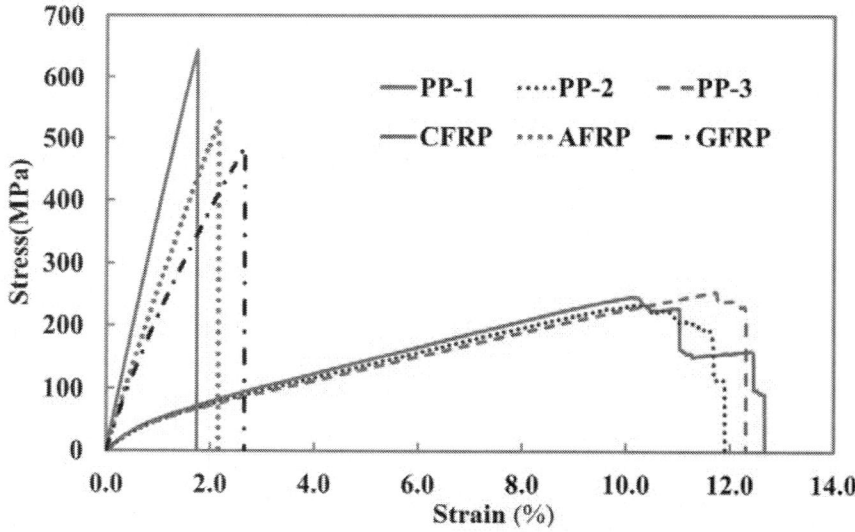

Figure 16. Stress-strain curves of CFRP, AFRP, GFRP and PP-band.

Table 1. Material properties of epoxy.

Material Property	Value	Units
Tensile Strength	20	MPa
Tensile shear strength	9.6	MPa
Compressive shear bond strength	21	MPa
Modulus of elasticity	1.5	GPa

5.2.2. Properties of Masonry

Material properties of brick, mortar, and masonry were determined by performing compression, shear and bond tests on bricks, mortar and masonry prisms. Material testing results are summarized in Table 2. Compression test on clay burnt bricks was carried out according to ASTM C67-03a [42]. Average compressive strength of brick samples was determined by performing compression test on three brick samples of 75 mm × 38 mm × 50 mm. Brick work for masonry walls was conducted using a cement lime mortar. Two different types of mortar were used for diagonal compression test and three point bending test. Weight mixed proportion of cement, lime and sand (140 g:1110 g:2800 g) with 0.14 water/cement ratio was used for in-plane diagonal compression test and (250 g:1000 g:2800g) with 0.25 water/cement was used for out-of-plane load bending test. Compressive strength of mortar for in-plane and out-of-plane load tests was determined by performing a direct compression test using 50 mm × 50 mm × 50 mm mortar cubes according to ASTM C109-02 [43]. Compressive strength of masonry was determined by performing compression test using masonry prisms of five bricks joined together with 5 mm mortar thickness according to ASTM C1314-03a [44]. Average values of brick, mortar and masonry compressive strength for in-plane and out-of-plane load tests are given in Table 2. Shear strength and bond strength of brick mortar assembly was determined by performing shear test and bond test using the test setup shown in Figure 17a,b.

Table 2. Material properties of masonry.

Material Property	Diagonal Compression Test	Out-of-Plane Load Test	Units
Average compressive strength of bricks	26.10	26.10	MPa
Average compressive strength of mortar	1.03	1.16	MPa
Average compressive strength of masonry	13.42	13.60	MPa
Average Shear strength of mortar	0.023	0.025	MPa
Average bond strength of mortar	0.0032	0.0043	MPa

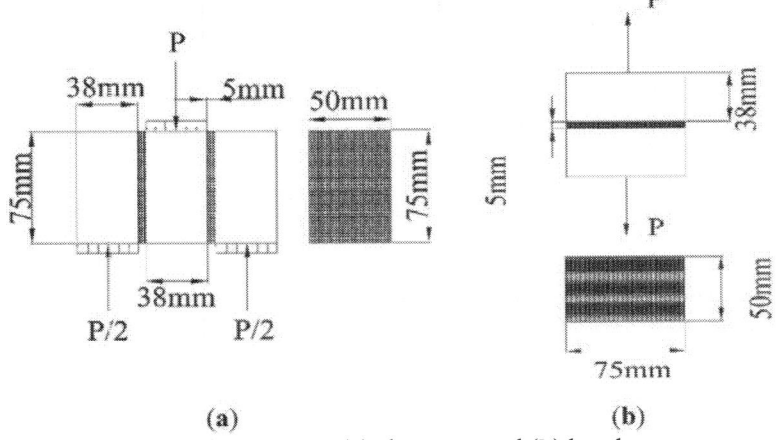

(a)   (b)

Figure 17. Test setup for masonry (a) shear test and (b) bond test.

6. CONCLUSIONS

Current experimental study gives a good basis for comparison and suitability of different retrofitting methods. FRP is an expensive material with high tensile strength but with low tensile failure strain ranging from 2%–4%, whereas PP-band is a low cost material with low tensile strength and higher tensile failure strain. Composite of FRP + PP-band has not only increased the initial strength and deformation capacity but also the residual strength of masonry wall panels. FRP and URM masonry panel has shown a brittle failure as compared to PP-band and FRP + PP-band retrofitted masonry panels. In the case of only FRP retrofitted masonry panels, once the FRP debonding started, it did not stop until complete detachment form the wall surface. This resulted in sudden drop in shear strength and breaking of masonry panels into several pieces. In the case of FRP + PP-band retrofitted panels, FRP was not completely detached because of the hindrance provided by PP-band at certain locations. This effect kept FRP attached on some parts of the masonry panel, which resulted in higher values of residual strength compared to only PP-band retrofitted masonry panels. Due to the holding effect of PP-band, the panels were capable of withstanding much bigger displacements as compared to only FRP retrofitted masonry panels.

In out-of-plane bending tests it is seen that application of PP-band along with the CFRP has changed the failure mode of FRP retrofitted masonry panels as initial drop in strength and separation of brick layers were not observed in FRP + PP-band retrofitted masonry panels. In FRP retrofitted panels, panels were broken in the form of longitudinal strips, but FRP + PP-band retrofitted masonry panels remained as single units during the entire loading history. In the case of FRP + PP-band retrofitted masonry panels, the initial failure of masonry panel was just like FRP retrofitted masonry panels due to debonding of FRP from the brick surface, but panels had sufficient residual strength and long deformation capacity.

In this study, different types of FRPs were selected to find the most suitable and appropriate type of FRP to be used with PP-band. Experimental results showed that in all types of FRP retrofitted panels, full strength of FRP was not utilized and higher strength and stiffness values of FRP have not played any important role as all types of FRP has shown almost similar increase in initial strength. Most suitable type of FRP is that which has minimum cost. Failure occurs either due to detachment of FRP from the brick surface or due to debonding of FRP and epoxy from the brick surface, which forces not using high quality and higher volumes of FRP, but to use strong epoxy or glue and bricks with good surface qualities. Current study has proven that FRP + PP-band composite is a high performance composite solution for seismic retrofitting of masonry structures and proposed retrofitting technique, using composite of FRP and PP-band is a very viable solution for seismic retrofitting of URM masonry structures.

ACKNOWLEDGMENTS

The authors thankfully acknowledge the technical and financial support mainly provided by Meguro Laboratory, Institute of Industrial Science, University of Tokyo Japan and this work was also partially supported by the Deanship of Scientific Research at King Faisal University through its 13th annual project 13011.

AUTHOR CONTRIBUTIONS

All authors tried their best to contribute effectively to perform and analyze this experimental work. Saleem Muhammad Umair and Kimiro Meguro planned experimental scheme and methods. Models were constructed and retrofitted by Saleem Muhammad Umair. Muneyoshi Numada and Kimiro Meguro supervised the experimental work and Muhammad Nasir Amin contributed for the analysis of results and failure patterns.

REFERENCES

1. Shrive, N.G. The use of fiber reinforced polymers to improve seismic resistance of masonry. *Constr. Build. Mater.* **2006**,*20*, 269–277.
2. Yoshimura, M.; Meguro, K. Proposal of retrofitting promotion system for low earthquake resistant structures in earthquake prone countries. In Proceedings of the 13th World Conference on Earthquake Engineering, Vancouver, BC, Canada, 1–6 August 2004.
3. Bruneau, M. State-of-the-art report on seismic performance of unreinforced masonry buildings. *J. Struct. Eng.* **1994**,*120*, 230–251.
4. Lotfi, H.R.; Shing, P.B. An interface model applied to fracture of masonry structures. *J. Struct. Eng.* **1994**, *120*, 63–80.
5. Calvi, G.M.; Bolognini, D. Seismic response of reinforced concrete frames infilled with masonry panels weakly reinforced. *J. Earthq. Eng.* **2001**, *5*, 153–185.
6. Magenes, G.; Calvi, G.M. In-plane seismic response of brick masonry walls. *Earthq. Eng. Struct. Dyn.* **1997**, *26*, 1091–1112.
7. Velazquez-Dimas, J.I.; Ehsani, M.R. Modeling out-of-plane behavior of URM walls retrofitted with fiber composites. *J. Compos. Constr.* **2000**, *4*, 172–181.
8. Elgawady, M.; Lestuzzi, P.; Badoux, M. A review of conventional seismic retrofitting for URM. In Proceedings of the 13th Internatonal Brick/Block Masonry Conference, Amsterdam, The Netherlands, 4–7 July 2004.

9. Schwegler, G. Masonry construction strengthened with fiber composites in seismically endangered zones. In Proceedings of the 10th European Conference on Earthquake Engineering, Vienna, Austria, 28 August–2 September 1994; pp. 467–476.

10. Gilstrap, J.M.; Dolan, C.W. Out-of-plane bending of FRP reinforced masonry walls. *Compos. Sci. Technol.* **1998**, *58*, 1277–1284.

11. Ehsani, M.R.; Saadatmanesh, H.; Velazquez-Dimas, J.I. Behavior of retrofitted URM walls under simulated earthquake loading. *J. Compos. Constr.* **1999**, *3*, 134–142.

12. Valluzzi, M.R.; Tinazzi, D.; Modena, C. Shear behavior of masonry panels strengthened by FRP laminates. *Constr. Build. Mater.* **2002**, *16*, 409–416.

13. Zhao, T.; Zhang, C.J.; Xie, J. Experimental study on earthquake strengthening of brick walls with continuous carbon fibre sheet. *Mason. Int.* **2003**, *16*, 21–25.

14. Stratford, T.; Pascale, G.; Manfroni, O.; Bonfiglioli, B. Shear strengthening masonry panels with sheet GFRP. *J. Compos. Constr.* **2004**, *8*, 434–443.

15. Tan, K.H.; Patoary, M.K.H. Strengthening of masonry walls against out-of plane loads using fiber-reinforced polymer reinforcement. *J. Compos. Constr.* **2004**, *8*, 79–87.

16. ElGawady, M.A.; Lestuzzi, P.; Badoux, M. In-plane seismic response of URM walls upgraded with FRP. *J. Compos. Constr.* **2005**, *9*, 524–534.

17. Prota, A.; Marcari, G.; Fabbrocino, G.; Manfredi, G.; Aldea, C. Experimental in-plane behavior of tuff masonry strengthened with cementitious matrix-grid composites. *J. Compos. Constr.* **2006**, *10*, 223–233.

18. Marcari, G.; Manfredi, G.; Prota, A.; Pecce, M. In-plane shear performance of masonry panels strengthened with FRP. *Composites* **2007**, *38*, 887–901.

19. Alcaino, P.; Santa-Maria, H. Experimental response of externally retrofitted masonry walls subjected to shear loading. *J. Compos. Constr.* **2008**, *12*, 489–498.

20. Willis, C.R.; Yang, Q.; Seracino, R.; Griffith, M.C. Damaged masonry walls in two-way bending retrofitted with vertical FRP strips. *Constr. Build. Mater.* **2009**, *23*, 1591–1604.

21. Roca, P.; Araiza, G. Shear response of brick masonry small assemblages strengthened with bonded FRP laminates for in-plane reinforcement. *Constr. Build. Mater.* **2010**, *24*, 1372–1384.

22. Santa-Maria, H.; Alcaino, P. Repair of in-plane shear damaged masonry walls with external FRP. *Constr. Build. Mater.* **2011**, *25*, 1172–1180.

23. Luciano, R.; Sacco, E. Damage of masonry panels reinforced by FRP sheets. *Int. J. Solids Struct.* **1998**, *35*, 1723–1741.

24. Mayorca, P.; Meguro, K. Efficiency of polypropylene bands for the strengthening of masonry structures in developing countries. In Proceedings of the 5th International Summer Symposium, Japan Society of Civil Engineers (JSCE), Tokyo, Japan, 26 July 2003; pp. 125–128.

25. Drozdov, A.D.; Al-Mulla, A.; Gupta, R.K. Structure-property relations for polymer melts: Comparison of linear low-density polyethylene and isotactic polypropylene. *Adv. Mater. Res.* **2012**, *1*, 245–268.

26. Sathiparan, N.; Mayorca, P.; Nesheli, K.; Ramesh, G.; Meguro, K. In-plane and out-of-plane behavior of PP-band retrofitted masonry panels. In Proceedings of the 4th International Symposium on New Technologies for Urban Safety of Mega Cities in Asia, Singapore, 18–19 October 2005; pp. 231–240.

27. Meguro, K.; Mayorca, P.; Sathiparan, N. Shaking table test on timber masonry house models retrofitted with PP-band meshes. In Proceedings of the 7th Intertional Symposium on New Technologies for Urban Safety of Mega Cities in Asia, Beijing, China, 21–22 October 2008.

28. Dar, A.M.; Saleem, M.U.; Numada, M.; Meguro, K. Experiment study on reduction of PP-band mesh connectivity for retrofitting of masonry structure. *Bull. Earthq. Resist. Struct.* **2014**, *47*, 67–80.

29. Sathiparan, N.; Meguro, K. Seismic behavior of low earthquake-resistant arch-shaped roof masonry houses retrofitted by PP-band meshes. *ASCE Pract. Period. Struct. Des. Constr.* **2012**, *17*, 54–64.

30. Sathiparan, N.; Mayorca, P.; Meguro, K. Shake table tests on one-quarter scale models of masonry houses retrofitted with PP-band mesh. *Earthq. Spectr.* **2012**, *28*, 277–299.

31. Turco, V.; Secondin, S.; Morbin, A.; Valluzzi, M.R.; Modena, C. Flexural and shear strengthening of un-reinforced masonry with FRP bars. *Compos. Sci. Technol.* **2006**, *66*, 289–296.

32. Ghiassi, B.; Xavier, J.; Oliveira, D.V.; Kwiecien, A.; Lourenço, P.B.; Zajac, B. Evaluation of the bond performance in FRP-brick components re-bonded after initial delamination. *Compos. Struct.* **2015**, *123*, 271–281.

33. Ghiassi, B.; Xavier, J.; Oliveira, D.V.; Lourenço, P.B. Application of digital image correlation in investigating the bond between FRP and masonry. *Compos. Struct.* **2013**, *106*, 340–349.

34. Ghorbani, R.; Matta, F.; Sutton, M.A. Full-field deformation measurement and crack mapping on confined masonry walls using digital image correlation. *Exp. Mech.* **2015**, *55*, 227–243.

35. Santa-Maria, H.; Alcaino, P.; Luders, C. Experimental response of masonry walls externally reinforced with carbon fibers. In Proceedings of the 8th US National Conference on Earthquake Engineering, San Francisco, CA, USA, 18–22 April 2006.

36. Mahmood, H.; Russell, A.P.; Ingham, J.M. Laboratory testing of unreinforced masonry walls retrofitted with glass FRP sheets. In Proceedings of the 14th International Brick/Block Masonry Conference, Sydney, Australia, 17–20 February 2008.

37. Nardone, F.; Protra, A.; Mafredi, G. Design criteria for FRP seismic strengthening of masonry walls. In Proceedings of the 14th World Conference on Earthquake Engineering, Beijing, China, 12–17 October 2008.

38. Mahmood, H.; Ingham, J.M. Diagonal compression testing of FRP retrofitted unreinforced clay brick masonry panels. *J. Compos. Constr.* **2011**, *15*, 810–820.

39. Mayorca, P.; Meguro, K. Proposal of a methodology to design PP-band meshes to retrofit low earthquake resistant houses. In Proceedings of the 6th International Symposium on New Technologies for Urban Safety of Megacities in Asia, Dhaka, Bangladesh, 9–10 December 2007.

40. Sinan, A.; Anil, O.; Emin, M.K.; Mustafa, K. An experimental study on strengthening of masonry infilled RC frames using diagonal CFRP strips. *Composites* **2008**, *39*, 680–693.

41. Ozsayin, B.; Yilmaz, E.; Ispir, M.; Ozkaynak, H.; Yuksel, E.; Ilki, A. Characteristics of CFRP retrofitted hollow brick infill walls of reinforced concrete frames. *Constr. Build. Mater.* **2011**, *25*, 4017–4024.

42. *Standard Test Methods for Sampling and Testing Brick and Structural Clay Tile*; ASTM C67–03a; ASTM International: West Conshohocken, PA, USA, 2003.

43. *Standard Test Method for Compressive Strength of Hydraulic Cement Mortars (Using 2-in or [50-mm] Cube Specimen)*; ASTM C109/C109M-02; ASTM International: West Conshohocken, PA, USA, 2002.

44. *Standard Test Methods for Compressive Strength of Masonry Prisms*; ASTM C1314–03a; ASTM International: West Conshohocken, PA, USA, 2003.

CHAPTER 10

Ferrimagnetism and Ferroelectricity of the Composite Matrix: $SrBi_2Nb_2O_9(SBN)_x$-$BaFe_{12}O_{19}(BFO)_{100-x}$

Marta Jussara Souza da Rocha[1], Mucio Costa Campos Filho[1,2], Klara Rhaissa Burlamaqui Theophilo[1], Juliano Casagrande Denardin[3], Igor Frota de Vasconcelos[4], Eudes Borges de Araujo[5], Antonio Sergio Bezerra Sombra[1]*

[1]Laboratório de Telecomunicações e Ciência e Engenharia de Materiais, Fortaleza, Brazil;
[2]Laboratório de Espalhamento de Luz, Departamento de Física, UFC, Campus do Pici, Fortaleza, Brazil;
[3]Universidade de Santiago do Chile (USACH), Estación Central, Santiago, Chile;
[4]Departamento de Engenharia Metalúrgica e de Materiais, UFC, Fortaleza, Brazil;
[5]Universidade Estadual Paulista Júlio de Mesquita Filho, Faculdade de Engenharia de Ilha Solteira, Departamento de Física e Química, São Paulo, Brazil.

ABSTRACT

A study of the dielectric, magnetic and structural properties of composites based on M-type barium hexaferrite BFO ($BaFe_{12}O_{19}$) and SBN ($SrBi_2Nb_2O_9$) is presented. The magneto-dielectric matrix composite $(SrBi_2Nb_2O_9)_x(BaFe_{12}O_{19})_{100-x}$, (x = 0, 25, 50, 75 and 100 wt%) were prepared by using a new procedure based in the solid state method. X-ray powder diffraction patterns, Raman and Infrared spectroscopy, Mössbauer spectroscopy and scanning electron microscopy (SEM) were carried out for better understanding of the microstructural, dielectric and magnetic properties. Radiofrequency (RF) dielectric permittivity, dielectric loss measurements and magnetic and electric hysteresis loops properties are also discussed throughout this paper. The hysteresis loops showed that composite samples preserve the ferrimagnetism and ferroelectricity for hexaferrite when SBN is added to the composite (BFO25P), although they become less coercive. In addition, the effects of organic binders group (TEOS, PVA and glycerin) on structural properties were also investigated.

Keywords: Composites; Magneto-Dielectric Materials

1. INTRODUCTION

The ceramics composite matrix are uniform, multiphases materials that have been strongly studied in recent works because of the possibility to obtain materials with desirable properties, specially for application in electronics devices. This properties wouldn't be easily observed for a single phase material, requesting extreme conditions, for example, low temperatures, to its observation thus limiting its applications.

In this work, our main goal is to develop a dielectric material that is able to respond to both electric and magnetic stimulus, i.e. that is ferroelectric and ferromagnetic. To do so, we use the Aurivillius ceramic $SrBi_2Nb_2O_9$ and the Hexaferrite $BaFe_{12}O_{19}$. Such a material could be applied in the same way that common dielectrics (as dielectric resonator antennas, for example) but opening a wide range of possibilities to make the application of ceramics to electronic devices, memories and telecommunications more useful and powerful.

The Aurivillius family of compounds that can be represented by the general formula $[(Bi_2O_2)^{2+}(A_{m-1}B_mO_{3m+1})^{2-}]$ in which A can be a monovalent, divalent or trivalent cation or a combination of those in suitable proportions, B can be a tetravalent or pentavalent cation with m having values of 2, 3, 4… etc. Their structures comprise a stacking of n peroviskite units of normal composition $A_{m-1}B_mO_{3m+1}$ interleaved with Bi_2O_3 layers along the pseuedotetragonal c-axis known as bismuth layered ceramics (BLFC).

They are potential materials for microelectronic applications when desirable properties are reached such as high dielectric permittivity, high Curie temperature, high fatigue resistance [1-3]. However, dielectric loss is still too high for practical applications for electronic devices. Fortunately, the structure of peroviskite allows easy replacement of cations and combinations among phases as well.

Recently, some researchers reported additions of impurities in bismuth-layered composites [1-3]. This work focused on a new alternative for a composite matrix by using barium hexaferrite.

M-type ferrite has a hexagonal structure and has been largely used for ferrofluids and magnetic devices. It has some interesting properties such as high saturation magnetization and high Curie temperature and has attracted interest for many years. By combining ferroelectric SBN ceramics and M-type ferrite, it was possible to produce the magneto-dielectric composite $SrBi_2Nb_2O_9(SBN)_x$-$BaFe_{12}O_{19}(BFO)_{100-x}$. By bringing these two materials together, it is expected to develop a bi-phase material, which would be ferroelectric and ferromagnetic. To do so, it would require SBN not to interact with BFO changing their characteristic properties, and the original phases can remain in the composite. The microstructure investigation is able to reveal whether the materials interact with each other forming new phases or not. If there is no undesirable chemical interactions between SBN and BFO, the new bi-phase ceramic will be formed and it will probably be ferroelectric and ferromagnetic.

The behavior of this new class of ceramics is not clearly known so far. Considering such context a precise investigation should be carried out for each composite combination. These matrix composites have interactions with to magnetic and electric fields which are not completely understood because of cumbersome procedures for getting both ferroelectric and ferromagnetic properties under simple physical conditions. By using a composite matrix it is possible to mitigate those drawbacks and two non-interacting ferromagnetic and ferroelectric ceramics would provide magneto-dielectric properties. In this work, we have studied the microstructure, dielectric and magnetic properties of $SrBi_2Nb_2O_9(SBN)_x-BaFe_{12}O_{19}$ $(BFO)_{100-x}$ matrix.

2. EXPERIMENTAL PROCEDURE

SBN $(SrBi_2Nb_2O_9)$ were synthesized from stoichiometric quantities of Bi_2O_3 (99.99%), Nb_2O_5(99.99%), $SrCO_3$ (99.99%) and BFO $(BaFe_{12}O_{19})$ from BaO (99.99%) and Fe_2O_3 (99.99%) derived from solid-state reaction method. Both reactions of the uncalcined powders are shown below:

$$Bi_2O_3 + Nb_2O_5 + SrCO_3 \rightarrow SrBi_2Nb_2O_9 + CO_2$$
$$BaO + 6Fe_2O_3 \rightarrow BaFe_{12}O_{19}$$

Bi_2O_3, Nb_2O_5, $SrCO_3$ for SBN have been mixed and ball-milled with zirconium balls in a polyethylene container by using a planetary mill for 8 hours, in order to lower it's grain size and enlarge it's surface area aiming to achieve more reactive powders. The same calcination conditions were carried out for ferrite for 2 hours. SBN and BFO samples were calcinated at 800°C for 5 hours and at 1000°C for 24 hours, respectively.

SBN and BFO powders were mixed as the following rule SBN_xBFO_{100-x} (x = 0, 25, 50, 75 and 100 wt%). Three series of samples were produced for different binders, namely polyvinyl alcohol (PVA), glycerin and tetraethyl orthosilicate (TEOS). The two powders in the defined proportions are mixed with 3 wt% of one of the binders. The powders were used to make cylindrical pellets under an uniaxial pressure of about 0.1 MPa for 5 minutes using a hydraulic press.

The bulks were sinterized at 1050°C for 5 hours. Samples were named in proportion of BFO and SBN (BFO100 -100% BFO, BFO75-75% BFO + 25% SBN, BFO50- 50% BFO + 50% SBN, BFO25-25% BFO + 75% SBN and SBN100-100% SBN) and also depending on the binder's name used during their formation (P, T, G, polyvinyl alcohol, tetraethyl orthosilicate and glycerin respectively). The sintered pellets were polished with fine emery paper in order to make both the surfaces flat and parallel and were electroded with high-purity silver paste for dielectric and electrical measurements to ensure a good electrical contact by using an impedance analyzer to cover the frequency range of interest.

The samples powders were analysed using Mössbauer spectroscopy and X-ray diffraction, both at room temperature. The Rietveld analysis was performed by the refinement program DBWS. The bulk samples were analized with Raman and infrared spectroscopy and Scanning electron microscope, for investigation of the structure and radiofrequency measurements, as well as, magnetic and electric hyteresis for investigation of the electromagnetic properties.

2.1. X-Ray Diffraction

The X-ray diffractograms were obtained at room temperature, by using non-sinterized powder samples and a Siemens D501 graphite-monochromator diffractometer using a Cu-K radiation detector (k = 0.1542 nm) with a Bragg-Brentano-geometry scintillation counter operating at 40 kV and 25mA. The ferrite powders were scanned through the 2θ angle ranging from $20°$ to $85°$ and the SBN powder ranging from $20°$ to $70°$ during five seconds for each step of counting time. The Rietveld analysis was performed by the refinement program DBWS. A Modified Thompson-Cox-Hasting pseudo-Voigt profile function was used to fit the calculated curve into the experimental diffraction data. The obtained density with such refinement was used for calculating the relative density (%) of the samples by the Archimedes method.

2.2. Raman and Infrared Spectroscopy

Infrared spectroscopy was obtained with a Mattson 7000 (FTIR) spectrometer, while Raman spectrum was obtained with a FTIR spectrometer VERTEX 70 equipped with RAM II Bruker FT-Raman module adapted for the rough surfaces (diffusion reflection spectroscope EasyDiff), using bulk sintered samples in both cases.

2.3. SEM

Scanning electron microscope (SEM) images for microstructure observation of the sintered pellets were obtained with a scanning electron microscope Vega XMUT/Tescan, Bruker.

2.4. Radiofrequency Measurements

Both surfaces of pellets were painted with silver paste for impedance at radiofrequency measurements. The permittivity and loss tangent were measured by using Agilent E4991A RF impedance/material analyzer connected to a personal computer, for frequency ranging from 1 MHz to 3 GHz, at room temperature.

2.5. Magnetic Hysteresis Measurements

For the measurement of the magnetic moment (M) versus the applied magnetic field (H) a VSM (vibrating sample magnetometer) was used. All measurements were done at room temperature. Magnetization was normalized by the mass of each sample in order to provide M in emu/g. The samples were measured over a maximum applied field of 8000 Oersted.

2.6. Electric Hysteresis Measurements

The electric hysteresis were obtained from a system composed by a Sawyer-Tower circuit connected to a Agilent 54622A digital oscilloscope, a Agilent 33220A function generator and a Trek 610E high-voltage font. All hysteresis loops were done at room temperature with disc shaped samples.

2.7. Mössbauer Spectroscopy

Mössbauer measurements were performed at room temperature by a FAST (ConTec) Mössbauer Systems spectrometer through transmission geometry and radioactive source of ^{57}Co in Rh matrix and isometric shifts are related to metallic iron α-Fe. The NORMOS program was used for adjusting iron sites in Barium hexaferrite structure and for determining hyperfine parameters by using powdered samples.

3. RESULTS AND DISCUSSION

3.1. X-Ray Diffraction

The X-ray diffraction patterns for BFO100, BFO50 and SBN100 are shown in **Figure 1**. The diffraction peaks presented by BFO100 sample were identified by JPCDS (file 78-0133), and powder diffraction patterns of $BaFe_2O_4$ located at 33, 16° and 46, 61° were identified by JPCDS (file 20-0132). Some low-intensity peaks observed at $2\theta < 30°$ were attributed to magnetite identified by JPCDS (file 76-0958). Mali and Ataie [4] suggest that the formation of barium hexaferrite comes from the reaction of intermediate phases of iron oxide and a spinel monoferite $BaFe_2O_4$, and the percentage of crystalline $BaFe_{12}O_{19}$ grows monotonically as increasing the temperature. This would explain the residual phases and a pure hexaferrite would be obtained for higher calcination temperatures. The presence of residual phases for barium hexaferrite was also reported by other authors [5]. The crystal structure found has hexagonal symmetry belonging to the P63/mmc space group. The average crystallite size was calculated by using the Debye-Scherrer's equation around 37, 16 nm.

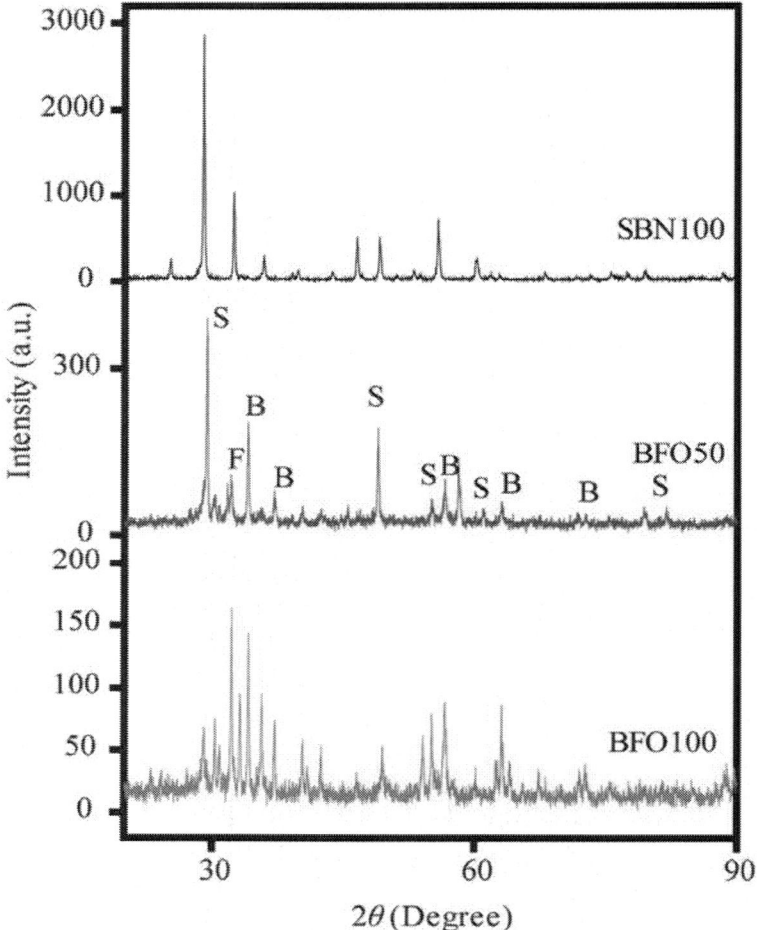

Figure 1. X-ray diffractograms for SBN100, BFO50, BFO100 samples.

For the SBN100 sample all the diffraction peaks were indexed according to JCPDFS (file 49-0607), without residual phase evidences. The crystal structure found is orthorhombic belonging to the A21am space group with crystalline size estimated about 45 nm.

The diffraction peaks for the present composite shows only a combination of the diffraction of the two base compounds. The representative peaks for SBN100 and BFO100 and spinel monoferrite were indexed by using S, B and F letters respectively.

3.2. Rietveld Refinement

Rietveld Refinement was performed only for the SBN100 and BFO100 samples.

The standard R-factors R_p, R_{wp}, R_{exp}, and goodness-offit parameter S, as well as the lattice parameters (a, b, c), the unit cell volume (V) and the density are given in Table 1. A low percentage of R_{wp} and S value close to unit (<1.3 normally) indicates that the refinement was successful. Better results were obtained to SBN100 refinement. The density and volume of unitary cell were close to values reported [3] and also from PDF database.

To the BFO, the presence of a small amount (~0.04%) of secondary phases was not considered in the refinement which increases the refinement parameters values. The density, unit cell and volume presented here are in good agreement with the expected results (JPCDS file peaks number 78-0133) already reported [6,7].

3.3. Archimedes Method

The relative densities of the present samples were measured by Archimedes method and are shown in **Table 2**. The expected density was achieved and better densification is related to TEOS binder use. The only difference between using TEOS, Glycerin and PVA as binders is the densification factor of the samples, since the organic material should vanish around 500°C and it is supposed not to interfere with the material atomic arrangement. Therefore, no structural differences should be observed, even though electrical and magnetic properties are influenced if any consistent change is observed as causing undesirable effects.

Table 1. Rietveld refinement parameters.

Sample	SBN100	BFO100
a (nm)	0.5515	0.5868
b (nm)	0.5513	0.5868
c (nm)	2.5024	2.3106
Density (g/cm³)	7.293	5.358
Volume (nm³)	0.761062	0.689074
R_p	10.74%	27.43%
R_{wp}	14.7%	34.96%
R_{exp}	11.6%	22.73%
S	1.27	1.54

3.4. Raman Spectroscopy

Figure 2 shows the Raman spectra for the composites, SBN100T, and BFO25T, BFO50T, BFO75T and BFO100T, respectively. The characteristic bands for SBN were all in agreement with data from the literature [8]. It is usual to all Aurivillius ceramics, once they are peroviskite like materials, a vibration mode over the 800 cm^{-1} - 900 cm^{-1} frequency range (associated to an octahedral sub-structure). The 835.95 cm^{-1} mode was identified to be associated to NbO$_6$ octahedral stretching mode A$_{1g}$. The 585 cm^{-1} mode is associated to E$_g$ modes related to the stretching of oxygen bonds in the octahedron. The lower frequency modes are probably related to lattice vibrations or backscattered photons.

Table 2. Relative density of the samples obtained from the Archimedes method.

Binder	Sample	Relative Density
TEOS	BFO100T	93.83%
	BFO75T	83.09%
	BFO50T	91.88%
	BFO25T	87.19%
	SBN100T	82.70%
PVA	BFO100P	83.09%
	BFO75P	80.77%
	BFO50P	85.60%
	BFO25P	78.11%
	SBN100P	67.92%
Glycerin	BFO100G	66.04%
	BFO75G	81.02%
	BFO50G	83.61%
	BFO25G	83.32%
	SBN100G	80.30%

A detailed analysis of hexaferrite vibration modes was carried out by Kreisel [9]. According to his work, 683.7 cm^{-1} mode was clearly observed, which is related to the symmetric stretching of bi-pyramid shaped sub-structure of FeO_5. This mode is commonly between 670 - 710 cm^{-1} for M-type ferrite. This is the strongest band in ferrites containing such structure. No other Raman active modes are expected for M-type ferrite above 800 cm^{-1} frequency, thus 845.3 cm^{-1} band should be related to $BaFe_2O_4$ monoferrite. Some expected bands for $BaFe_{12}O_{19}$ are not observed in spectrum [9], this fact is a conesquence of strong noise in the spectrum, hiding lower intensity bands. Opaque and dark samples usually show this sort of spectrum in FT-Raman measurements.

The composite spectra show no new band, indicating that the vibration modes remain invariant when the two materials are mixed. An apparent displacement of the 585 cm^{-1} SBN band in composite samples may be a result of insufficient resolution to identify this band on SBN from 683.7 cm^{-1} band on BFO. The bands are located at the same frequency, just varying their intensities proportionally to the sample composition, showing that both are non-interacting materials and that they form a bi-phase ceramics.

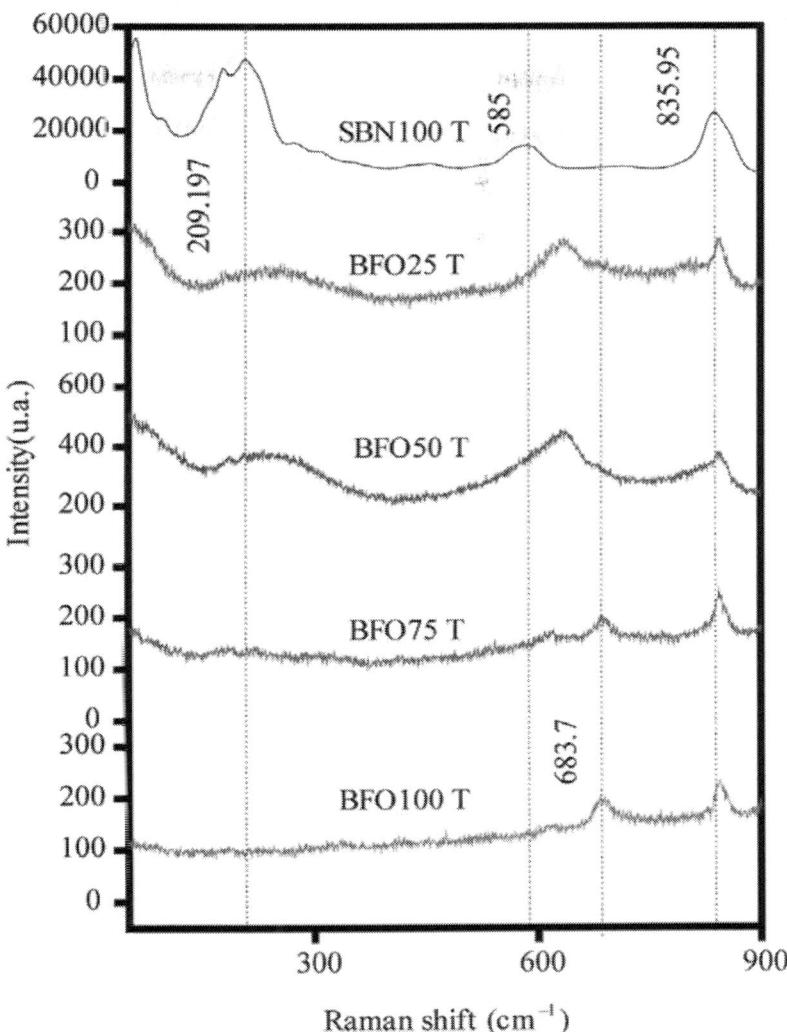

Figure 2. Raman spectra for the SBN100T, BFO75T, BFO50T, BFO25T, BFO100T samples.

3.5. Infrared Spectroscopy

Figure 3 shows the IR spectra. The NbO_6 structure could be related to the 600 - 640 cm^{-1} bands which are expected for octahedral stretching. The bismuth layers bands are hidden once they would be located about the same frequency range where NbO_6 octahedral stretching bands is found, and are wide enough to hide bismuthlayer bands.

Similar results were already reported for the ferrite spectra [10,11-13]. The small displacement of the bands could be related to different methods of sample preparation [11-13]. The identified bands for barium hexaferrite in this present work are 609.4 cm⁻¹, 430 cm⁻¹ and 543.89 cm⁻¹, which are all in good agreement with the characteristic bands for crystalline $BaFe_{12}O_{19}$ (552, 434 and 583 cm⁻¹) [11-13]. These bands are related to Fe-O vibration bonds. Again, the composites did not apparently show new bands.

3.6. Mössbauer Spectroscopy

The Mössbauer Spectroscopy was carried out only for BFO100, since there is no iron atoms in SBN and the presence of strontium atoms caused great difficulties to obtain the spectra when this material is added into ferrite.

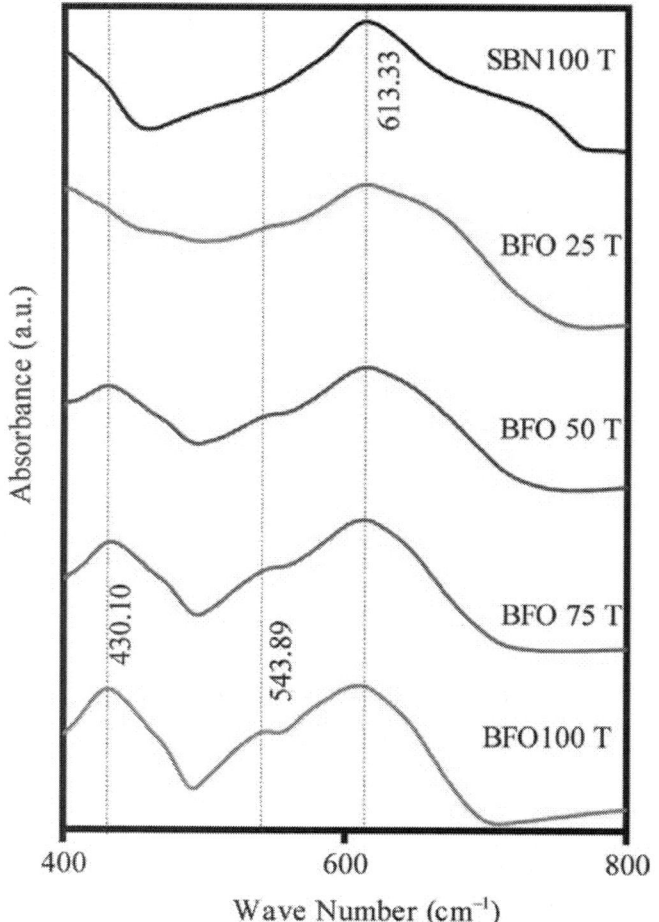

Figure 3. IR spectra for the SBN100T, BFO75T, BFO50T, BFO25T, BFO100T samples.

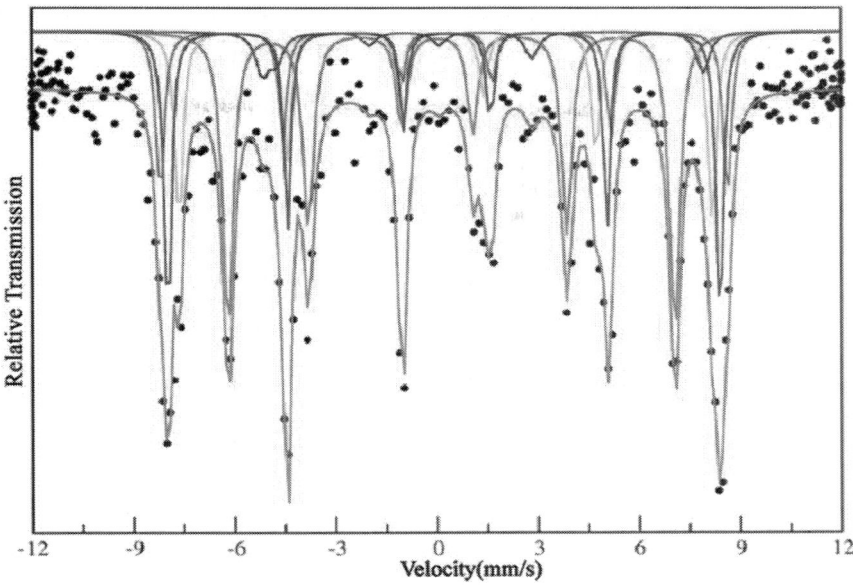

Figure 4. Mössbauer spectrum for the BFO100 sample.

Figure 4 shows five observed sextets related to a characteristic complex structure of five iron sites. The sites occupied by iron atoms are 12k, 4f1, 4f2, 2a and 2b, located in the spinel (4f1 and 12k) or hexagonal blocks (4f2, 2a and 2b) which were found in the structure. The hyperfine parameters are given in **Table 3** and a fitting method was applied to irons sites based on the same arguments used by Sankaranarayanan and Shaman [14] since our spectrum results show all iron sites formed in the same profile.

The strongest hyperfine field sextet (52, 16T) is associated to 4f2-octahedra site of barium hexaferrite, located on the hexagonal block. The hyperfine fields of 2b (40.65T) and 12k (41.09T) do have close values, being the lowest ones. Although the high quadrupolar splitting associated to 2b makes it possible to identify which one is related to the 2b or 12k site, since 2b site, which has a trigonal bipyramidal symmetry, is highly distorted and is expected to show extremely large quadruple-pole spliting values (2.310 mm/s) and relative intensity (6%) lower than 12k site, which has large relative intensity (>35%) for samples treated at high temperatures (>900°C). Other two sextets have showed high hyperfine fields and they are corresponding to 4f1 (49.00T) tetrahedral and 2a (50.79T) octahedral sites.

Also, the isomer shift is useful for determining valence states, and electron shielding. The oxidation states in which iron form compounds is +3 and +2 and the configuration of the valence electrons is [Ar]3d^5 and [Ar]3d^6. Due to the screening effect of 3d electrons, this electron density at the nucleus is higher for Fe^{3+} than in the Fe^{2+}, thus ferrous ions have larger positive isomer shifts than ferric ions. The isomer shift observed in our ferrite is between 0.32 - 0.38 mm/s which is in great accordance with the ferric isomer shifts reported. [13,15].

3.7. SEM

The SEM micrographs are shown in Figures 5(a)-(c). Note that for BFO100 sample a dense ceramic is shown, although some pores can be observed, with heterogeneous distribution of grain sizes and shapes, showing sharp edged grains and also some completely asymmetric agglomerates. The grain size is ranging approximately from 0.8 to 3.0 µm, which allows the existence of single-domain grains, since some of the grains are smaller than typical domain-wall dimensions, to this ferrite (1 µm) [16]. Nevertheless most of the grains exhibit multi-domain morphology. $BaFe_{12}O_{19}$ has been reported to have non-isometric crystallite size, which would explain such a wide range of size variation [11].

For the SBN100 the micrographs reveal a much more porous sample, but still dense. The grain size distribution is homogenous and the agglomerate shapes are more symmetric than found in BFO100 samples. The grain size varies between 0.4 - 1 µm, which was also observed by Shrivastava and Jah [2]. The composite sample shows good density with few vacancies. The distribution of ferrite and SBN is homogeneous and each grain can be easily identified. The addiction of SBN leads to an increase of the ferrite grain size, which can reach more than 5 µm in composite samples. This change in morphology would probably induce a modification in the magnetic properties.

3.8. Radiofrequency Measurements

As it is shown in **Table 4**, the values of permittivity (ε') and loss (tanδ) changes according to different binders. The change in relative density of the sample is related to this effect. The permittivity of BFO samples for different binders is associated to the densification of the sample due to the presence of more vacancies in the lower density samples [17]. The value of the permittivity of this ferrite is slowly decreasing with frequency, because the electronic jump between ferric ions is not able to follow the changes in polarization up to high frequencies [17,18]. SBN is more likely to vary its permittivity as a function of density since the presence of vacancies will induce more significant decreasing in ε'. This is related to the fact that SBN is an insulator, with relative permittivity values going up to 10^2 [2,19].

The differences in density of samples is an important factor for the composite samples. TEOS shows higher densification for the majority of the samples and so higher permittivity values have been observed for BFO75T and BFO50T samples. **Figure 6** shows the frequency dependence of permittivity.

Another factor which influences the values of both permittivity and loss is the grain size. Since the addition of SBN is leading to an increase of the ferrite grain size, thus it leads to an easier polarization [20]. The increase of these parameters is observed for composite samples with higher percentage of BFO (BFO50 and BFO75).

Table 3. Hyperfine parameters of the Mössbauer measurements.

Sample	Sites	Coordination	IS (mm/s)	QS (mm/s)	H_{hf} (T)	R_A (%)
	12k	octahedral	0.351	0.401	41.09	36%
	4f1	tetrahedral	0.334	0.097	49.00	18%
BFO100	4f2	octahedral	0.372	−0.099	52.16	15%
	2a	octahedral	0.371	−0.098	50.79	24%
	2b	trigonal bipyramidal	0.326	2.310	40.65	6%

(a)

(b)

(c)

Figure 5. SEM micrographs for (a) BFO100; (b) BFO50; (c) SBN100.

Table 4. Values of permittivity and loss in the RF range.

Samples	100 MHz		500 MHz		1 GHz	
	ε'	$\tan\delta$	ε'	$\tan\delta$	ε'	$\tan\delta$
BFO100G	12.38	0.0072	12.58	0.02	13.15	0.040
BFO75G	23.17	0.0374	22.80	0.023	23.56	0.032
BFO50G	69.75	0.167	48.71	0.212	45.340	0.235
BFO25G	36.12	0.004	36.84	0.117	41.81	0.202
SBN100G	35.46	0.048	37.51	0.246	43.03	0.428
BFO100P	13.19	0.0061	13.43	0.036	14.13	0.050
BFO75P	77.04	0.0484	78.03	0.068	96.91	0.117
BFO50P	33.10	0.0854	31.07	0.0217	33.15	0.00052
BFO25P	42.42	0.0115	44.02	0.174	51.70	0.308
SBN100P	27.89	0.0037	28.03	0.0027	28.90	0.0007
BFO100T	8.04	0.0055	8.05	0.002	8.10	0.003
BFO75T	96.68	0.0046	104.8	0.238	143.3	0.490
BFO50T	108.94	0.128	91.44	0.231	88.94	0.282
BFO25T	35.35	0.0055	36.47	0.146	42.16	0.255
SBN100T	39.05	0.0086	39.19	0.008	41.18	0.006

The dielectric loss shows similar values to TEOS and PVA series and higher values to Glycerin. **Figure 6** shows the frequency dependence of loss tangent ($\tan\delta$). The observed behavior for composite samples indicates the existence of resonance absorption at higher frequencies [21]. At low frequency, a weak oscillatory behavior is observed for loss values of SBN. This could be explained by the stronger influence of DC conductivity and charge carriers over this frequency range since no loss peak is observed [22]. The ferrite shows a dispersive behavior for the loss tangent, which decreases as frequency increases. This kind of behavior was studied by Koops [23] and is related to the existence of different regions inside the samples with different values of conductivity. The samples with higher losses were BFO50T and BFO75T.

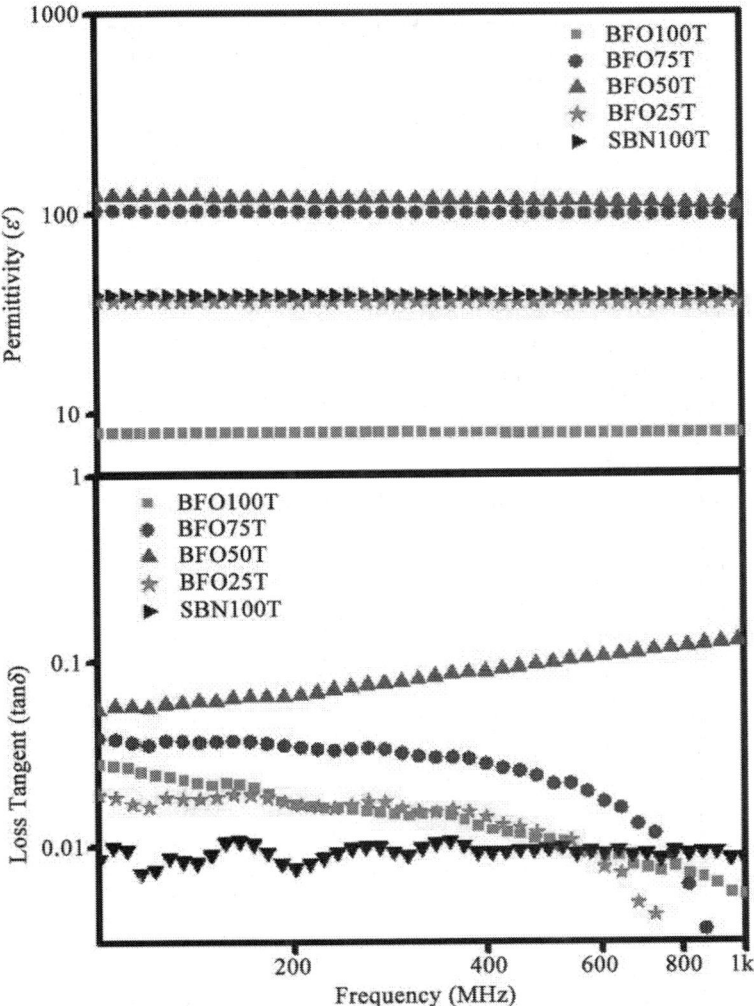

Figure 6. RF measurements of (a) permittivity and (b) loss tangent for TEOS samples.

3.9. Magnetic Hysteresis

Figure 7 shows the hysteresis loops for TEOS and glycerin series. A little depression in the hysteresis loop for pure ferrite samples was observed. This behavior is a consequence of coupling between the barium hexaferrite and the spinel ferrite, which is found for a small percentage and is anti-ferromagnetic. The presence of this extra phase can induce a lowering in the coercivity of the material. This anomalous shape is better observed in the glycerin series loops.

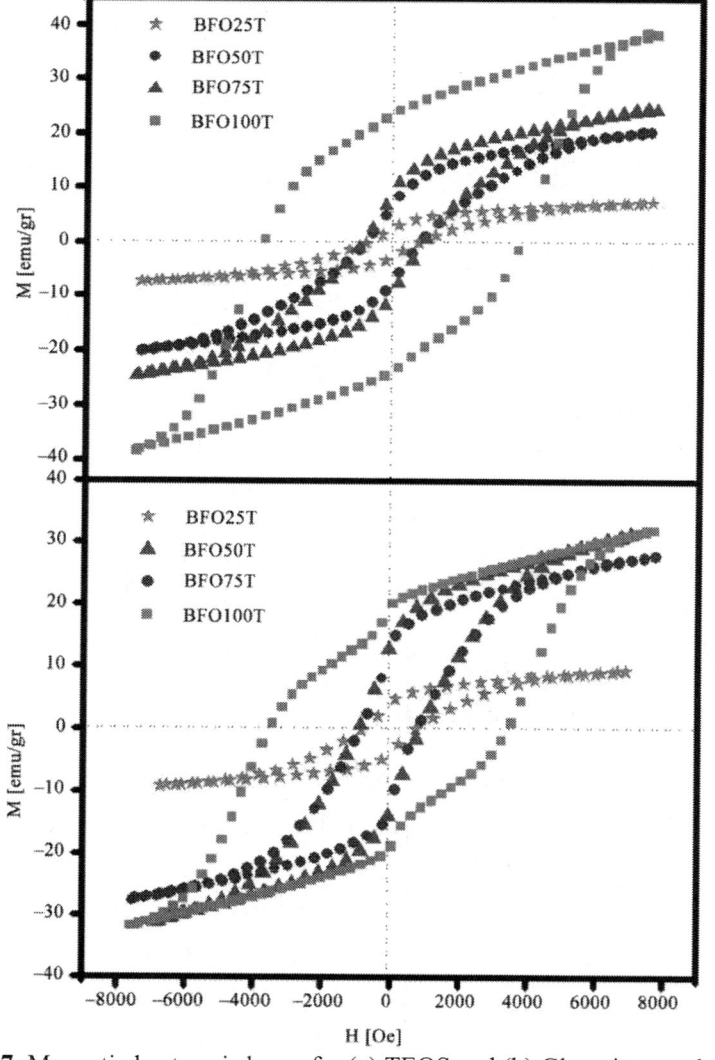

Figure 7. Magnetic hysteresis loops for (a) TEOS and (b) Glycerin samples.

The hysteresis loops show a narrowing behavior by adding SBN. The present material coercivity is reduced as a function of SBN in the composite. And this is coherent because SBN is not magnetic, therefore it is not able to respond to any magnetic stimulation. No response by SBN to magnetic fields causes the lowering in all magnetization parameters. Another important fact is the increase of the ferrite grain size for the composite samples. Depending on the size of the grain, only multi-domains microstructure will be formed and this will lead to a decrease in coercivity and coercive fields, due to the easier domain-wall motion in the multi-domain environment [6,24]. The values of reminisce magnetization, saturation magnetization and coercive fields are shown in **Table 5**. We can see that no strong deviation was observed from TEOS to PVA series. Glycerin series show a little quantitative difference, but follow the same qualitative behavior. Once again, samples density plays an important role of defining the denser samples with higher magnetization parameters and coercive fields. Similar results were reported by Ogasawara and Oliveira [25].

It is known that a good crystalline structure leads to a better magnetization where temperature and dwell time are parameters that strongly influence the enhancement of crystalline structure [6]. Although some works reports high coercicivity and coercive fields (>5 KOe) for barium hexaferrite [16,26], this was not observed in this present work because high synthesis temperature used here allows a growth of crystallite size and the formation of multi-domain structures [25]. For the pure ferrite, the higher coercive field observed was 3.7 KOe and the higher magnetization of saturation was 38.3 emu/g, which is in good agreement with similar works [6,18,25].

3.10. Electrical Hysteresis

Figure 8 shows P-E hysteresis loops, recorded at room temperature, at 1 Hz frequency, for the selected SBN100P, BFO25P, BFO50P and BFO75P composite samples. The hysteresis loops data are characteristic of a lossy linear capacitor, rather than a ferroelectric material. For the SBN100P the electric field amplitude was gradually increased to show that the polarization saturation tendency is not observed for this sample and as a consequence each hysteresis loop present essentially the same shape, as shown **Figure 8**(a).

The nonlinear P-E nature of a typical ferroelectric material is characterized by a hysteresis effect, in which the polarization trends to a saturation value for high electric field and the direction of the spontaneous polarization can be reversed by an external applied electric field. On the other hand, a P-E linear behavior is typically observed for normal dielectric or paraelectric materials. As the SBN100P sample was prepared from pressed SBN powders, the P-E behavior observed in **Figure 8**(a) can be due to typical defects of a non sintered sample. In contrast, the BFO addition at the composite composition is expected to increase the conductivity and consequently a linear P-E behavior could be observed as shown **Figure 8**(b). Thus, the spontaneous polarization values shown in **Figure 8** do not correspond to the absolute values of the studied samples.

Table 5. Magnetic Hysteresis parameters of the samples.

Samples	Remanent Magnetization (emu/g)	Coercive Field (Oe)	Saturation Magnetization (emu/g)	Maximum Field (Oe)
BFO100T	23.98	3744.4	34.97	7680
BFO75T	10.07	932.8	20.39	7676.4
BFO50T	7.60	813.6	17.39	7539
BFO25T	2.34	725	5.99	7388
BFO100P	21.13	3699.4	31.33	7578
BFO75P	15.84	1110	27.34	7418.4
BFO50P	8.73	633.8	18.13	7629
BFO50P	3.76	923	6.46	7648.6
BFO100G	19.99	3460.4	28.90	7692.8
BFO75G	13.86	833	26.24	7731.6
BFO50G	14.10	868	22.31	7010.4
BFO25G	4.01	747.4	7.33	6888.2

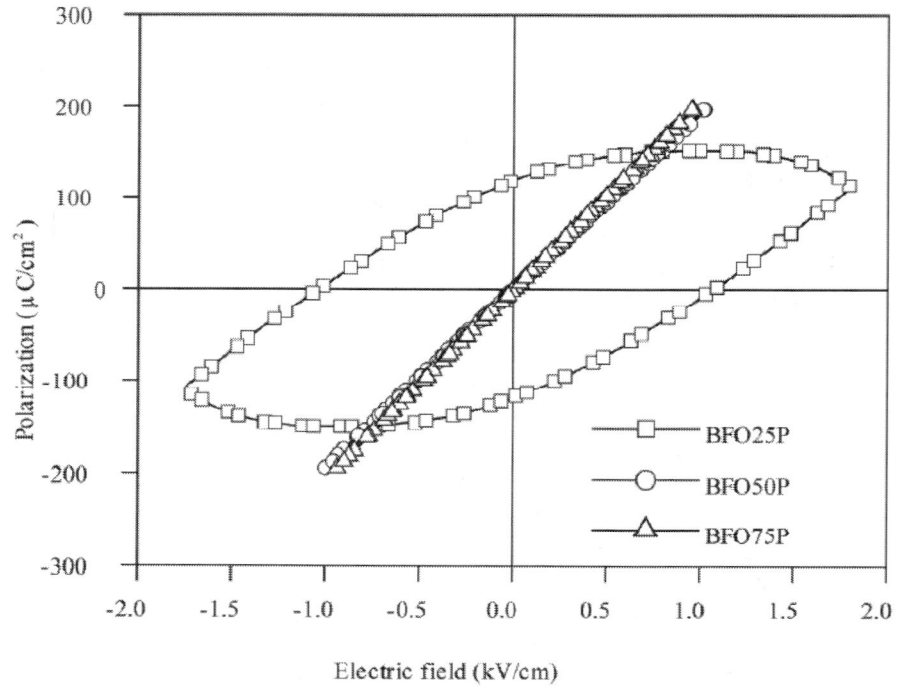

(b)

Figure 8. Electric hysteresis loops recorded at 1 Hz frequency for (a) SBN100T sample and (b) BFO25P, BFO50P and BFO75P composites.

4. CONCLUSIONS

In this paper a study of the magnetic and dielectric properties of composites based on M-type barium hexaferrite BFO (BaFe$_{12}$O$_{19}$) and SBN (SrBi$_2$Nb$_2$O$_9$) is presented. The magneto-dielectric matrix composite (SrBi$_2$Nb$_2$O$_9$)$_x$

$(BaFe_{12}O_{19})_{100-x}$, (x = 0,25,50,75 and 100 wt%) were prepared by a new procedure using the solid state reaction method.

A study of the effects of an organic group of binders (TEOS, PVA and glycerin) on the structural properties of the ceramic-matrix composites was also investigated.

SEM micrographics have revealed a good densification and a grain size ordered microns. The relative densities achieved with TEOS as a binder were above 80% for all samples, showing that TEOS is a better binder to this ceramics. It was also observed that density and grain size strongly influences the magnetic and electrical properties, in general, denser samples show better electric and magnetic properties.

A complex behavior was observed for loss tangent over the radiofrequency range, which means that less lossy samples could not help keeping this characteristic over entire frequency range. The magnetic hysteresis loops showed that composite samples preserve the ferrimagnetism for hexaferrite when SBN is added to the composite, although they become less coercive. For electric hysteresis the density of the samples are not high enough to define the true behavior of ferroelectricity in composite samples. For further works, the properties over microwave frequency range, thermal influences on the dielectric properties will be investigated for possible applications of the specimen.

5. ACKNOWLEDGEMENTS

This work was partly sponsored by CAPES, CNPq, FUNCAP (Brazilian agencies) and the U. S. Air Force Office of Scientific Research (AFOSR) (FA9550-11-1- 0095). The support from Fondecyt 1080164; Núcleo Milenio Magnetismo Básico y Aplicado; and CONICYT under Proyecto BASAL FB0807, is gratefully acknowledged.

REFERENCES

1. V. Shrivastava, A. K. Jha and R. G. Mendiratta, "Dielectric Studies of La and Pb Doped $SrBi_2Nb_2O_9$ Ferroelectric Ceramic," Materials Letters, Vol. 60, No. 12, 2006, pp. 1459-1462.

2. V. Shrivastava, A. K. Jha and R. G. Mendiratta, "Structural and Electrical Studies in La-Substituted $SrBi_2Nb_2O_9$ Ferroelectric Ceramics," Physica B: Condensed Matter, Vol. 371, No. 2, 2006, pp. 337-342.

3. R. R. Das, P. B., W. Pérez and R. S. Katiyar, "Effect of Ca on Structural and Ferroelectric Properties of $SrBi_2Ta_2$- O_9 and $SrBi_2Nb_2O_9$ Thin Films," Ceramics International, Vol. 30, No. 7, 2004, pp. 1175-1179.

4. A. Mali and A. Ataie, "Structural Characterization of NanoCrystalline $BaFe_{12}O_{19}$ Powders Synthesized by Sol-Gel Combustion Route," Scripta Materialia, Vol. 53, No. 9, 2005, pp. 1065-1070.

5. S. S. Fortes, J. G. S. Duque and M. A. Macedo, "Nanocrystals of $BaFe_{12}O_{19}$ Obtained by the Proteic Sol-Gel Process," Physica B: Condensed Matter, Vol. 384, No. 1-2, 2006, pp. 88-90.

6. U. Topal, H. Ozkan and H. Sozeri, "Synthesis and Characterization of Nanocrystalline $BaFe_{12}O_{19}$ Obtained at 850°C by Using Ammonium Nitrate Melt," Journal of Magnetism and Magnetic Materials, Vol. 284, 2004, pp. 416-422.

7. B. S. Zlatkov, M. V. Nikolic, O. Aleksic, H. Danninger and E. Halwax, "A Study of Magneto-Crystalline Alignment in Sintered Barium Hexaferrite Fabricated by Powder Injection Molding," Journal of Magnetism and Magnetic Materials, Vol. 321, No. 4, 2009, pp. 330-335.

8. N. Ortega, P. Bhattacharya and R. S. Katiyar, "Enhanced Ferroelectric Properties of Multilayer $SrBi_2Ta_2O_9/SrBi_2$- Nb_2O_9 Thin Films for NVRAM Applications," Materials Science and Engineering B, Vol. 130, No. 1-3, 2006, pp. 36-40.

9. J. Kreisel, G. Lucazeau and H. Vincent, "Raman Spectra and Vibrational Analysis of $BaFe_{12}O_{19}$ Hexagonal Ferrite," Journal of Solid State Chemistry, Vol. 137, No. 1, 1998, pp. 127-137.

10. L. B. Kong, T. S. Zhang, J. Ma and F. Boey, "Progress in Synthesis of Ferroelectric Ceramic Materials via HighEnergy Mechanochemical Technique," Progress in Materials Science, Vol. 53, No. 2, 2008, pp. 207-322.

11. H.-F. Yu and H.-Y. Lin, "Preparation and Thermal Behavior of Aerosol-Derived $BaFe_{12}O_{19}$Nanoparticles," Journal of Magnetism and Magnetic Materials, Vol. 283, No. 2-3, 2004, pp. 190-198.

12. T. González-Carreño, M. P. Morales and C. J. Serna, "Barium Ferrite Nanoparticles Prepared Directly by Aerosol Pyrolysis," Materials Letters, Vol. 43, No. 3, 2000, pp. 97-101.

13. F. M. M. Pereira, C. A. R. Junior, M. R. P. Santos, R. S. T. M. Sohn, F. N. A. Freire, J. M. Sasaki, J. A. C. de Paiva and A. S. B. Sombra, "Structural and Dielectric Spectroscopy Studies of the M-Type Barium Strontium Hexaferrite Alloys (Ba x Sr_{1-x} $Fe_{12}O_{19}$)," Journal of Materials Science: Materials in Electronics, Vol. 19, No. 7, 2008, pp. 627-638.

14. V. K. Sankaranarayanan and D. C. Khan, "Mechanism of the Formation of Nanoscale M-Type Barium Hexaferrite in the Citrate Precursor Method," Journal of Magnetism and Magnetic Materials, Vol. 153, No. 3, 1996, pp. 337- 346.

15. J. Li, W. Sturhahn, J. M. Jackson, V. V. Struzhkin, J. F. Lin, J. Zhao, H. K. Mao and G. Shen, "Pressure Effect on the Electronic Structure of Iron in $(Mg,Fe)(Al,Si)O_3$ Perovskite: A Combined Synchrotron Mössbauer and XRay Emission Spectroscopy Study up to 100 GPa," Physics and Chemistry of Minerals, Vol. 33, No. 8-9, 2006, pp. 575-585.

16. G. Mendoza-Suárez, J. A. Matutes-Aquino, J. I. EscalanteGarcía, H. Mancha-Molinar, D. Ríos-Jara and K. K. Johal, "Magnetic Properties and Microstructure of Baferrite Powders Prepared by Ball Milling," Journal of Magnetism and Magnetic Materials, Vol. 223, No. 1, 2001, pp. 55-62.

17. K. K. Mallick, P. Shepherd and R. J. Green, "Dielectric Properties of M-Type Barium Hexaferrite Prepared by Co-Precipitation," Journal of the European Ceramic Society, Vol. 27, No. 4, 2007, pp. 2045-2052.

18. F. M. M. Pereira, M. R. P. Santos, R. S. T. M. Sohn, J. S. Almeida, A. M. L. Medeiros, M. M. Costa and A. S. B. Sombra, "Magnetic and Dielectric Properties of the MType Barium Strontium Hexaferrite (Ba x Sr_{1-x} $Fe_{12}O_{19}$) in the RF and Microwave (MW) Frequency Range," Journal of Materials Science: Materials in Electronics, Vol. 20, No. 5, 2009, pp. 408-417.

19. B. Harihara Venkataraman and K. B. R. Varma, "Frequency-Dependent Dielectric Characteristics of Ferroelectric $SrBi_2Nb_2O_9$ Ceramics," Solid State Ionics, Vol. 167, No. 1-2, 2004, pp. 197-202.

20. C. Singh, S. B. Narang, I. S. Hudiara, K. Sudheendran and K. C. J. Raju, "Complex Permittivity and Complex Permeability of Sr Ions Substituted Ba Ferrite at X-Band," Journal of Magnetism and Magnetic Materials, Vol. 320, No. 10, 2008, pp. 1657-1665.

21. A. L. Ruoff, "Material Science," Prentice Hall, Englewood Cliffs, 1973.

22. A. K. Jonscher, "Dielectric Relaxation in Solids," Journal of Physics D: Applied Physics, Vol. 32, No. 14, 1999, pp. 57-70.

23. C. G. Koops, "On the Dispersion of Resistivity and Dielectric Constant of Some Semiconductors at Audiofrequencies," Physical Review, Vol. 83, No. 1, 1951, pp. 121- 124.

24. J.-Y. Yu, S.-L. Tang, L. Zhai, Y.-G. Shi and Y.-W. Du, "Synthesis and Magnetic Properties of Single-Crystalline $BaFe_{12}O_{19}$ Nanoparticles," Physica B: Condensed Matter, Vol. 404, No. 21, 2009, pp. 4253-4256.

25. T. Ogasawara and M. A. S. Oliveira, "Microstructure and Hysteresis Curves of the Barium Hexaferrite from CoPrecipitation by Organic Agent," Journal of Magnetism and Magnetic Materials, Vol. 217, No. 1-3, 2000, pp. 147-154.

26. J. Ding, D. Maurice, W. F. Miao, P. G. McCormick and R. Street, "Hexaferrite Magnetic Materials Prepared by Mechanical Alloying," Journal of Magnetism and Magnetic Materials, Vol. 150, No. 3, 1995, pp. 417-420.

27. Y.-Q. Qu, A.-D. Li, Q.-Y. Shao, Y.-F. Tang, D. Wu, C. L. Mak, K. H. Wong and N.-B. Ming, "Structure and Electrical Properties of Strontium Barium Niobate Ceramics," Materials Research Bulletin, Vol. 37, No. 3, 2002, pp. 503-513.

CHAPTER 11

Carbon Fibre Sensor: Theory and Application

Alexander Horoschenkoff[1] and Christian Christner[2]

[1] Munich University of Applied Sciences, Germany
[2] Universität der Bundeswehr München, Germany

1. INTRODUCTION

The piezoresistive[1] - carbon fibre sensor (CFS) consists of a single carbon fibre roving with electrical connected endings embedded in a sensor carrier (patch) for electrical insulation. Depending on the requirements of the application different patch types (e.g. glass fibre reinforced plastic (GFRP), polyester film, neat epoxy resin) are used. In terms of the mechanical properties GFRP is a particularly suitable patch material. CFSs with a GFRP patch exhibits an improved linearity of the signal due to the supportive effect of the glass fibres to the carbon sensor fibre especially in the case of compression loading. In [6] the ex-PAN fibre T300B was identified as one suitable carbon fibre for CFSs because of the excellent linear piezoresistive behaviour, the high specific resistivity and a high breaking elongation of the fibre. Figure 1 shows a CFS with a single layer GFRP patch (UD prepreg EG/913).

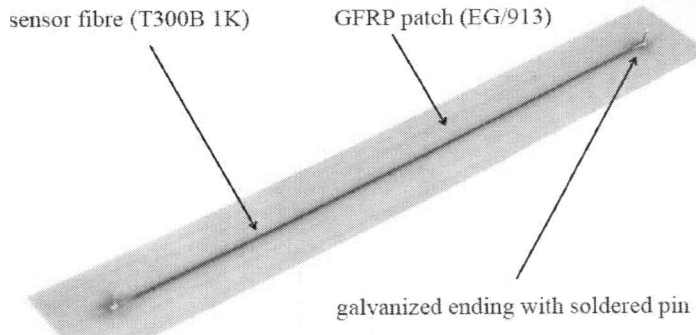

sensor fibre (T300B 1K) GFRP patch (EG/913)

galvanized ending with soldered pin

Figure 1. Carbon fibre sensor with an ex-Pan sensor fibre and a GFRP patch

The manufacturing process of a CFS includes three basic steps: Pre-curing of the carbon fibre, preparation of the electrical connection and embedding of the sensor fibre into the sensor carrier.

The pre-curing process is used to stabilize the carbon fibre roving and to align the filaments of the roving. For this purpose the twisted carbon fibre roving is impregnated by a resin with low viscosity and cured by using a special tooling. Good results for the impregnation of the carbon fibre roving T300B 1K were obtained by using the epoxy resin EP301 S (HBM) and a twist of 20 turns per meter. Spring elements provided a constant tension force along the roving during the curing process at 180°C for 1.5 hours.

For preparation of electrical connections a galvanic process is applied based on a nickel electrolyte. In order to attain a homogeneous nickel coating of the filaments the resin must be removed at the fibre endings[2] - . An applied current of 40 mA for 30 seconds leads to an excellent nickel coating. Depending on the application of the CFS (surface application or structural integration) the ends of the sensor fibre can be provided with soldered pins.

The embedding process of the sensor fibre depends on the used patch type and patch material. Figure 2 shows a micro section of a carbon fibre sensor in longitudinal and transverse direction.

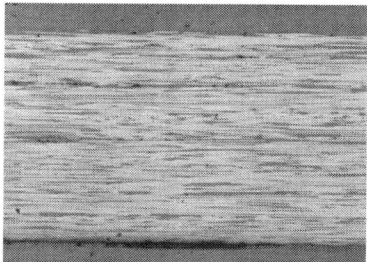
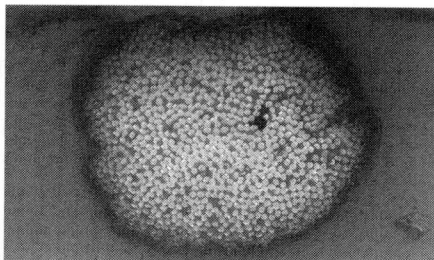

(a) Longitudinal direction   (b) Transverse direction

Figure 2. Microsection of a carbon fibre sensor in longitudinal and transverse direction Sensorfibre: 1K roving of the carbon fibre T300B Patch: single layer UD prepreg EG/913

2. ELECTROMECHANICAL PROPERTIES OF CARBON FIBRES

Referring to Chung [2] and Dresselhaus [3] crystalline carbon fibres have the same crystal structure as graphite. The layered and planar structure of crystalline carbon fibres is shown in Figure 3. In each layer the sp2sp2 hybridized carbon atoms are arranged in a hexagonal lattice. Within a layer (xx-yy plane) the carbon atoms are bonded by three covalent bonds (overlapping sp2sp2 orbitals), and a metallic bonding is provided by the delocalization of the pzpz orbitals. This delocalization of the fourth valence electron leads to a good electrical conductivity of the carbon fibre. The individual layers are held together by weak van der Waals forces (zz-direction). These different types of bonds within and between the layers result in the anisotropy of the mechanical, thermal and electrical material properties of carbon fibres.

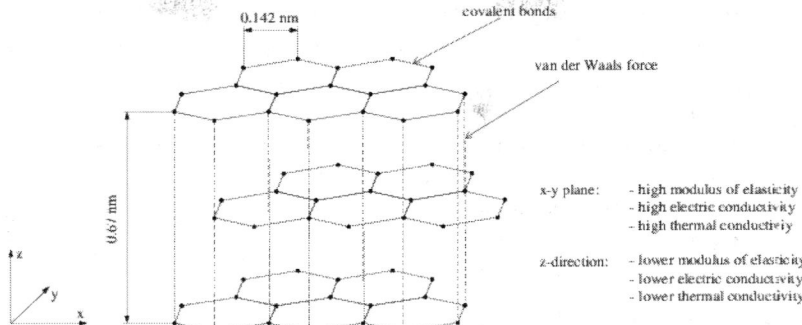

Figure 3. Graphite structure of crystalline carbon fibres

2.1. Specific Resistivity

The degree of crystallinity and the micro structure of carbon fibres are controlled by the carbonisation process and determine the mechanical and electromechanical properties. Concerning ex-PAN fibres the Young's modulus ranges from 200 GPa (HT fibres) to 600 GPa (HM fibres). Ex-pitch fibres have a higher modulus up to 900 GPa. Figure 4 shows the correlation between the Young's modulus and the specific resistivity of different carbon fibre types.

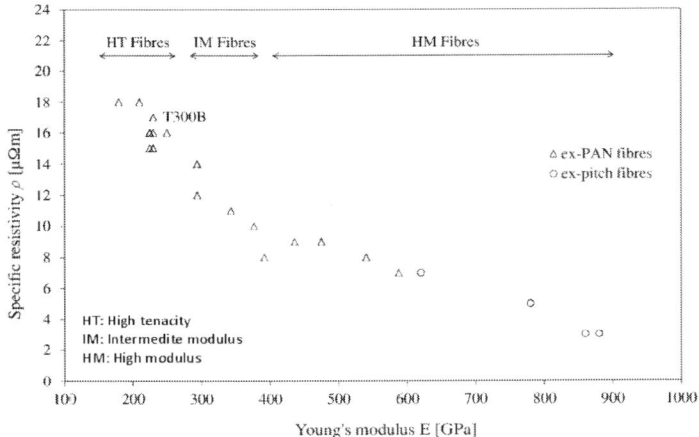

Figure 4. Correlation between the specific resistance $\rho\rho$ and the Young's modulus EE of different ex-PAN and ex-pitch carbon fibres

The specific resistivity of HT fibres (high tenacity) is in the range of 15 $\mu\Omega\mu\Omega$m up to 18 $\mu\Omega\mu\Omega$m. Higher orientated fibres (IM fibres and HM fibres) show a specific resistivity below 14 $\mu\Omega\mu\Omega$m.

The specific resistance of carbon fibres is strongly temperature dependent [15]. In consequence, temperature effects have a high influence on the signals of CFSs and must be considered by an appropriate temperature compensation (see section 3).

2.2. Piezoresistivity

The electrical resistance RR of a carbon fibre is given by:

$$R = \rho \frac{L}{r^2 \pi}$$

where$\rho\rho$.. is the specific resistivity, LLthe length and rr the radius of the fibre. The total differential of $R=f(\rho,L,r)R=f(\rho,L,r)$ is then yielded by the following Equation (2).

$$\mathrm{d}R = \frac{\partial R}{\partial \rho}\mathrm{d}\rho + \frac{\partial R}{\partial L}\mathrm{d}L + \frac{\partial R}{\partial r}\mathrm{d}r = \frac{L}{r^2 \pi}\mathrm{d}\rho + \rho\frac{1}{r^2 \pi}\mathrm{d}L - 2\rho\frac{L}{r^3 \pi}\mathrm{d}r$$

Using Equation (1) and Equation (2) the relative change in resistance of a carbon fibre (dR/R0)(dR/R0) can be expressed as:

$$\left(\frac{\mathrm{d}R}{R_0}\right) = \frac{\mathrm{d}\rho}{\rho} + \varepsilon(1 + 2\nu) = k\varepsilon$$

$$\text{with } \varepsilon = \frac{\mathrm{d}L}{L} \text{ and } \nu = -\frac{\mathrm{d}r/r}{\mathrm{d}L/L}$$

The term dρ/ρdρ/ρ denotes the piezoresistive effect (material effect) and the term $\varepsilon(1+2\nu)\varepsilon(1+2\nu)$ represents the geometric effects. The strain sensitivity kk covers both effects. It should be noted that the strain sensitivity kk depends on the effective Poisson ratio $\nu\nu$ [3] - Therefore, the strain sensitivity (dRR)=klεl+ktεt(dRR)=klεl+ktεt must be defined for a corresponding Poisson ratio [4] - .

For some applications of CFSs it can be useful to split the strain sensitivity k=1.71k=1.71 into the longitudinal strain sensitivity $\nu\nu$ and the transverse strain sensitivityklkl. The relative change in resistance is then given by:

$$\left(\frac{\mathrm{d}R}{R}\right) = k_l \varepsilon_l + k_t \varepsilon_t$$

The strain sensitivity µµ of a T300B 1K fibre was determined as vv for a corresponding Poisson ratio vfvfof 0.28. The sensor fibre exhibits a longitudinal strain sensitivity vpvp in the range of 1.72 to 1.78 and a transverse strain sensitivity −εy/εx−εy/εx in the range of 0.37 to 0.41. The piezoresistivity of the ex-PAN fibre is linear up to a strain level of approximately 6000 v=0.285v=0.285m/m [1, 6]. Problems may occur at the metallized fibre endings. In order to avoid any influence of the electrical connections the strain level at the fibre endings should not exceed a level of 2500 μm/m.

This can be realized by using special tabs which exhibit a higher stiffness in the regions of the metallic connections. An example of such load relieving tabs is given in Figure 5. The lay-up of the tabs ensures a four times smaller strain level in the region of the electrical connections compared to the strain level which occurs in the testing area.

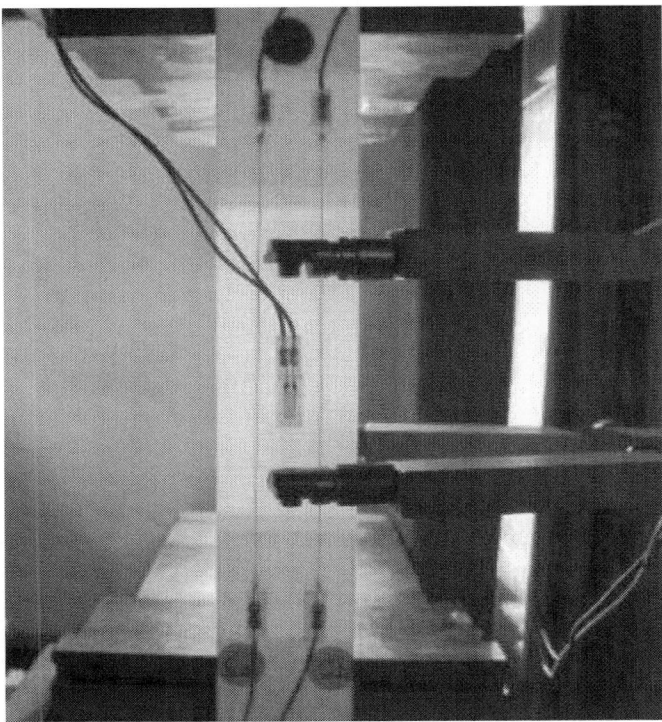

Figure 5. CFS patch with load relieving tabs Lay-up in the testing area: [0∘][0∘]Lay-up in the regions of the metallized fibre endings: [0∘5][05∘]

Figure 6 shows the excellent linear piezoresistive behaviour of the ex-PAN fibre T300B 1K up to a strain level of 6000 μm/m (loading and unloading). Bending tests were performed to investigate the compression behaviour of CFSs. These investigations are not completed. First results show that the linearity of the signal depends significant on the carrier material.

Ex-pitch fibres show a nonlinear piezoresistive behaviour.

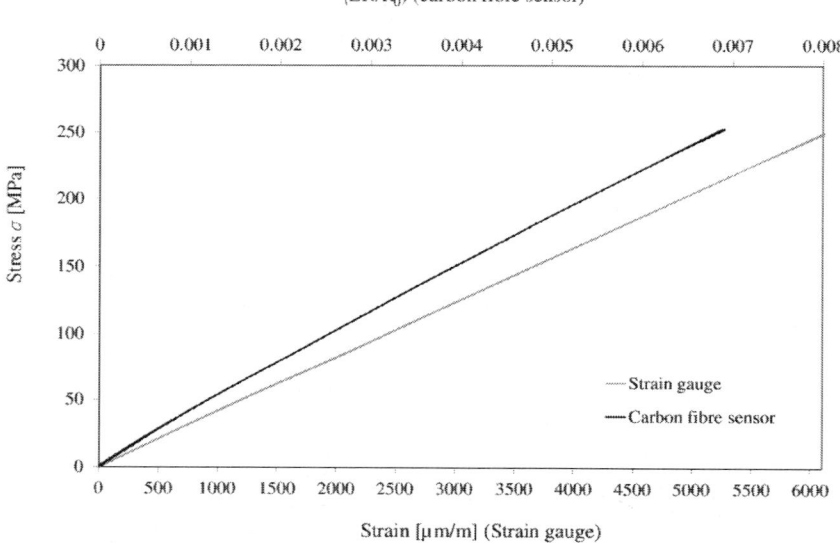

Figure 6. Characterization of the linear piezoresistivity up to a strain level of 6000 $\mu\mu$m/m (loading and unloading) by means of a CFS patch with load relieving tabs at the endings Sensor fibre: T300B 1K (ex-PAN fibre)Lay-up in the testing area: [0∘8][08∘]Lay-up of the load relieving tabs: [±45∘,0∘3] sym[±45∘,03∘]sym

3. THE WHEATSTONE BRIDGE

The change in resistance of a CFS due to an applied strain is usually small. The Wheatstone bridge is an electrical circuit which allows the determination of very small changes in electrical resistance with great accuracy. Furthermore, the Wheatstone bridge minimizes the high influence of temperature changes on the CFS signal. The measurement circuit, illustrated in Figure 7, consists of four resistances, a supply voltage and the output voltage of the bridge.

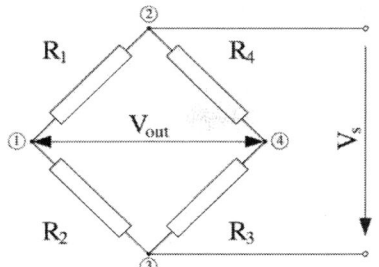

$R_1 \ldots R_4$:   resistances, bridge arms

V_s: bridge supply voltage

V_{out}:   bridge output voltage

$$\frac{V_{out}}{V_s} = \frac{1}{4}\left(\frac{\Delta R_1}{R_1} - \frac{\Delta R_2}{R_2} + \frac{\Delta R_3}{R_3} - \frac{\Delta R_4}{R_4}\right)$$

Figure 7. General Wheatstone bridge circuit

The relative change in resistance can be determined by the ratio of output voltage to input voltageVout/VsVout/Vs. There are two configurations of the Wheatstone bridge which are of special importance for the use of CFSs. These configurations of the Wheatstone bridge are known as "half bridge" configuration and "full bridge" configuration.

In the case of a half bridge circuit (Figure 7), the bridge is formed by two CFSs (R1R1andR2R2) and two completion resistors (R3R3andR4R4). The full bridge configuration (Figure 7) is formed by four CFSs and needs no additional resistors. Both circuits will enable a compensation of temperature effects (thermal dependency of the specific resistance and thermal expansion) if the thermal conditions of the connected CFSs are identical. In the case of a half bridge circuit with one active CFS (R1R1) the output voltage VoutVoutis given by:

$$V_{out} = \frac{V_s}{4}\left(\frac{\Delta R_{1,mech} + \Delta R_{1,therm}}{R_1} - \frac{\Delta R_{2,therm}}{R_2}\right)$$

Equation (5) shows that the thermal effect is compensated by the mechanically unloaded carbon fibre sensor (R2R2). Detailed information about the principle and use of the Wheatstone bridge can be found in classical textbooks on strain gauge techniques (e.g. [5, 13])

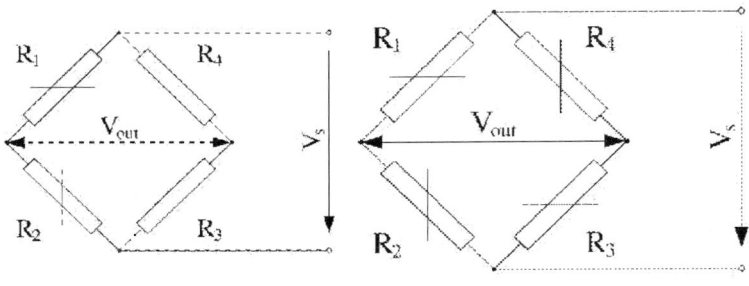

(a) Half bridge circuit   (b) Full bridge circuit

Figure 8. Half bridge and full bridge configuration of the Wheatstone bridge

4. INTEGRAL STRAIN MEASUREMENT OF THE CARBON FIBRE SENSOR

Equation (3) describes the relative change in electrical resistance $(\Delta R/R0)(\Delta R/R0)$ of a carbon fibre sensor due to an applied elastic strain ε. It is important to understand that a CFS measures the strain integrally along its whole fibre length. Thus, Equation (3) can be written as:

$$\left(\frac{\Delta R}{R_0}\right) = k\frac{1}{L}\int_0^L \varepsilon(x)\,\mathrm{d}x$$

Equation (6) shows that a CFS measures the displacement between the terminal points of the sensor fibre.

$$\left(\frac{\Delta R}{R_0}\right)\frac{L}{k} = \int_0^L \varepsilon(x)\,\mathrm{d}x = [u(x = L) - u(x = 0)]$$

This integral strain measurement of CFSs in accordance to Equation (7) can be used to create carbon fibre sensor meshes (CFS meshes). Such a CFS mesh allows the determination of the two dimensional (2D) state of strain and state of deformation of a whole structure or larger areas of a structure.

5. CARBON FIBRE SENSOR MESHES

The strain and deformation analysis by means of CFS meshes are based on linear or higher order displacement approximations. The displacement functions depend on the used element type such as 3-node triangle elements, 6-node triangle elements or quadrilateral elements. The basic theory of strain analysis with CFS meshes using 3-node triangle elements with a linear displacement approximation is presented below.

A triangular CFS element defined by its vertices 1, 2, 3 and its local coordinate system is given in Figure 9.

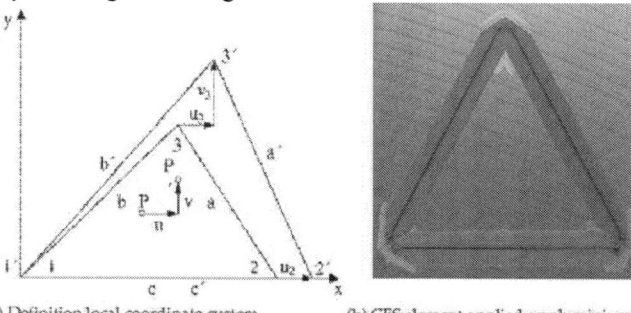

(a) Definition local coordinate system   (b) CFS element applied on aluminium

Figure 9. Triangle CFS element with 3 nodes

The displacement u(x,y) and v(x,y) of an inner point PP can be determined by the displacements of the vertex in using a displacement function.

Assuming a linear displacement function the displacement u(x,y) and v(x,y) within the element can be calculated in accordance to Equation (8) and Equation (9).

$$u(x,y) = N_1 u_1 + N_2 u_2 + N_3 u_3$$

$$v(x,y) = N_1 v_1 + N_2 v_2 + N_3 v_3$$

Hereby u_i and v_i denote the displacements of the vertex 1, 2, 3 in the x- and y-direction while N_i represents the shape functions of the element. The shape functions N_i of a 3-node triangle can be found in classical books on the theory of finite element analysis (e.g. [16]). The engineering strains ε_x, ε_y and γ_{xy} are defined by:

$$\varepsilon_x = \frac{\delta u}{\delta x}; \qquad \varepsilon_y = \frac{\delta v}{\delta y}; \qquad \gamma_{xy} = \frac{\delta u}{\delta y} + \frac{\delta v}{\delta x}$$

Considering the local coordinate system (u1=0u1=0, v1=0v1=0,v2=0v2=0) the strains within the triangle can be calculated by:

$$\varepsilon_x = \frac{u_2}{x_2}; \qquad \varepsilon_y = \frac{v_3}{y_3}; \qquad \gamma_{xy} = \frac{x_2 u_3 - x_3 u_2}{x_2 y_3}$$

In consequence of the linear displacement approximation the strains are independent of the coordinates (x and y) and thus constant within the element. The unknown displacements u_2 , u_3 and v_3 of the vertex 2 and 3 can be determined by the signals of the carbon fibre sensors.

In order to verify this linear approach an experimental investigation of a CFS mesh applied on a 1000 mm x 1000 mm x 5 mm PMMA plate was performed. The simply supported plate was loaded with a single static force at the center. In addition to the experiment a finite element analysis (FEA) was performed. In [9] the results of this experiment and of the corresponding finite element simulations are presented in detail [5] . Figure 10 shows the PMMA plate and the applied CFS mesh.

Figure 10. Carbon fibre sensor mesh applied on a 1 m x 1 m PMMA plate. Each sensor has a length of 300mm. [9]

Figure 11 shows the determined strain εxεx for each element of the mesh. There was a good correlation between the measured and the calculated strain levels. The accuracy was in the range of±5%±5%.

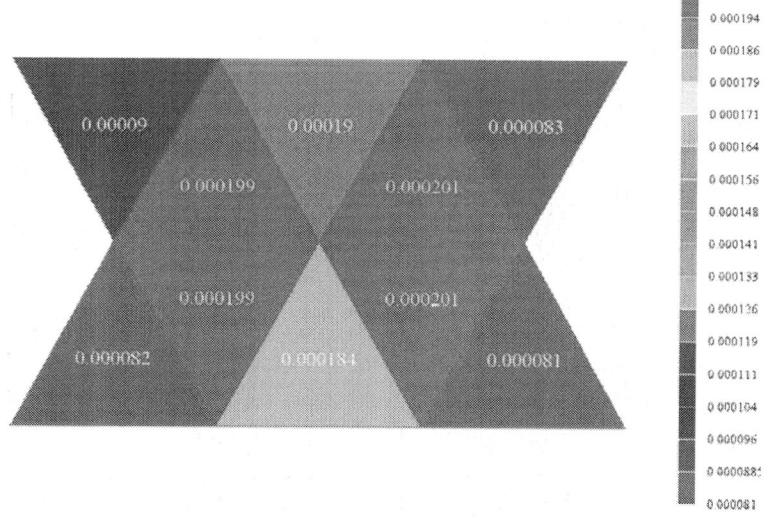

Figure 11. Strain εxεx measured by the carbon fibre sensor mesh

The results of the performed investigation show that CFS meshes are a reliable instrument to determine the strain fields and principle strains of lightweight structures.

The principle strains (strain level and direction) are of particular interest in case of structures made of composite materials. For example, tailored fibre placement (TFP) is an advanced textile manufacturing process for CFRP structures in which the carbon fibre rovings are placed in accordance to the direction of principal stresses.

The finite element simulation is a powerful tool to analyse the stress fields and principle directions of lightweight structures. At the design and optimization processes the FEA is almost the only way to evaluate the structural load.

However, there is a lack of techniques to review the results of the FEA. CFS meshes offer a high potential to verify the results of the finite element analysis.

6. MICRO CRACK DETECTION

A major failure mode of multidirectional reinforced laminates is transverse matrix cracking. Matrix cracks will reduce the effective stiffness of the laminate and will result in local stress concentrations at the crack tip. Furthermore, interlaminar crack growth and local delamination can occur. Due to its integral strain measurement method the CFS has a high potential to detect matrix crack initiation and monitor crack growth. Figure 12 shows a thin GFRP laminate (Lay-up:[90∘2,0∘,90∘2][902∘,0∘,902∘]) which has two embedded CFSs. Matrix cracks along the CFS will affect the sensor signal which will give a clear indication of the crack density. A study was performed to investigate the influence of cracks on the sensor signal [10].

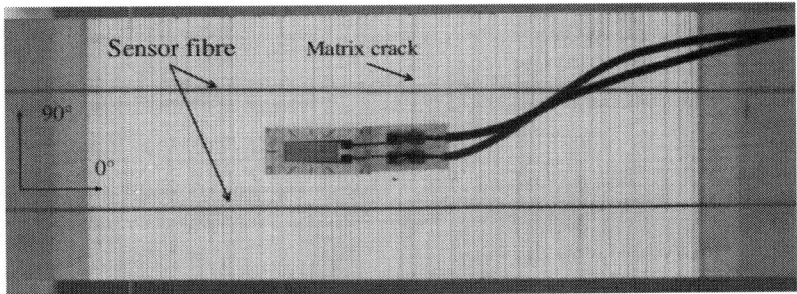

Figure 12. Multidirectional reinforced GFRP laminate with two embedded CFSs. Transverse matrix cracking will affect the sensor signal.Lay-up: [90∘2,0∘,90∘2][902∘,0∘,902∘]

The study was performed on the GFRP laminate [0∘,90∘5,0∘,90∘5,0∘][0∘,905∘,0∘,905∘,0∘]. Three CFSs were embedded in the mid-plane 0∘0∘-layer. At a strain level higher than 3000 μm/m matrix cracks appeared in the 90∘90∘-layers. The following techniques were applied to characterize the influence of damages on the sensor signal:

- Acoustic emission analysis in combination with pattern recognition technique
- Microscopy and micrographs
- Analytical calculations
- Finite element analysis

Figure 13 shows the correlation between the crack density, the CFS signals, the acoustic emission energy (AE-energy), the strain level measured by a strain gauge and the according stress level. It can be seen that the matrix crack initiation causes first AE-signals and a change of the slope of the CFS signals.

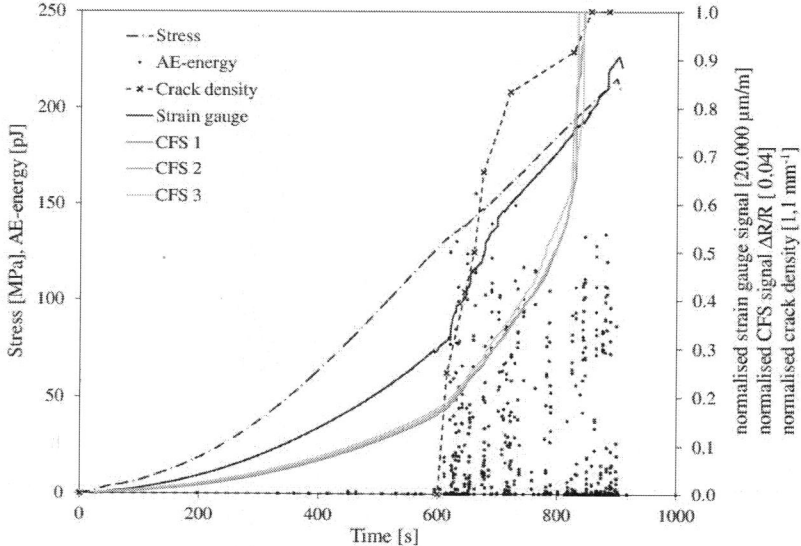

Figure 13. Acoustic emission and CFS signals measured on an GFRP laminate under uniaxial tensile load [10]Lay-up: [0°,90°5,0°,90°5,0°] [0°,905°,0°,905°,0°]

Furthermore, a good correlation between the sensor signal ($\Delta R/R0$)($\Delta R/R0$) and the crack density can be observed. After having reached a crack density of 0.8 mm^{-1} the signals of the embedded carbon fibre sensors increase disproportionately. The analytical approach of Garret and Bailey [4, 11, 12] and a FEA were applied to calculate the reduction of the stiffness of the laminate. Müller showed that the global stiffness loss of the laminate due to the matrix cracking can be measured by means of the CFS. However, at high strain levels (>70%>70%ofεultimateεultimate) there is a strong influence of high local stresses at the crack tip on the CFS signal. These stress concentrations may result in filament breakage and in extremely high signal levels.

Based on this result CFSs can be used for damage monitoring and for the prediction of the lifetime of damaged structures if the damage level can be characterized by stiffness loss. Figure 14 shows the cyclic loading of a laminate to measure the stiffness loss and to determine Ladeveze's material parameters [8]. Based on this calibration the lifetime of a structure, e.g. pressure vessel, can be predicted.

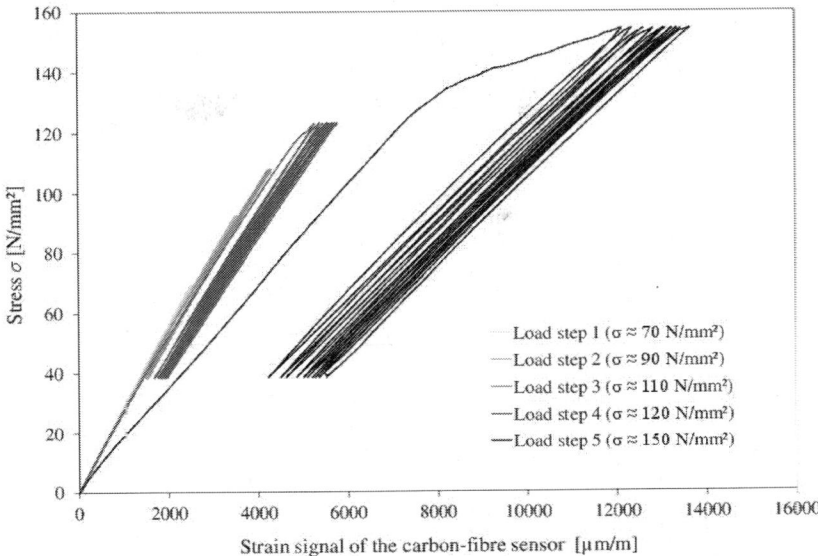

Figure 14. Damage measurement and determination of damage variables like the energy release rate Material: GFRP, EG/913 Lay-up: [0∘,90∘5,0∘,90∘5,0∘] [0∘,905∘,0∘,905∘,0∘]

7. IMPACT AND DELAMINATION DETECTION

An impact loading can cause small damages inside the composite material which may not be found by visible inspection. One major concern is delamination damages or the disbonding of interfaces. These damages result in sublaminates having lower buckling resistance and compression strength. Although a small delamination-damage does not necessarily constitute failure, the damaged area may undergo a time-dependent growth and may attend a critical size.

The principles for achieving damage tolerant primary composite structures were established by the aircraft companies [14]. Maintenance intervals and inspection plans are determined in such a way that readily detectable damages will be repaired before damage growth can affect the fatigue strength of the structure. The influence of undetectable damages is covered by the so called barely visible impact damage (BVID) which defines the damage that establishes the strength values to be used in analysis to demonstrate compliance with the load requirements. One method to determine the influence of the BVID on the mechanical performance is the compression after impact test procedure (CAI-test procedure, i.e. Boeing BSS 7260). A 4 mm thick quasi-isotropic test specimen (150 x 100 mm) is damaged by a dropped weight impact testing machine. An impact level of about 3 J/mm will cause the BVID. This means that only a small remaining indentation is visible on the surface of the specimen, but delamination may be found inside. The strength of the damaged specimen is reduced by 10 to 20% compared to the undamaged material. This shows that in many cases the exploitation of material performance is limited, since the skin thicknesses of a composite structure are designed to absorb an impact.

For aircraft structures health monitoring systems have an extremely high potential to improve the efficiency of composite structures. Based on a monitoring system the material design values can be increased and the inspection intervals can be enlarged. Both aspects can result in a weight reduction of the structure up to 10%. A preliminary study was performed to investigate the use of CFSs as a sensor system to detect the BVID. The CAI specimen was used to integrate a rectangular CFS mesh (Figure 15 and Figure 16).

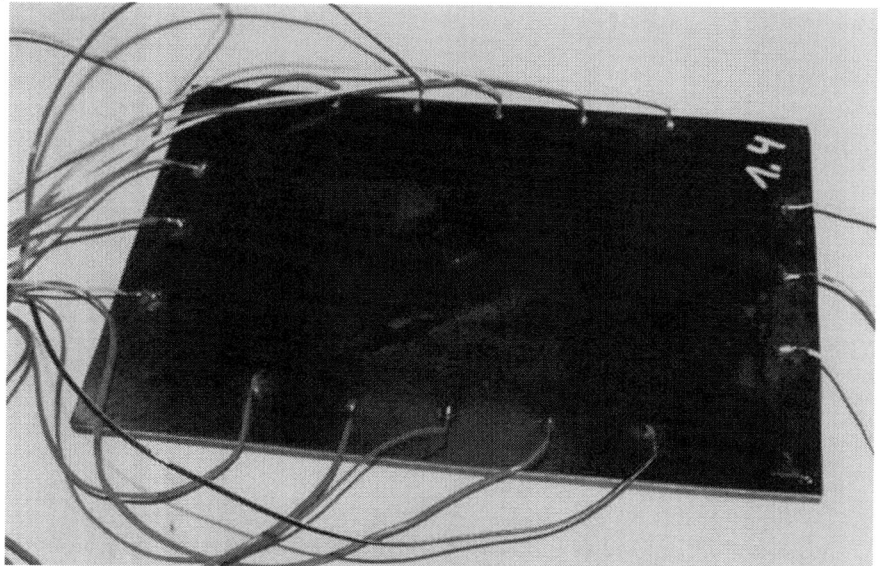

Figure 15. Compression after impact (CAI)-specimen with integrated CFS mesh to detect damages below the barely visible impact damage level (BVID)

The distance between the CFSs varied from 30 to 100 mm. Three methods were investigated to detect the impact damage:

- Online measurements of the resistivity during the impact test
- Offline measurements, comparison of the resistivity before and after the impact
- Active thermography, CFSs used as heating element

It has been shown that all three procedures are suitable to detect impact damages. A distance of 50 mm between the CFSs is necessary to detect even small damages below the BVID. In a second step the use of CFSs will be investigated to detect debonding of skin and stringer. A health monitoring system for a complex aircraft structure will be based on several technologies like ultrasonic inspection sensors and strain sensors. The CFS technology will complement the established sensors due to its specific and simple integral measurement principle.

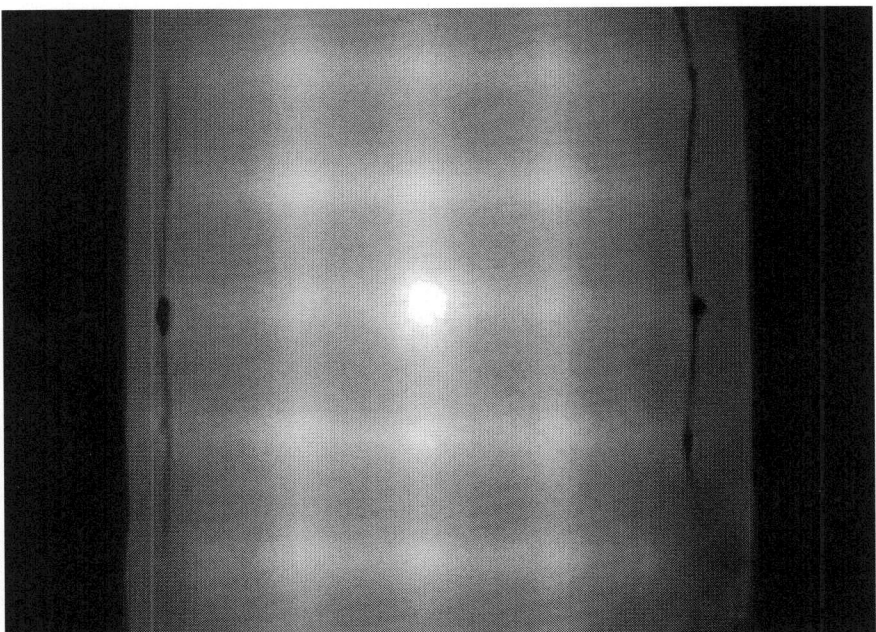

Figure 16. Compression after impact (CAI)-specimen with integrated CFS mesh. The CFSs are used as heating element for active thermography. A small delamination damage is visible (BVID)

8. APPLICATION

CFSs offer a high potential to be used as a sensor element for composite materials for stress analysis, damage detection and the monitoring of manufacturing processes. Two industrial applications have been selected to demonstrate this.

8.1. Tabletop of A Ct-Scanner

Carbon fibre reinforced plastics (CFRP) are used for tabletops of computer tomography (CT) scanners, since CFRP fulfills the X-Ray transparency which is necessary to get the picture quality sufficient for medical diagnosis.

CFSs can be embedded in the tabletop of a CT scanner to measure its deflection. Based on the measured deflection the CT images can be readjusted to result in an improved medical attendance [7]. Figure 17shows the tabletop of a CT scanner with ten u-shaped CFSs applied. For this application u-shaped CFSs are used to avoid that any metal wiring is part of the scan plane for all operation positions.

Figure 17. CT tabletop with u-shaped CFSs to measure the deflection. The metallic wiring is attached to the clamping support.

The lengths of the u-shaped CFSs vary from 250 to 1250 mm. The determination of the beam deflection is based on the integral strain measurement of CFSs (Equation (6)). Considering small deformations the relation between the elastic strain of the outer fibre $\varepsilon \hat{\varepsilon}^{\wedge}$ and the beam deflection vv is given by:

$$v = \iint v'' dx dx = \iint \frac{\widehat{\varepsilon}}{e_y} dx dx$$

where e_y denotes the distance from the neutral axis. Assuming a cantilever beam (see Figure 18) the slope v'v' of the beam can be determined directly by the signals of the CFSs. Assuming that the CFS starts at the clamping support $(x=0x=0)$ the slope v' at the end of the applied CFS $(x=l_i)$ becomes:

$$v'(x = l_i) = \frac{l_i}{ke_y} \left(\frac{\Delta R}{R} \right)_i$$

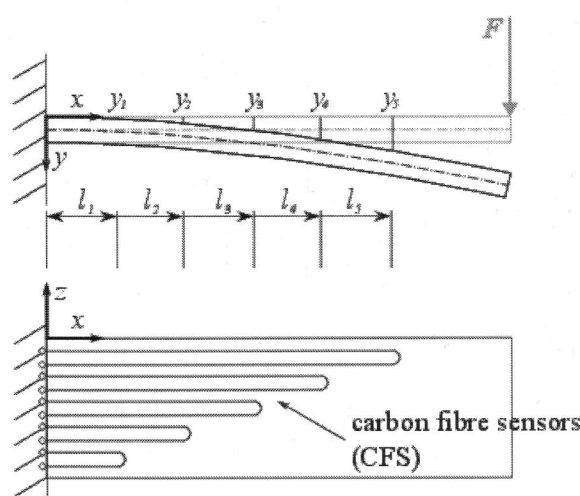

Figure 18. Side and top view of a cantilever beam with five u-shaped CFS

The deflection y of the beam can be calculated by numerical integration.

$$v(x = l_i) = \sum_{n=1}^{i} \left[v'(x_n) - \frac{v'(x_n) - v'(x_{n-1})}{2} \right] (l_n - l_{n-1})$$

The index i denotes the number of CFSs applied. The quality of the approximation in accordance to Equation (14) depends on the complexity of the loading, the number of applied CFSs and the used sensor configuration.

In the case of the CT table (Figure 17) the deflection of the table can be determined with an accuracy of $\pm 0.3 \pm 0.3$ mm for different operation positions. Figure 19 shows a typical measurement. The operating position of the

CT tabletop varies stepwise from 0 mm to 2000 mm and back to 0 mm. The weight of the used dummy is 100 kg and electrical half bridges are used for temperature compensation.

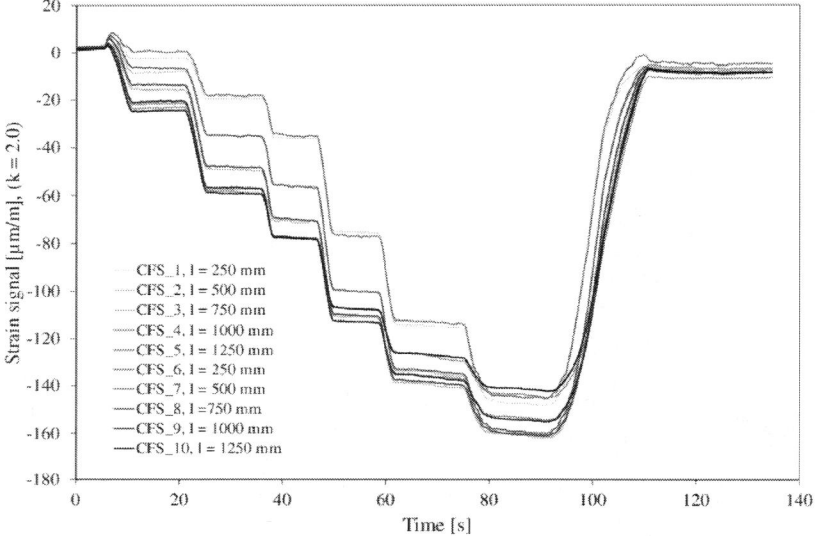

Figure 19. CFS signals of the CT table for a weight of 100 kg and different operation positions, ranging from 0 to 2000 mm

8.2. Pressure Vessels

For thin walled assumptions the longitudinal stress σlσl and hoop stress σrσr in the cylindrical portion of a pressure vessel, away from the ends, are given by:

$$\sigma_l = \frac{pr}{2t}$$

$$\sigma_r = \frac{pr}{t}$$

where *t* is the thickness and r is the radius of the vessel. The relation σr/σl shows that the efficiency of pressure vessels can be increased if hoop wrapped vessels are used. A metal cylinder is reinforced by carbon fibres having a radial orientation (type II vessels). The strength of the vessel is increased remarkably.

There are two aspects for the use of CFSs for pressure vessels which can be easily integrated by means of the winding process:

- Determination of the pressure level of the vessel
- Monitoring of degradation processes due to fatigue or overloading of the radial fibre reinforcement

For such an application a sensor patch with four CFSs connected in full-bridge configuration is particularly suitable, since the full-bridge circuit minimizes the influences of thermal effects and shows an improved long term stability of the signal. Figure 20 shows such a hoop wrapped vessel with an embedded CFS patch.

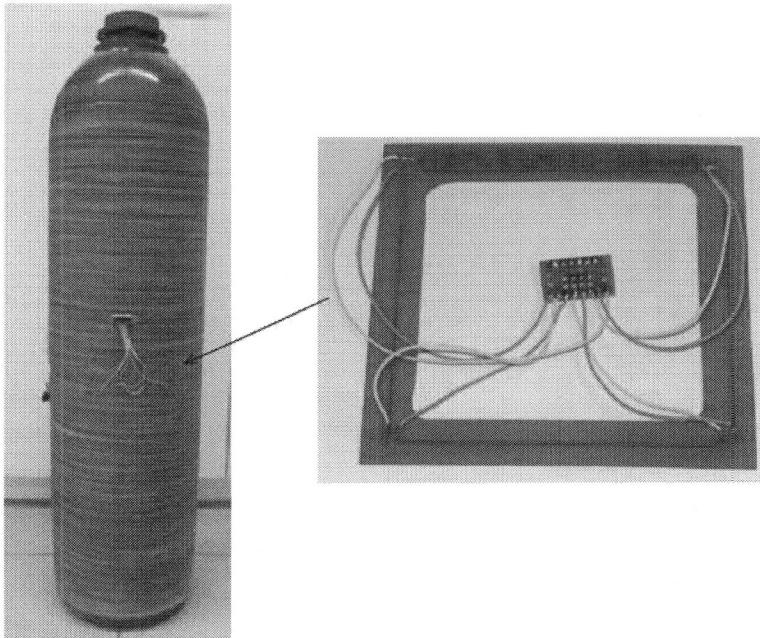

Figure 20. Hoop wrapped pressure vessel (type II vessel) with embedded CFSs

Experimental studies showed that the pressure of the vessel can be determined by using a CFS patch with a resolution of about 1 bar. A representative measurement of the performed test is shown in Figure 21. The pressure load of the vessel was increased stepwise up to a maximum pressure level of 100 bar. The subsequent pressure relief was performed in the same manner.

Micro cracks in the CFRP layers of the pressure vessel may occur as a result of mechanical or thermal overloading. The failure of the matrix causes a loss of stiffness of the CFRP layers and reduces the global stiffness of the vessel. Depending on the crack density and the crack growth the lifetime of the structure will decrease. Based on the damage master curve of the structure, an estimation of the remaining lifetime can be performed (see Figure 14).

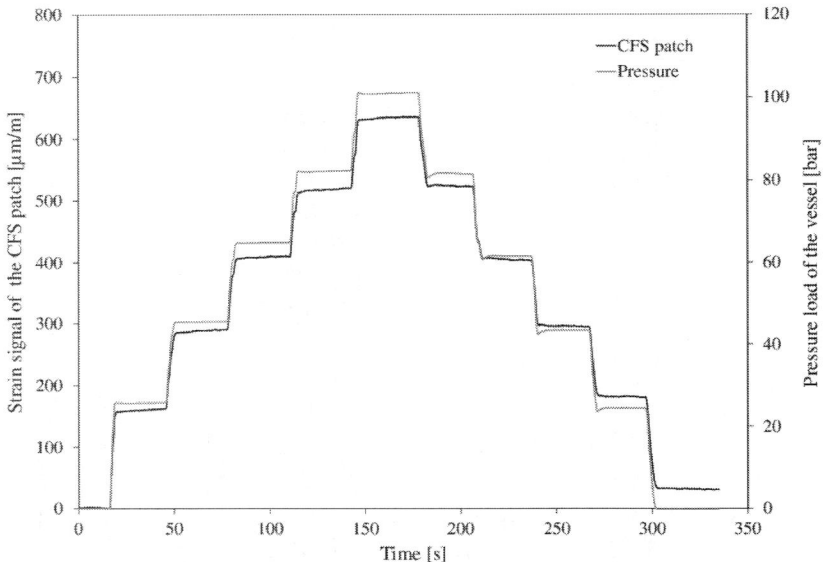

Figure 21. Strain signals of a CFS patch applied near the surface of a type II pressure vessel subjected to a pressure test

9. CONCLUSION

There are three main aspects which made carbon fibre sensors (CFSs) very interesting to be used for composite materials:

- Material-conformity
- Linear piezoresistivity up to high strain levels
- Integral strain measurement method

CFSs based on a T300B 1K ex-Pan fibre exhibit an excellent linear piezoresistivity up to a strain level of 6000 μm/m. The according strain sensitivity was determined as k=1.71 (related to a Possion ratio of $v = 0.28$). The longitudinal strain sensitivity kl of the CFS is in the range of 1.72 - 1.78. Transverse to the fibre direction CFSs exhibit a transverse strain sensitivity kt of approximately 0.4. This significant transverse strain sensitivity must be considered in praxis. Tension load was applied for all tests, the characterization of the compression behaviour is under examination.

A disadvantage of CFSs is the high influence of temperature on the signal which will be an aspect for future research. At the moment a long term stability of.±2 μm/m (T=const.) can be achieved for a half or full bridge circuit.

CFSs can be used for strain analysis, damage monitoring and the control of manufacturing processes. Concerning strain and stress analysis an approach for CFS meshes was developed based on triangular elements with linear displacement approximation. A good correlation was found between the measurement and the finite element calculation. The use of CFS meshes can be a new approach to complement finite element analysis from the experimental side.

Concerning monitoring aspects the influence of material damages on the sensor signal were studied. It has been demonstrated that based on the integral strain measurement method the CFS is an excellent sensor to detect delaminations and matrix cracks in multidirectional reinforced laminates. Therefore, the CFS offers unique features for fracture mechanics: Measurement of strain levels and detection of matrix cracks. By means of two examples the CFS technologies could be demonstrated successfully:

- Determination of the deflection of a tabletop of a CT-Scanner
- Determination of the strain level and the crack density of a pressure vessel
- Strain measurement and matrix crack detection are in the focus of safe and damage tolerant composite structures making the CFS technology a complement of established sensors.

NOTES

1. Piezoresitivity describes the change in electrical resistance of a conductor due to an applied strain.
2. For example, the epoxy resin EP 310 S can be burned off or removed using acid.
3. The effective Poisson ratio kk depends on the Poisson ratio of the sensor fibrekk, the Poisson ratio of the sensor patch klkl and the strain ratio ktkt of the structure.
4. Conventional strain gauges exhibits the same behaviour. The strain sensitivity of strain gauges is usually defined for a corresponding Poisson ratio kk (steel).
5. A quadratic displacement approach for a 3-node triangle is also presented and verified in [9]. However, the quadratic displacement approach of a 3-node triangle requires additional strain gauges at the nodes to determine the 12 unknown coefficients of the shape function.

REFERENCES

1. C. Christner, A. Horoschenkoff, H. Rapp, 2012Longitudinal and transverse strain sensitivity of embedded carbon-fibre sensors, Journal of Composite Materials Online First: DOI:

2. D. Chung, 1994Carbon Fiber Composites, Butterworth-Heinemann.
3. M. Dresselhaus, G. Dresselhaus, K. Sugihara, I. Spain, H. Goldberg, 1988Graphite Fibers and Filaments, Springer.
4. K. Garrett, J. Bailey, 1977Multiple transverse fracture in 90° cross-ply laminates of a glass fibre-reinforced polyester, Journal of Material Science 12
5. K. Hoffmann, 1987An introduction to measurements using strain gages, Hottinger Baldwin Messtechnik.
6. A. Horoschenkoff, T. Müller, A. Kröll, 2009On the charaterization of the piezoresistivity of embedded carbon fibres, 17th International Conference on Composite Materials, Edinburgh.
7. A. Horoschenkoff, T. Müller, C. Strössner, K. Farmbauer, 2011Use of carbon-fibre sensors to determine the deflection of composite-beams, 18th International Conference on Composite Materials, International Conference on Composite Materials, Jeju.
8. P. Ladeveze, E. L. Dantec, 1992Damage modelling of the elementary ply for laminated composites, Composite Science and Technology 43
9. T. Matzies, C. Christner, T. Müller, A. Horoschenkoff, H. Rapp, 2011Carbon-fibre sensor meshes: Simulation and experiment, 21st International Workshop on Computational Mechanics of Materials, Limerick.
10. T. Müller, A. Horoschenkoff, H. Rapp, M. Sause, S. Horn, 2010Einfluss von zwischenfaserbrüchen in 0/90laminaten auf die elektrische widerstandsänderung von eingebetteten carbonfasern, 59th Deutscher Luft- und Raumfahrkongress.
11. J. Nairn, 1989The strain energy relase rate of composite microcracking: A variational approach, Journal of Composite Materials 23
12. J. Nairn, S. Hu, 1994Matrix microcracking, Damage Mechanics of Composite Materials 9
13. C. Perry, H. Lissner, 1955The Strain Gauge Primer, Mc Gram Hill.
14. H. Razi, S. Ward, 1996Principles for achieving damage tolerant primary composite aircraft structures, 11th DoD/FAA/NASA Conference on Fibrous Composites in Structural Design.
15. I. Spain, K. , V. , H. Goldberg, I. Kalnin, 1982Unusual electrical resistivity behavior of carbon fibres, Solid State Communications 817-819
16. O. Zienkiewicz, R. Taylor, 2000Finite Element Method 1The Basis, Elsevier.

CHAPTER 12

Composite Material and Optical Fibres

Antonio C. de Oliveira [1] and Ligia S. de Oliveira[1]

[1] Laboratório Nacional de Astrofísica, Ministério da Ciência Tecnologia e Inovação, Brazil

1. INTRODUCTION

For ease of handling and polishing, optical fibres are generally mounted in some form of rigid structure. A schematic of a typical fibre termination used in astronomical instruments (typical of multiple-fibre spectrographs) is show in Fig. 1. The fibre is first placed within a flexible tube (polyimide or similar), often referred to as the strain relief tube. The fibre and tube are placed within a rigid ferrule. Adhesive is applied to the fibre and tubing to fix them in place. The ferrule can then be easily manipulated for polishing, or mounting within an instrument, without risk of damage to the fibre. The role of the strain relief tube is to prevent stresses occurring at the point where the fibre enters the ferrule given bending at this point may lead to breakages. For coupling an array of optical fibres to a microlens array, such as in IFU (Integral Field Unity), a brass plate with an array of drilled microholes may be used. Optical fibres positioned in an array of accurately drilled holes, as illustrated schematically in Fig. 2.

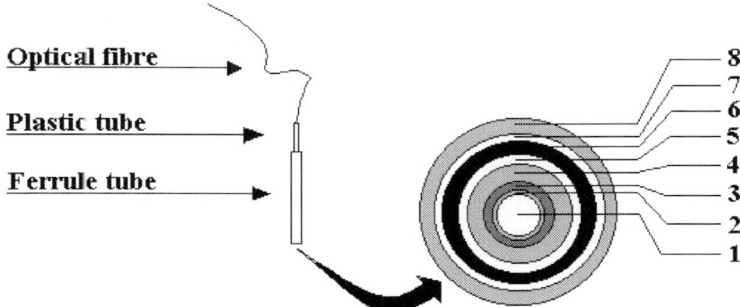

Figure 1. Schematic diagram of a single fibre mounting assembly: 1, core; 2, cladding; 3, polyamide buffer; 4, acrylate buffer; 5, epoxy; 6, plastic tube; 7, epoxy; 8, ferrule steel tube.

This system, called microholes array, contains a grid of holes spaced by the pitch of the microlens array. The holes are machined using custom made drills with two different diameters. This produces a stepped hole, with the smaller diameter hole used for fibre positioning while the larger hole is used to accommodate a ferrule. The small holes are approximately 10 μm larger than the fibre diameter to allow sufficient space for a glue to penetrate. Using a stepped hole also allows a greater depth of material to be machined than by using a small drill alone. This permits a thicker, hence more robust, piece of material to be used. A support plate is also used, positioned above the fibre-positioning array with spacers, to maintain accurate angular alignment of the ferrules with respect to the microlens optical axis. Each ferrule contains a polyimide strain relief tube to prevent mechanical stress, occurring at the point where the fibre enters the ferrule. To secure the fibres, ferrules, and polyamide tubes in place the whole input assembly is immersed in a container of EPOTEK 301-2 adhesive. This epoxy is a natural choice due to its excellent wicking properties and low shrinkage upon curing. After curing, which takes approximately three days at room temperature; any excess glue is removed prior to optical polishing.

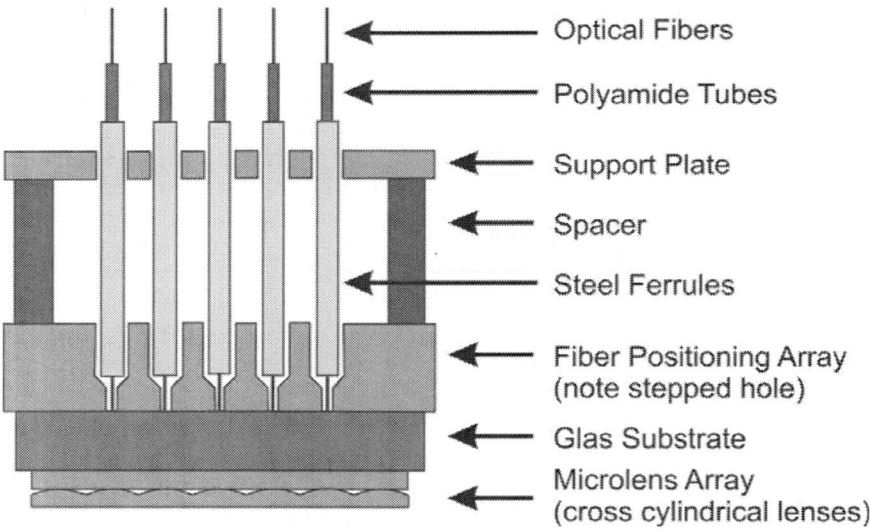

Figure 2. Schematic micro lens array and fibre positioning array used to constructed IFUs

The problem here may be an increase of FRD (Focal Ratio Degradation) caused by contraction of the metal ferrule or brass plate at low temperature causing stress on the fibres and consequent loss of throughput. Astronomic instruments like that in general work in environments with significant thermal gradients, a common characteristic of ground-based observatories. An interesting alternative to the conventional steel ferrule may be a quartz tube. Quartz material has no problems of contraction in the temperature gradients experienced in that places, -10 °C to 20 °C, but is very expensive and difficult to obtain. The ideal condition requires a material with elasticity controlled so as not to cause stress or shift the positioning of optical fibre under temperature gradients. For just such purposes, we have developed a special composite formed from a mixture of EPO-TEK 301-2 and some refractory material oxide in nano-particle form, cured and submitted to a customized thermal treatment. To avoid bubbles and points of stress, this mixture is prepared in a separate receptacle inside a vacuum chamber. The resulting material is more resistant and harder than EPO-TEK 301-2 and is found to be well suited to the fabrication of optical fibre arrays. An important secondary characteristic is the ease with which it can be polished. This feature is a result of the micro particles, which keep the polished surface very homogeneous during the final polishing procedure. The resulting composite combines the beneficial characteristics of both the epoxy and the oxide. main factor its coefficient of thermal expansion is significantly lower than simple solidified epoxy; the exact value depending on the relative concentrations. While the characteristics of this particular composite are still under study, it is clearly possible deploy this material in the construction of devices for several fibre instruments.

2. NEW MATERIALS TO SUPPORT OPTICAL FIBRES

Similar micrcholes arrays and the support plates of the Fig. 2, used to construct Eucalyptus IFU, were made with toolmakers brass (de Oliveira et al., 2002). The problem here is that differential expansion between the metal array and the glass microlens substrate may lead to the bond between them failing at low temperatures. Although the coefficient of thermal expansion of epoxy is much greater than those of steel, brass or glass, the elasticity of the epoxy accommodates the dimensional changes without breakage: however, this can introduce a small amount of stress build-up.

It is well known that mechanical deformation causes focal ratio degradation (FRD) by the formation of microbends in the fibre (Clayton 1989). FRD is a non-conservation of *étendue* such that the focal ratio is broadened by propagation in the fibre. When mounting the fibre, the appropriate epoxy and tubing should be selected, and general care must be taken to minimize mechanical stress and avoid additional FRD (de Oliveira et al., 2005). This is straightforward at room temperature, but greater care must be taken in the choice of materials for use at low temperatures. When the fibre assembly is cooled to temperatures around -10 °C, the epoxy, tubing and ferrule will all shrink differentially, and this may cause the level of FRD to increase. There is other problem when the fibre assembly is warmed to around 20 °C and cooled to around -10 °C. The UV epoxy, currently used used to cement the metal and the glass may be damaged if the system will be submitted several times at big changes of temperature. In this case the system may detach in places due the thermal gradient.

Microhcles array and support plate made with solidified EPOTEK represent a first step to change the metallic base for a polymeric base. The choice of the EPOTEK 301-2 is appropriate due to its excellent wicking properties and low shrinkage upon curing. This properties are very good to use in optical fibres system, so that, if it is possible to get plates adequate to machine with this epoxy we can get total compatible in the construction of the system.

2.1. Epoxy Solidified

The epoxy EPO-TEK 301-2 has low viscosity and requires a container to constrain its flow until it is solidified. Generally aluminium has been used to make these containers but it is possible to use brass, plastic or acrylic. The complete curing process takes approximately three days at room temperature and when it is dry, it is transparent to visible light. To avoid bubbles and points of stress, the epoxy is prepared in a separate container inside a vacuum chamber. The correct amount is allowed to set in the container, which is placed inside a dry environment. Once cured, some thermal treatment may be necessary. Several kinds of blocks, cylinders and plates can be made to test the polishing qualities of these test pieces. The results are very encouraging with the hardness similar to acrylic resin. The Fig. 3and 4 shows steps to obtain samples machined to manufacture blocks of epoxy solidified.

Figure 3. EPO-TEK 301-2 solidified, during the machining procedure

Machining quality is important as burrs inside the microholes may prevent the fibres from being threaded into the holes, or cause stress, or breakage, of the fibres. The quality of such microhole arrays inspected visually using a microscope, give very encouraging results displaying a minimum of burring; scarf, remaining in the holes, may be readily removed by cleaning in an ultrasonic bath.

There are several advantages to the use of solidified epoxy as compared to brass, for example, in the fabrication and use of fibre support devices. Ease of machining and compatibility with other epoxies used to attach glass or silica, may be the most important of these advantages. Although the coefficient of thermal expansion of epoxy is much greater than that of steel or glass, Tab. 1, its elasticity accommodate thermally induced dimensional changes without breakage. This also avoids excessive stress associated with increases in FRD but, in principle, could be deleterious in compromising the critical positioning stability of optical fibres as the temperature varies.

Figure 4. Plate of EPO-TEK 301-2 solidified and machined

Table 1. Coefficient of Thermal Expansion of some materials

Material	α CTE at 20 °C	Units
Brass	19	$10^6 /$ °C
Carbon Steel	10.8	$10^6 /$ °C
In ox Steel	17.3	$10^6 /$ °C
Quartz	0.59	$10^6 /$ °C
EPO-TEK 301-2	55 at 61	$10^6 /$ °C

Machining quality is important as burrs inside the microholes could prevent the fibres from entering the hole, cause stress, or breakage, of the optical fibres. The quality of the microholes array may be inspected visually using a binocular microscope and the results are often very satisfactory with minimal burring present. Swarf present in the holes is readily removed by cleaning in an ultrasonic bath. TheFig. 5 shows a sample of epoxy solidified with a microholes array.

Figure 5. Photograph of the microholes array in a sample of the EPO-TEK 301-2 solidified

After machined, this sample has a diameter of 48 mm and a thickness of 3 mm. This microholes array is matrix 30x30 holes spaced on a 1.0 mm pitch. The holes were machined using custom made drills with diameters of 0.60 mm and 0.21 mm. This produces a stepped hole, with the smaller diameter hole used for fibre positioning while the larger hole is used to accommodate a ferrule. The small holes are approximately 10 μm larger than the fibre diameter to allow sufficient space for glue penetrates. The machining error in the position of the small holes was measured to be approximately 2 μm.

2.2. Experimental Stress Analyses

It is possible to use a very simple experiment of Photo elasticity method to evaluate the static stress in the plates of epoxy solidified. Classical two-dimensional photo elasticity is an optical experimental technique for determining stress fields in solids bodies. For a given analysis, polarized light is passed through a transparent sample or the body in question, and stress-induced or static stress changes in the light result in an interference-like pattern, which may be analysed to determine the principal stresses at each point within the body.

The basis for photo elastic measurement is a phenomenon of double refraction (also called artificial birefringence). There are two situations that may cause this phenomenon. Certain plastics exhibit the first situation when the sample of this type is subjected to an applied load, the resulting stress /strain field causes the molecules within the transparent material to have a preferred alignment. The second situation is exhibited by certain epoxies after dried and in this case the stress may be called static stress. In both situations the light wave vibrations have two preferred directions within the material and a wave of linearly polarised light entering the field is split into two waves which are linearly polarised at right angles to each other and which propagate with different velocities. That is, two rays travel along each an original line of propagation, and their electric vectors are mutually perpendicular. In fact, each vibration is collinear with one of the principal stress directions. (See Fig. 6.) Also, since the two waves travel at different velocities, a phase difference develops between then and by using certain optical elements.

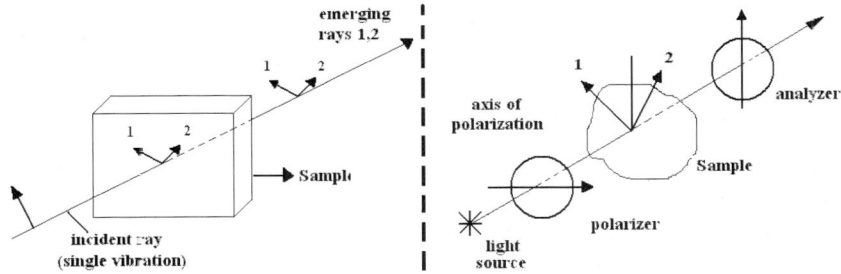

Figure 6. Left: propagation of light through photo elastic models. Right: plane polariscope used to get information of the stress inside of the sample by analyse of the artificial birefringence.

A very simple system to get qualitative images of the samples may be adapted with a transmission Polariscope, as shown in Fig. 6. The system uses a CCD camera to take the images. Stresses within solidified EPO-TEK 301-2 may be investigated with the use of the photo elasticity method whereby stress-induced birefringence is measured using polarized light. Static stress can be recorded as an interference pattern, which may be analysed to determine the principal stresses at each point within the material. Indeed, significant stress induced birefringence is detected in such samples, as is shown inFig.7 side left.

This stress can be alleviated through thermal shock induced by warming the sample to 80 °C for 30 min.Fig. 7 side right demonstrates the reduction in stress as the material is returned to room temperature. Of course, such experiments are only viable for transparent materials but they do give a warning that care must be taken in analysing the effects of thermally induced stress through measurement of fibre displacement in arrangements and FRD stress-induced.

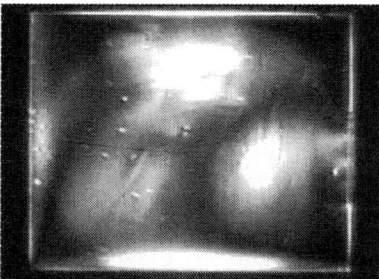 

Figure 7. Left: sample before thermal shock. Right: same sample after thermal shock

2.3. Composite

It is possible create composites using a mix of epoxy and several types of oxides in micro or nano-particle form. To avoid stress points and heterogeneous regions, the composite needs to be prepared using mixers of high speed. Ultrasonic chamber can be useful to ensure more uniformity to the mixture. Before the cure, this composite requires be subjected to a vacuum of 10^{-3} Torr to reduce bubbles inside of material. The mixture of EPO-TEK 301-2 with refractory material oxide in nano-powder, cured and submitted to a thermal treatment around 400 °C, produce a very interesting option instead simple epoxy solidified. The resulting material is more resistant and harder than EPO-TEK 301-2 and is found to be well suited to the fabrication of optical fibre arrays. Several different refractory material oxides in nano-powder may be used to produce different characteristics in this type of composite. So it is possible combine Zirconium oxide, Barium oxide, Silica oxide, Cerium oxide and others, Fig. 8, to obtain a material optimized to specific applications. The solidified mixture combines the beneficial characteristics of both the epoxy and the oxide; main factor its coefficient of thermal expansion is significantly lower than simple solidified epoxy; the exact value depending on the relative concentrations.

There are two important factors that consolidate the structure of this composite. The first is the process of cure of the liquid mixture. The second is the process of heating of the solid material obtained after the cure. The chemical reactions during the first process are limited by the time to reach the complete cure of the epoxy. Anyway, chemical analysis showed no evidence of endothermic or exothermic chemical reactions between oxides and epoxy. In fact, the materials involved in the mixture appear quite neutral. However, the heating procedure in temperatures around 400 C with slow cooling during 24 hours induces slight shrinkage on the material. Although the study still lacks depth, it is fairly simple to conclude that the structure undergoes some type of molecular rearrangement with some material loss and subsequent compaction. In fact this process carbonizes the external side of the solid material. To avoid total and destructive carbonization, both, the heating and cooling is done with the composite inside a steel container with refractory sand. After this procedure, the external part carbonized can be removed by machining process leaving the sample completely clean. The material thus obtained proves to be quite stable and resistant even though it has some degree of slow oxidation on its surface. This oxidation is evident from the slight colour change after a few weeks of exposure and manipulation, but still remains a high physical stability.

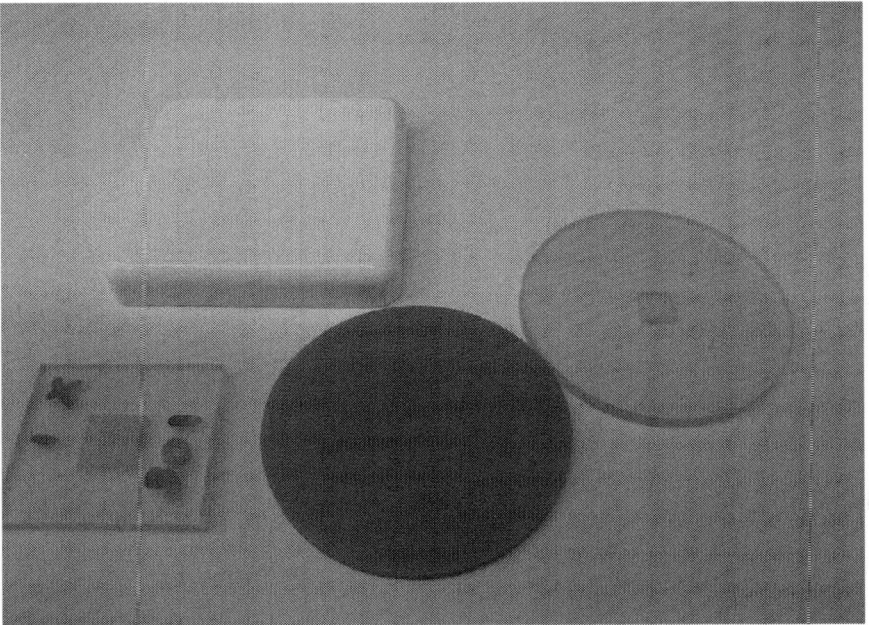

Figure 8. Samples of different composites at the centre and epoxy solidified at the borders

This composite has two physical characteristics very interesting for the construction of optical fibres holders. The first feature is its ability to sustain their polishing, with minimum quantities of abrasives during this procedure. In other words, when the composite is subjected to a polishing of high performance, the detachment of the refractory oxide nanoparticles reinforces gently the polishing process and increasing the efficiency of this procedure. The surface roughness measured in several samples, after high performance polishing was about 0.01 microns. Furthermore, the time for obtaining a polished surface with this quality is about 10 times less than the time required to polish a surface of brass of the same size.

3. SIMPLE COMPOSITE FERRULE

Mechanical deformation is a change of geometry of the optical fibre away from a straight cylinder. Large-scale bending, or macrobending is where the radius of the curvature of the bend is very large in comparison to the core diameter. On the other hand microbends are deformations of the cylindrical core shape, which are small, compared to the fibre diameter (Ransey 1988). It is well known that mechanical deformation causes FRD by the formations of microbends in the fibre (Clayton 1989). When mounting the fibre, the appropriate epoxy and tubing should be selected, and general care must be taken to minimize mechanical stress and avoid additional FRD. Currently steel ferrules tubes are used to prepare the extremities of the optical fibres for general purposes, in test lab or even as a part of some instrument. Although it is clear that inefficiencies can result in the use of metal ferrules submitted to low temperatures. Ferrules and inserts made with the composite described here, promises to be best option to handle the ends isolated of optical fibres.

3.1. Composite

While there is no direct evidence for the deterioration in Focal Ratio Degradation (FRD) of optical fibres in severe temperature gradients, the fibre ends inserted into metallic containment devices such as steel ferrules can be a source of stress, and hence increased FRD at low temperatures. In such conditions, instruments using optical fibres may suffer some increase in FRD and consequent loss of system throughput when they are working in environments with significant thermal gradients, a common characteristic of ground-based observatories. It is possible to use careful methodologies that give absolute measurements of FRD to quantify the advantages of using epoxy-based composites rather than metals as support structures for the fibre ends. This is shown to be especially important in minimizing thermally induced stresses in the fibre terminations. Furthermore, by impregnating the composites with small cerium oxide particles the composite materials supply their own fine polishing grit, Fig. 9, which aids significantly to the optical quality of the finished product. Different types of inserts are possible, Fig. 10, depending only on the precision of the machining.

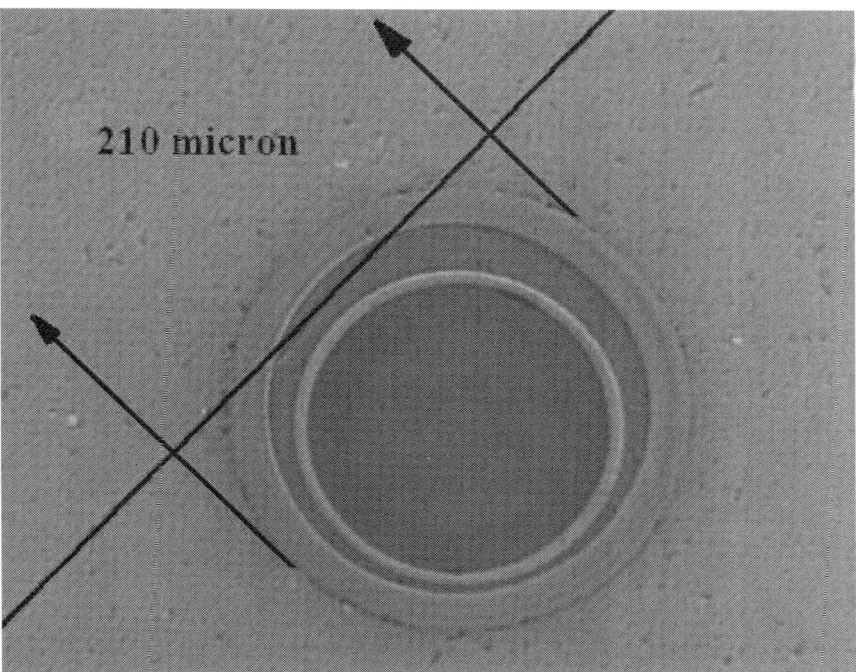

Figure 9. M:croscopic photo of optical fibre inserted in a composite ferrule, after polishing procedure.

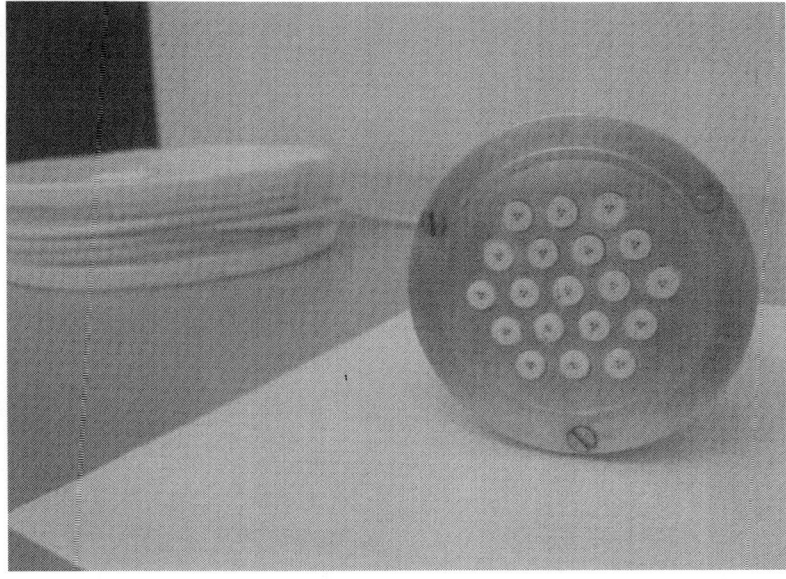

Figure 10. Inserts with optical fibres to be used in a fibres collector plate. Each insert can have several fibres.

4. MICROHOLES ARRAYS USING PLATES OF COMPOSITE

A system like that shown in section 1, Fig. 2, presented a problem in the past: This problem was the terrific facility to detach the glass substrate of the metal brass polished. Variations of temperature at long of time cause different expansion in the metal brass and the glass. After some time, the UV epoxy normally used to glue the microlens arrays with the microholes array, cannot support more the bonding between the metal brass plates due to the successive expansions and contractions caused by temperature variations. Experiments using plates made with epoxy solidified, Fig. 11 can resolve this kind of problem in the range of temperatures between −10 °C and 22 °C, typical of high altitude, ground-based observatories. Although the coefficient of thermal expansion of epoxy, around 60 x 10-6 in/in/°C, is much greater than that of brass metal, steel metal or glass, its elasticity accommodates dimensional changes thermally induced. This means less pressure on the optical fibre and consequently avoids increases in FRD associated with stress but, in principle, could be deleterious in compromising the critical positioning of fibres as the temperature varies. The ideal condition requires a material with elasticity controlled so as not to cause stress or shift the positioning of optical fibre under temperature gradients.

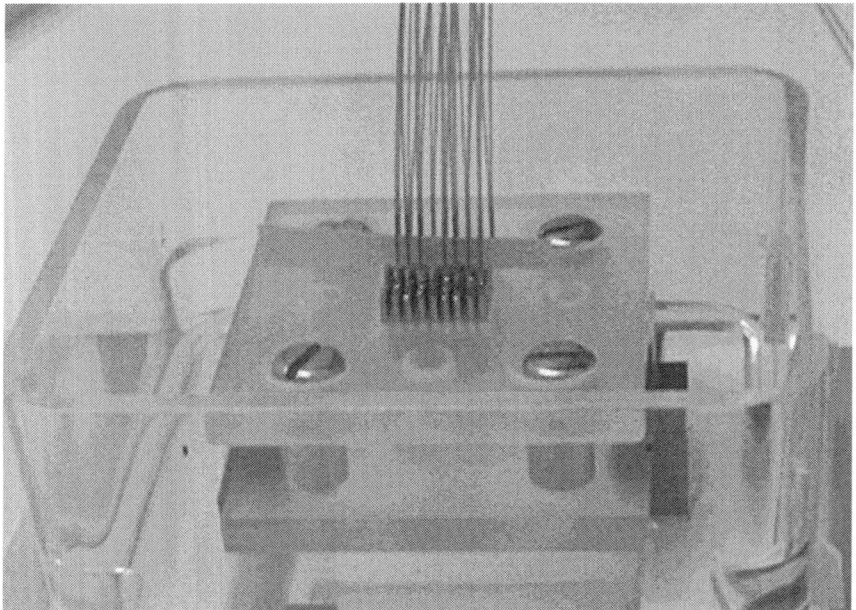

Figure 11. Microholes array device being prepared to be the input array of an IFU system

Notwithstanding the characteristics of this particular composite are still under study, this material was used successfully in the construction of devices for several fibres instruments. For example, we have used this composite to construct SIFS/IFU for the SOAR telescope in Chile, (de Oliveira et al. 2010) and FRODOspec/IFU for the Liverpool Telescope, (Macanhan et al. 2006).

4.1. Construction of Microholes Arrays Systems

To replace the brass metal or epoxy solidified and resolve the problems presented by both materials, we have used our composite to manufacture the parts of the microholes array device. The material composite obtained is less stressed and harder than EPOTEK 301-2 being a good choice to be used in optical fibres arrays. This material certainly has a combination of the characteristics from the epoxy and the refractory material oxides. The most important consequence of this combination is a coefficient of thermal expansion hither than metal brass but shorter than a simple epoxy solidified. The exactly number will depend of the relative concentration between the refractory material oxides and epoxy.

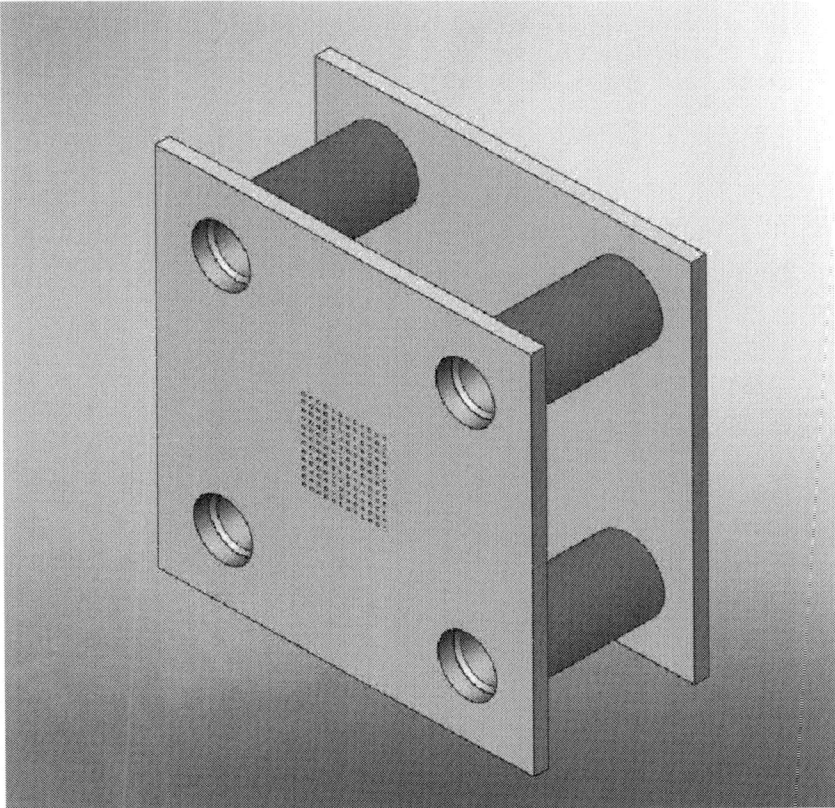

Figure 12. Schematic of the composite plates set to build the entrance device of lenslet IFU

In general, the schematic shown in Fig. 12 is the base of the entrance device of the lenslet IFU system. This device is much easier to be manufactured than the device described in section 1, Fig. 2. In fact, this new version does not require any precision in the holes confection on the composite plates. To obtain precision with the fibres position we have used a third plate called mask of precision, Fig. 13. This is a metal mask very thin obtained by a technique called electro formation. The mask obtained by this way may be configured to have holes with specifics diameters and pits, with error around 1 micron in the diameter and in the position of the holes. This technique may produce a metal nickel plate with 200 microns of thickness and the procedure is very cheap. Taking in account these facilities; the mask will define the precision of the fibres array. It is possible to obtain micro holes with the diameter exactly one or two microns larger than the diameter of the fibre used. For another hand, the diameter of the holes in the composite plates does not need to have any precision and may be much larger than the diameter of the fibre. Since that, the step holes with different diameters in the composite plates it is not more necessary, also will be not necessary to use ferrules and any kind of protection to the fibres. Eventually a device like as shown in the Fig. 13 need to be made under a microscope because the diameter of the fibre may be much small and the number of fibres involved at the assemble may be high.

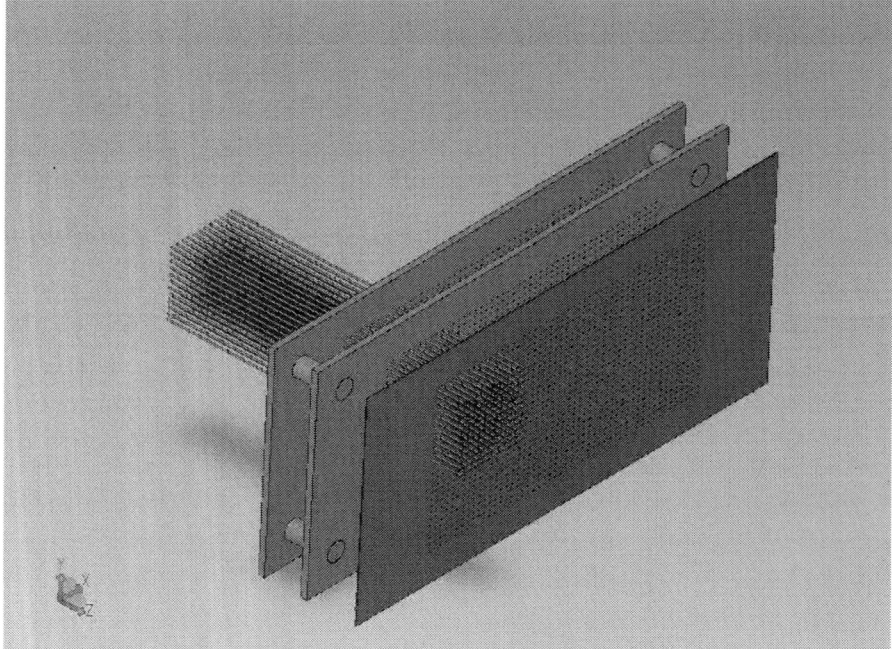

Figure 13. Entrance device during the assembling step, where the matrixes of holes in the composite plate set and in the precision mask are populated with the optical fibres terminations.

After assembled, the precision mask is glued against the composite plate and all set is immersed in EPOTEK 301-2 following the old procedure. To obtain the maximum throughput the surface of the fibres should be polished such that they are optically flat. This is the condition to attach the microlens array against the composite plate, Fig. 14 and Fig.15.

Figure 14. SIFS/IFU microlens glued

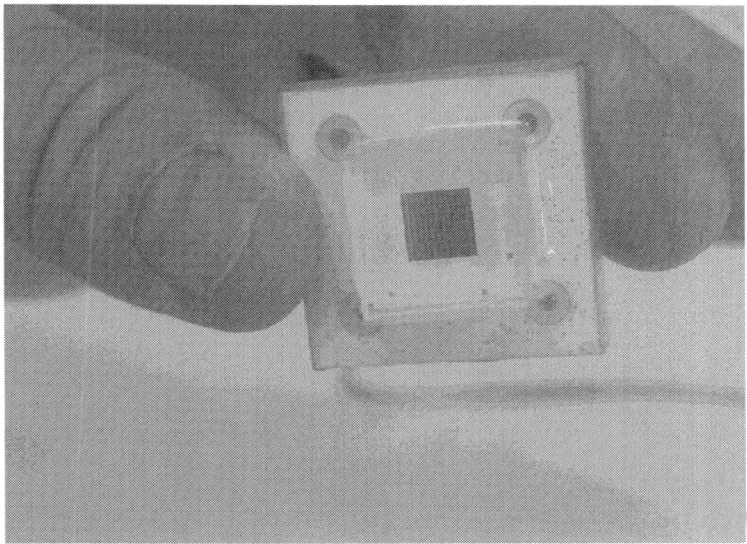

Figure 15. FRODOS IFU microlens glued

The pre-polishing process starts with the removal of excess glue with 2000 grit emery paper. Initial lapping with 6 μm diamond slurry on a copper plate and a second lapping with 1 μm diamond slurry on a tin-lead plate is used until the complete removal of the precision mask. Without the metal mask, the material of the composite plate is self-abrasive enough to produce a polishing of high performance of the optical fibres on a chemical cloth. This procedure is a basic condition to attach the microlens array against the composite plate of fibre terminations.

5. CHARACTERIZATION

The complete characterization of this composite may require several kinds of possible tests. However, applications with optical fibres in metrology involve analyses of displacement when the device is submitted at thermal gradients. More specifically, optical fibres arrays used in astronomic instruments need to resist low temperatures without displacement of the fibre position and without delamination problem between parts. Simples experiments show that the linear CTE assumes values between 20 and 40 x 10-6 / °C to 0 °C depending the concentration of the components. For example, a sample made with EPO-TEK 301-2, Barium oxide, Zircon oxide and Cerium oxide with proportions respectively 5:1:1:1, exhibits an α CTE around 30×10^{-6} $1/°C$.

Analysis of the Absolut Transmission in samples shows clearly that optical fibres inserted in brass or steel ferrules suffer increases in FRD when submitted to low temperatures. On the hand, the FRD increase in optical fibres inserted in ferrule made from EPO-TEK or in composite materials is minimised when submitted at the same negative variation of temperature. In fact, the result predicts a loss of around 10 per cent for the brass ferrules and around 3 per cent for the steel. Although it is clear that inefficiencies can result in the use of metal ferrules submitted to low temperatures, the losses are not easily quantifiable. The reason for this is that the ratio of the outer diameter of the fibre and the inside diameter of the ferrule defines the amount of epoxy between the ferrule and fibre. In the final analysis, this represents more or less compression in the fibre when the metal is compressed during the reduction of temperature.

5.1. Tests on Fibres in an Array

It is possible to do an experiment to observe the displacement of the fibres in an array of fibres constructed in the plates described in the section 4.2. An experimental array of 10 x 8 optical fibres is chosen as a representative test since it matches the base of the input array of IFUs as described before. A displacement is likely to occur with variations in temperature since the support material may suffer from some type of mechanical distortion. The experimental assembly consists of a support to hold the input array and to control thermal dissipation. A relay lens is used to project the image of the fibre array onto a CCD as a shown in the Fig. 16. The support to hold the array under examination (Fig. 17 and 18) is made of brass and had a canal for the introduction of liquid nitrogen. A continuous flux of dry nitrogen needs to be directed towards the surface of the input array to avoid condensation. Four small temperature sensors are cemented inside holes in the tested plate, close to the optical fibres. These sensors are necessary in order to test if the temperature along the plate reach thermalized state. A digital thermometer can be used to collect information from the sensors.

In our experiment, another sensor was installed inside the brass support together with a special electrical resistor to allow for temperature control. The temperature of the input array was controlled over a range between 23 °C and -10 °C. Images of the illuminated optical fibres can be obtained for a set of temperatures within the allocated range. With these images it is possible to obtain information regarding the change in the position of each fibre in the array. A simple algorithm may be used to process these results.

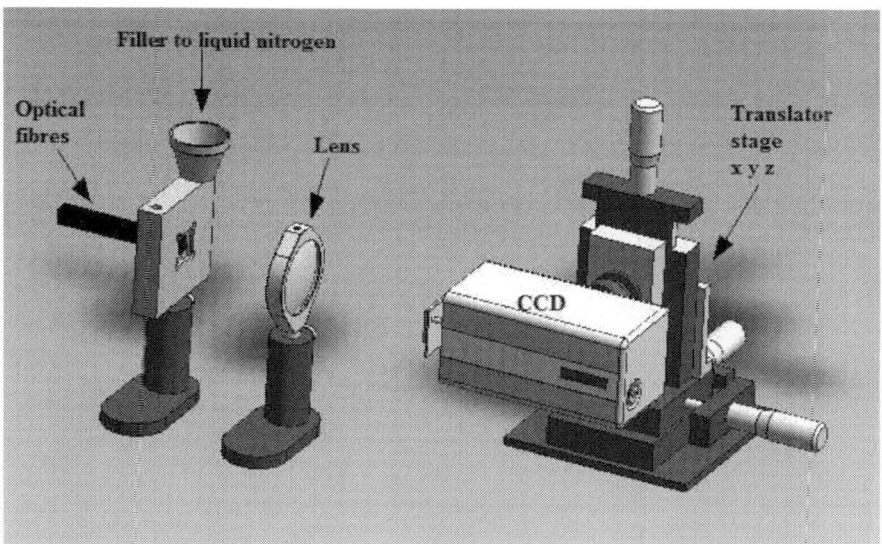

Figure 16. Diagram of the experimental set up to take images from the optical fibre array. The CCD is installed in a translation stage to put the image of the optical fibre illuminated exactly in the centre of the CCD plate. The holder support is used to keep the fibre plate array fixed and to control thermal dissipation.

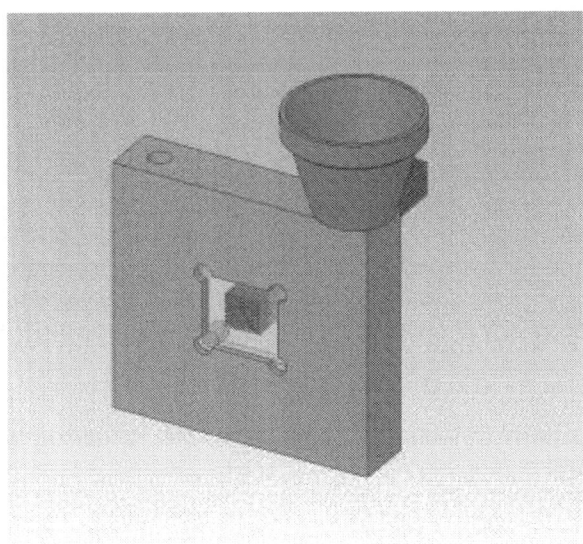

Figure 17. Schematic diagram of the holder where is fixed the fibre plate array.

5.2. Analyses of Displacements of Fibres in the Array

An image analysis by software then determines the centroid of each bright spotlight projected by each fibre from the array on the CCD. Several images like the sample shown in the Fig. 19 are used to determine an average value in the position of each bright spot light. This associates position vectors connecting each bright spotlight with the origin. The first fibre at the top/left is used as a position reference. The variations of vector's modulus during the temperature gradient are computed to produce a graph with sub pixel precision (Neal et al. 1997). In this section we present results of tests using fibre arrays submitted to negative temperature gradients. The purpose of these tests is to analyse how much the fibres in the array may be displaced from their original position as a function of the expansion of the material during the change of temperature. Figs. 20, 21 and 22 show the behaviour of arrays with 80 optical fibres when submitted to four temperature gradients. The accuracy calculated for this experiment was less than 0.2μm and the continue curve represents a fitted function.

Figure 18. Schematic diagram of the holder showing the canal to flow the liquid nitrogen during the procedure to decrease the temperature. The holder is made of brass.

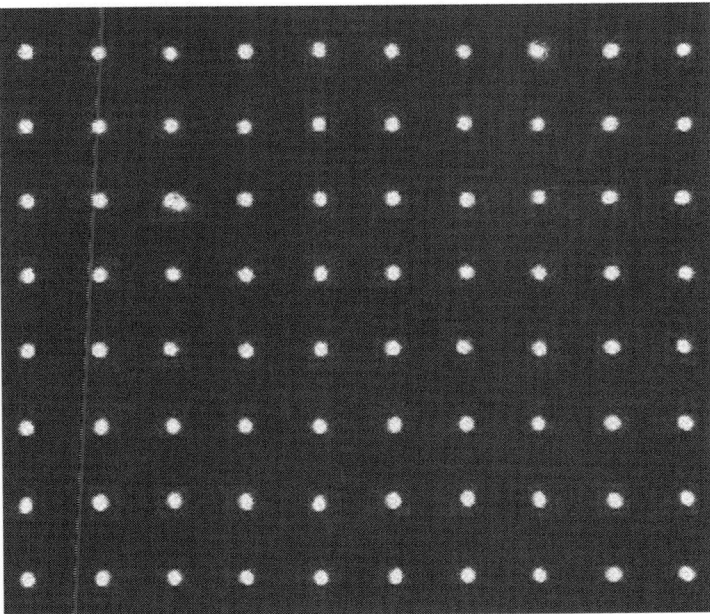

Figure 19. Image of the optical fibres matrix illuminated and projected on the CCD.

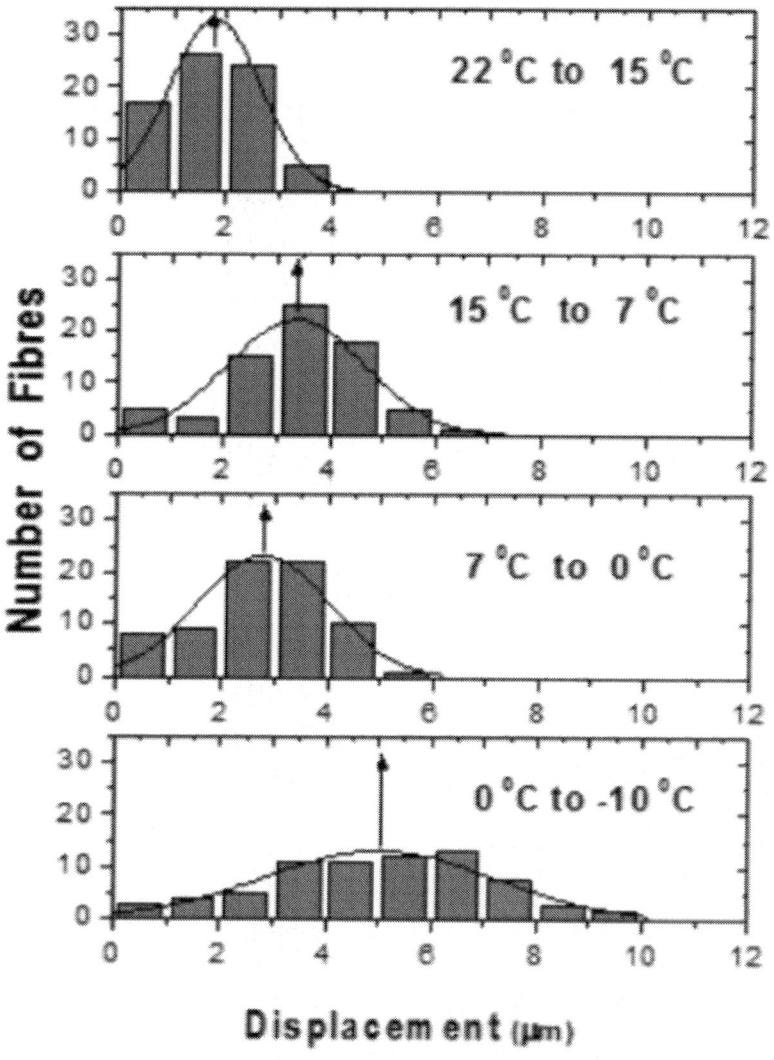

Figure 20. Distribution pattern of the optical fibres array constructed in metal brass plate submitted to four gradients of temperatures, 10 min each.

The bar graphs, demonstrates the distribution pattern of fibre positions as the temperature declines. It is possible to observe, the expansion of the EPO-TEK 301-2 epoxy material. The change in positions of the fibres amounts to ~12µm as the temperature approaches -10 °C. The array made of brass almost reaches this value despite having a totally different molecular structure to the epoxy. An interesting result was obtained with the optical fibre array in composite as may be observed in the Fig. 23. The behaviour of the composite array at low temperatures, represented in the bar graphs, is less noticeable than that obtained for the brass and epoxy arrays. In fact, when submitted to -10 °C the change of the positions at the fibres is less than 6µm. In the experimentation made with composite and brass plate samples, the temperature registered by all sensors on the plate was the same after some minutes and the error expected between the sensors would be around 0.2°C. However, we noted variations of ~1.2°C between the sensors in the experimentation with EPO-TEK plate. This may be explained by the fact that there are regions with different degrees of stress in the solidified EPO-TEK plate. These differences imply in a possible variation of thermal conductivity along the plate. As was shown in the section 2.3, stresses within solidified EPO-TEK 301-2 can be investigated with the use of the photo elasticity method whereby stress-induced birefringence is measured using polarised light. In fact it is quite common to observe static stress in plates of EPO-TEK solidified.

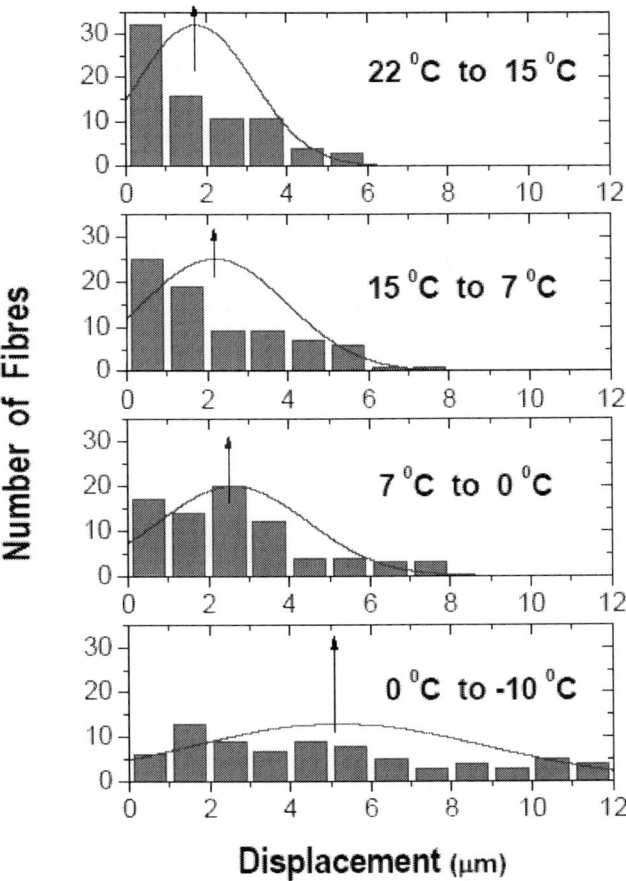

Figure 21. Distribution pattern of the optical fibres array constructed in epoxy EPO-TEK 301-2 submitted to four gradients of temperatures, 10 min each.

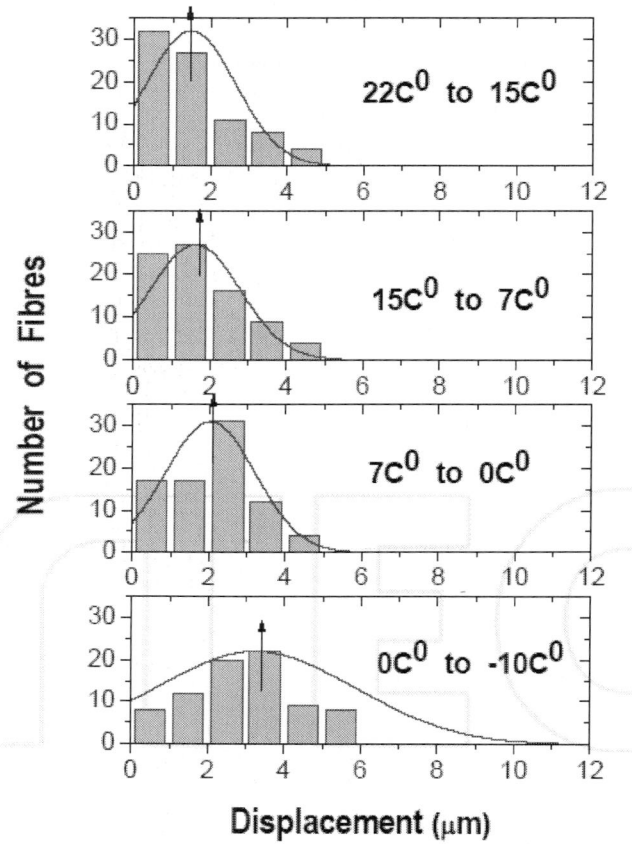

Figure 22. Distribution pattern of the fibres array in composite submitted to 4 gradients of temperatures, 10 min each.

The final conclusion for this experiment is that the composite epoxy material shows significant improvement and, in fact, has an even better performance than the brass or epoxy solidified. The chosen composite material (EPO-TEK 301-2 + zirconium oxide) retains the beneficial bonding properties of the epoxy while avoiding its thermal displacement properties.

5.3. FRD In Optical Fibres Samples

The mode dependent loss mechanisms are the causes of focal ratio degradation (FRD) in optical fibres, and are not often addressed by manufacturers. Mode dependent losses can be divided into two basic mechanisms. The first is waveguide scattering, which causes transfer of energy into loss modes by variations of the core diameter along the length of the fibre. The second is mechanical deformation. Mechanical deformation is a change of the geometry of the fibre away from a straight cylinder. Large scale bending, or macrobendings, is where the radius of curvature of the bend is very large in comparison to the core diameter. On the other hand, microbends are deformations of the cylindrical core shape, which are small, compared to the fibre diameter (Ransey 1988). It is well known that mechanical deformation causes FRD by the formation of microbends in the fibre (Clayton 1989). FRD is a non-conservation of *étendue* (or optical entropy) such that the focal ratio is broadened by propagation in the fibre. When mounting the fibre, the appropriate epoxy and, tubing should be selected and general care must be taken to minimise mechanical stress and avoid additional FRD.

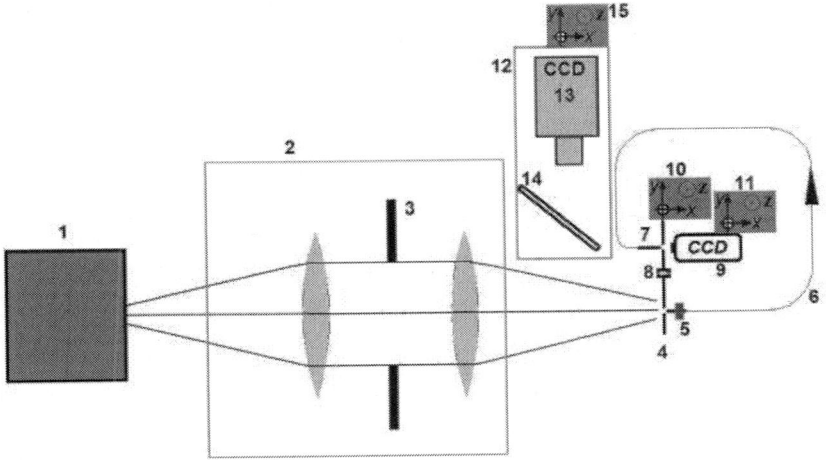

Figure 23. Diagram of the apparatus used to measure FRD – 1, light source, band pass filter and light diffuser; 2, telecentric optical system with unit magnification; 3, adjustable iris diaphragm; 4, alignment plate with a pinhole and both extremities of the tested fibre; 5, *peltier* device connected with the entrance of the optical fibre; 6, optical fibre; 7, exit of the optical fibre; 8, pinhole; 9, CCD; 10, xyz translation stage; 11, xyz translation stage; 12, microscope system; 13, CCD/lens; 14, beam splitter; 15, xyz translation stage

To measure the FRD properties of an optical fibre it is necessary to illuminate the test fibre with an input beam of known focal ratio. Then the output beam can be measured and compared with the input beam from a pinhole with the same diameter as the fibre core to determine the amount of FRD produced by the test fibre. The result is a plot of absolute transmission against output focal ratio. The experimental apparatus used to achieve this is illustrated in Fig. 23.

Illumination is provided by a 1-to-1 telecentric optical system that produces an image from an extensive uniformly illuminated source. This source is fed by a stabilized halogen lamp and has a band pass filter to provide light at 525nm, and filter's bandwidth of 100nm. An iris diaphragm placed in the collimated beam can be used to select the input focal ratio. A microscope with a CCD and beam splitter, monitored by a TV may be inserted between the pinhole/fibre plane, to be sure that the pinhole or the test fibre occupies the same position. To ensure accurate alignment of the fibre with the optical axis of the camera, the fibre is mounted in a tip-tilt translation stage. To begin the experiment, the pinhole device and the CCD are positioned to give us a reference image. In the test sequence, the pinhole is replaced with the entrance of the test fibre and the CCD is illuminated by the exit of the test fibre to give a projected image of the fibre. A distance of 9 mm between the CCD and the pinhole (or the entrance of the fibre in test) was determined as the best position to obtain images for optimal analysis. Background exposures are necessary for subtraction from the test exposures to remove the effects of hot pixels and stray light. In our experiments all fibres were tested at wavelength of 525nm, (defined using a Schott glass VG14 colour filter, ± 50 nm filter's bandwidth).

5.4. Reduction Software

We have developed a custom software package (DEGFOC 3.0) to reduce the fibre images and to obtain throughput energy curves. This software works with PC microcomputers in a WINDOWS environment. We found this to be an effective solution for use in the optical laboratory environment allowing for ease of analysis. The DEGFOC 3.0 package gives curves of enclosed energy as is shown in the Fig. 24 with the option to save the result in ASCII format to be used in any graphic software, (eg: ORIGIN).

Fibre throughputs are automatically determined as a function of output focal ratio. The first step is an estimation of the background level to be subtracted from the test exposures to remove the effects of hot pixels and stray light. The software then finds the image centre by calculating the weighted average of all pixels. It associates a radius with each pixel and calculates the eccentricity that, in the ideal case, should be zero. Our target here is to obtain the absolute transmission of the fibre at a particular input f-ratio. After establishing the distance between the fibre test and the CCD, the software defines concentric annuli centred on the fibre image. These are then used to define the efficiency over a range of f-numbers at the exit of the fibre, where each f-number value contains the summation of all energy emergent from the fibre. Each energy value is calculated by the number of counts within each annulus divided by total number of counts from the pinhole images. The limiting focal ratio that can propagate in the tested fibre is approximately f/2.2. Therefore we have defined f/2 to be the outer limit of the external annulus within which all of the light from the test fibre will be collected. The corresponding diameters of the annulus are converted to output focal ratios, multiplying them by the appropriate constant given by the distance between the fibre output end and the detector.

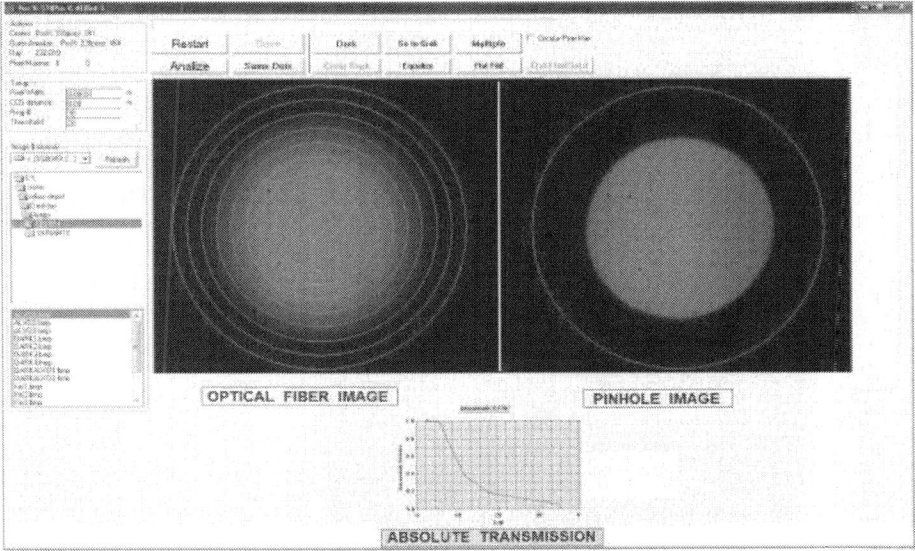

Figure 24. Print screen of the windows to the DEGFOC software.

5.5. Temperature Gradient & Frd In Optical Fibres

In this experiment we have controlled the temperature of samples between −10 °C and 22 °C, typical of high altitude, ground-based observatories. To achieve this variation we have used a *Peltier* device coupled with a temperature sensor connected to an electronic controller. The end of the test fibre is placed in contact with the *Peltier* plate by a support, and to avoid problems with water condensation at low temperatures, the test ferrule is installed inside a plastic container with a glass window. A positive pressure of nitrogen gas is maintained using a flexible tube from a gas source. With these experimental arrangements it is possible to obtain images of the optical fibres with one of extremities inside a ferrule experiencing low temperatures without water condensation. This avoids the formation of ice at the end of the fibre that could attenuate the light at its termination and contaminate the results. Our aim is to measure the effect of constriction of the ferrule on the optical fibre caused by the gradient in temperature.

Plots of absolute transmission versus output focal ratio for three samples in four configurations are presented here. We have plotted graphs with the extremes curves obtained at room temperature of 23 °C and at -10 °C after a time interval of 30 min. chosen to stabilize the thermal effects between one measurement and next. The throughput graph obtained from one fibre with brass ferrule, in dry atmosphere, is shown in Fig. 25. These results show an increase in FRD when the brass ferrule experiences a cold temperature. The total variation observed in the hatched area is very strong and diminishes as the output focal ratio of the fibre is increased. An analysis of the results demonstrates that the loss of light at F/2.3 would be around 10 per cent. This degradation is caused, presumably, by the contraction of the brass ferrule with decreasing temperature causing compressive stress of the ferrule on the fibre. The error bars, of ± 1 per cent, together the average curves, were defined after repeating each experimentation at least six times. Some experimental uncertainty in the control of temperature causing small variations in the compression force on the ferrule/fibre and consequently cause small variations in the throughput of the sample. However is evident the presence of some dissipative process, which changes the borderline of the stress during the variations of temperature. Taking in account that the experiment is made with input focal ratio around the Numerical Aperture of the fibre (F/2.27) there is the possibility that it is changing because the stress.

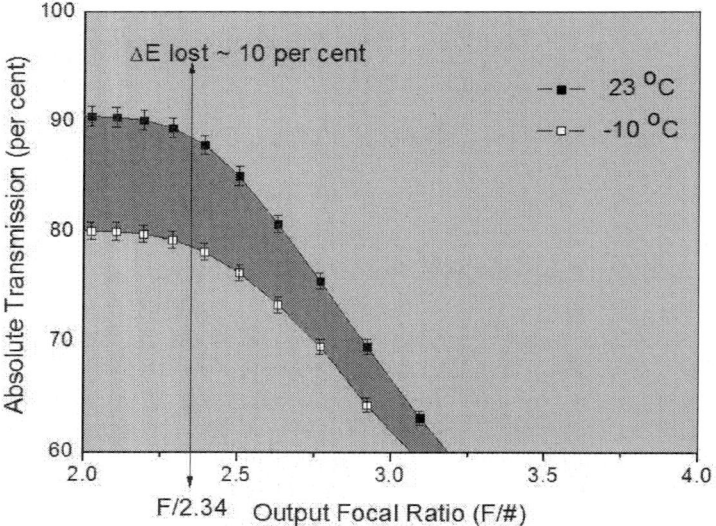

Figure 25. Performance of the optical fibre using brass ferrule. The ferrule was submitted to a negative temperature gradient of 23 °C in a dry atmosphere. The gradient was obtained, reducing the temperature, 23 °C to -10 °C, in 30 min of interval time. Two extremes curves were measured in this interval, producing the hatched area.

Such effects are critical to the design and implementation of fibre spectrographs. These results imply serious restrictions in the use of metal ferrules for optical fibres operating in ambient conditions that experience large changes of temperature typical of many observatories both during the night and throughout the year.

The throughput for samples with fibres inserted into epoxy ferrules and composite ferrules, in dry atmosphere, are shown for comparison in Figs. 26 and 27 using the same experimental procedures. Both graphs, present a very similar curves, with loss of light at F/2.34 around 2 per cent to the epoxy ferrule and 1 per cent to the composite ferrule. It seems that the loss of energy through stress-induced FRD effects is significantly less than that observed with the metal ferrule samples. The similarity of the epoxy and composite results imply that we are seeing similar effects due to the similar structure of both materials. In fact the composite material uses the same epoxy as a substrate. A natural compression happens during the cooling process, but does not produce a compressive stress of the brass ferrule on the fibre. The elastic properties of the epoxy may neutralize the mechanical stress on the fibre during the contraction process. The same error bars, of ± 1 per cent, were obtained after six repetitions of the experiment.

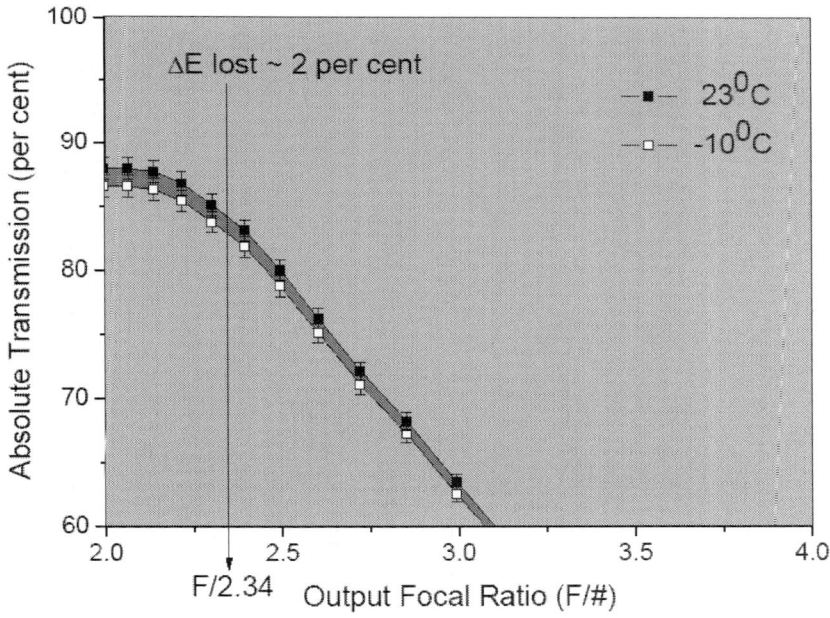

Figure 26. Performance of the optical fibre using epoxy ferrule. The ferrule was submitted to the negative temperature gradient following the same conditions of the experimentation using metal ferrule.

In general, the variation in the FRD results obtained with different samples from the same optical fibres is ±~1 per cent because the noise of the measurements. Analyse of the throughput curves obtained at room temperature from the epoxy and composite ferrules is ~ 3 per cent less on average when comparing the same curve obtained from the metal ferrules samples. The explanation for this difference may be in the aging process of the epoxy and composite ferrules. Both samples were submitted to six thermal cycles, between 50 °C and -20 °C after machining to avoid anomalous results during the experimentations. However, this procedure may increase the intrinsic FRD of the fibre given that the material structure of the ferrule may suffer accommodation pressing the fibre extremity. On the other hand, small differences of size in the hatched area of lost energy between similar samples could be expected. Differences like that would be explained by the difficulty to quantify the total length of the fibre immersed in epoxy inside the ferrule. The procedure of inserting fibre and epoxy into the ferrule is virtually handmade. There is no way to accurately control the amount of epoxy into opaque ferrules, because it is not possible to visualize the level of epoxy. Exception perhaps for polished quartz ferrules. Obviously, the length of fibre immersed in epoxy defines the length that would be submitted at the stress from the ferrule contraction.

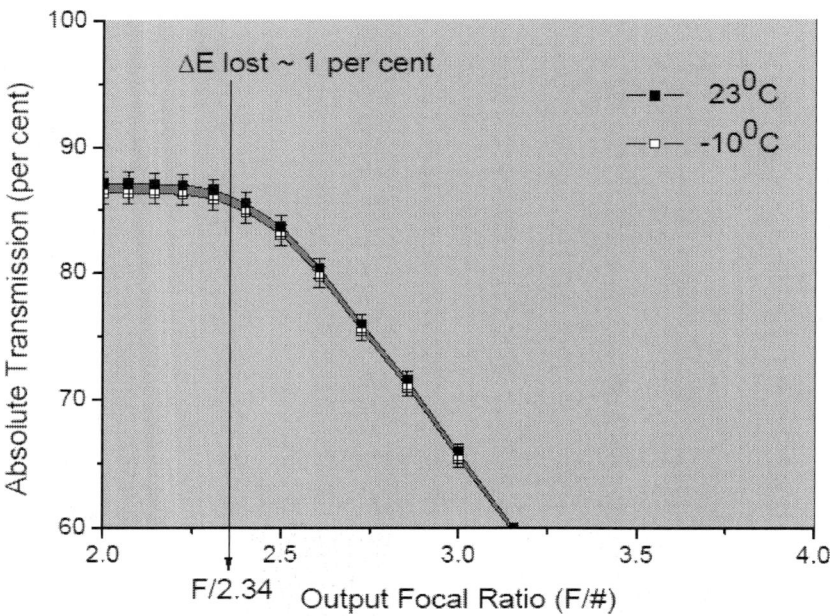

Figure 27. Performance of the optical fibre using composite ferrule. The ferrule was submitted to the same negative temperature gradient of the anterior experimentation using metal ferrule, steel ferrule and epoxy ferrule.

6. POLISHING SUBSTRATE

There are two ways of surface preparations in optical fibres, cleaving and polishing. In general applications directed to scientific instrumentation require optical fibres with extremities polished. This is the way where it is possible to optimize the spot light from the optical fibre. Furthermore, all fibre connectors require polishing and high performance may be reached with special machines and dedicated procedures. Currently, polishing procedures to optical fibres are based on very delicate glass paper or lapping discs soaked in abrasive liquid solutions. Often, this kind of liquid abrasive is very expensive taking in account your composition based in sophisticated chemistry keeping micro diamonds in suspension. Other options, considers abrasive silica and aluminium oxide mixed with oil solution or water solution. A very interesting application for the composite described here is its use as a high-performance abrasive disc to polish optical fibres.

6.1. Abrasive Discs of Composite To Polish Optical Fibres

It is possible to fabricate composite discs, controlling the abrasive capacity through the correct choice of oxide and quantity mixed with epoxy. There are several manufacturers of oxide refractory with high purity such that it is possible to compose a complete grid of polishing discs. We can consider two major advantages in the use of polishing discs manufactured with composite: The first lies in the fact that the entire polishing process can be done using distilled water only. The second advantage is that after the polishing procedure, the disc can be restored to its original flatness and completely cleaned by machining process. The efficiency of this composite disc to polish optical fibres is based in the fact that some of the oxides of the mixture are naturally abrasives. Materials like cerium oxide or silica oxide can be prepared in liquid solutions abrasives and has been used for a long time in polishing procedures of lenses and other optical devices. Discs of composite can be made in any size and can easily be adapted in the rotation device of polishing machine. Fig. 28 shows an array of optical fibres polished using discs of composite.

Figure 28. Microscopic photo of part of the optical fibres array, after polishing procedure, using discs of composite containing cerium oxide.

7. CONCLUSION

The motivation of this work was to test the performance of optical fibres inserted in ferrules made with different materials at low temperatures. The problem of finding a material best suited to securing fibres for astronomical spectrographs to cope with thermal stresses, FRD minimization and the need to achieve adequate polishing finish led us to investigate the use of composite materials. As has already been demonstrated, epoxies can be used not only as a means of holding fibres within structures (slit blocks, fibre arrays etc.) but also as a material to fabricate the structures themselves. The properties that require investigation in this context are CTE matching, machinability, bonding to glass and ease of polishing. In this context we have made several samples to evaluate FRD performance and position displacement of the inserted fibres when submitted to low temperatures.

ACKNOWLEDGEMENT

This work was financially supported by the FAPESP project no. 1999/03744-1 and CNPq project 62.0053/01-1- PADCT III/ Milenio. We wish to thank the staff of the Laboratório Nacional de Astrofísica/ MCT.

REFERENCES

1. C. A. Clayton, 1989The Implications of Image Scrambling and Focal Ratio Degradation in Fibre Optics on the Design of Astronomical Instrumentation, Astronomy and Astrophysics, 2131-2April 1989), 5025150004-6361
2. A. C. de Oliveira, et al.2002The Eucalyptus Spectrograph, Proceedings of SPIE Instrument Design and Performance for Optical/Infrared Ground-based Telescopes, 14171428Waikoloa, Hawaii, USA, August 25-28, 2002
3. A. C. de Oliveira, et al.2005Studying Focal Ratio Degradation of Optical Fibres with a Core Size of 50μm for Astronomy, Mon. Not. R. Astron. Soc., 3563October 2004), 107910870035-8711
4. A. C. de Oliveira, et al.2010The SOAR Integral Field Unit Spectrograph Optical Design and IFU Implementation, Proceedings of SPIE Modern Technologies in Space- and Ground-based Telescopes and Instrumentation, 77394S773941S-12, San Diego, California, USA, June 27, 2010
5. L. W. Ransey, 1988Focal Ratio Degradation in Optical Fibers of Astronomical Interest, In: Fiber Optics in Astronomy, Samuel C. Barden, 2640Astronomical Society of the Pacific, 0-93770-720-1Francisco, California, USA

6. V. B. P. Macanhan, et al.2006FRODOSPEC Integral Fibre Unit, Proceedings of SAB XXXII Reunião da Sociedade Astronômica Brasileira, 194195Atibaia, São Paulo, Brasil, August 3, 2006

Index